Analysis of sinter-based additive manufacturing for functional products

Vom Promotionsausschuss der

Technischen Universität Hamburg

zur Erlangung des akademischen Grades

Doktor-Ingenieur (Dr.-Ing.)

genehmigte Dissertation (kumulativ)

von

Mohammad Karim Asami

aus

Karlsruhe

2026

1. Gutachter: Prof. Dr.-Ing. Claus Emmelmann

2. Gutachter: Prof. Dr.-Ing. Ingomar Kelbassa

Tag der mündlichen Prüfung: 27. März 2026

Light Engineering für die Praxis

Reihe herausgegeben von

Claus Emmelmann, Laser Zentrum Nord GmbH, Hamburg, Germany

Technologie- und Wissenstransfer für die photonische Industrie ist der Inhalt dieser Buchreihe. Der Herausgeber leitet das Institut für Laser- und Anlagensystemtechnik (iLAS) an der Technischen Universität Hamburg (TUHH). Die Inhalte eröffnen den Lesern in der Forschung und in Unternehmen die Möglichkeit, innovative Produkte und Prozesse zu erkennen und so ihre Wettbewerbsfähigkeit nachhaltig zu stärken. Die Kenntnisse dienen der Weiterbildung von Ingenieuren und Multiplikatoren für die Produktentwicklung sowie die Produktions- und Lasertechnik, sie beinhalten die Entwicklung lasergestützter Produktionstechnologien und der Qualitätssicherung von Laserprozessen und Anlagen sowie Anleitungen für Beratungs- und Ausbildungsdienstleistungen für die Industrie.

Karim Asami

Analysis of Sinter-Based Additive Manufacturing for Functional Products

Edited by Claus Emmelmann

Springer

Karim Asami
Fraunhofer IAPT
Hamburg, Hamburg, Germany

ISSN 2522-8447 ISSN 2522-8455 (electronic)
Light Engineering für die Praxis
ISBN 978-3-032-28432-7 ISBN 978-3-032-28433-4 (eBook)
https://doi.org/10.1007/978-3-032-28433-4

This Springer imprint is published by the registered company Springer Nature Switzerland AG
The registered company address is: Gewerbestrasse 11, 6330 Cham, Switzerland

Abstract

The manufacture of functional products in a globalized environment today requires the integration of new approaches and methods in product development to optimize the production cycle. In recent years, additive manufacturing processes have emerged as a key technology for optimizing products. Sinter-based additive manufacturing processes represent a small subarea that can benefit from conventional sinter-based manufacturing processes in terms of existing knowledge. The material extrusion process for metals, in particular, has several challenges to overcome, such as achieving material standards and identifying potential applications. However, the process offers fundamental advantages as no powder handling is required during the manufacture, which ensures fast and simple processing. In addition, this process can be used to produce self-contained porous structures, which are not feasible with other metal-based additive manufacturing technologies. To overcome the challenges and create new optimization approaches, the integration of machine learning has proven to be a promising tool. With the aid of machine learning approaches, the experimental effort and, consequently, development time and costs can be significantly reduced. To be successful and reduce the experimental effort, however, a database must be created so that the models can make reliable predictions.

The generation of design guidelines is essential for the development of new products and for increasing industrial acceptance, and must be systematically developed for this process. In order to achieve this goal, this thesis identifies the process parameters and design restrictions along the entire process chain and develops design guidelines for the manufacture of functional products. The material used is AISI 316L stainless steel, which is suitable for a wide range of applications due to its excellent material properties. In addition, infill structures were simulated for a lightweight optimization strategy, and machine learning models were developed and implemented based on the outcomes in order to predict component properties without time-consuming and computationally expensive optimization iterations. By further integrating machine learning models for the prediction of quality parameters in the material extrusion process, suitable process parameters for production can be identified. An investigation of the density prediction demonstrates that a prediction accuracy of 97.44 % is achievable, and the surface roughness has been determined with an accuracy of 75.71 %.

The validation of the developed design guidelines involved different designs of bipolar plate demonstrators, which are applied in fuel cell and electrolysis cell applications. A geometric analysis of the uniform serpentine structure, which was additionally manufactured using the laser-based powder bed fusion as an additional additive manufacturing process, demonstrated that the average standard deviation in the geometry is 52 % lower. The findings from the lightweight design optimization have been applied to an injection mold demonstrator that can be used for the production of these bipolar plates. By integrating infill structures, a weight reduction of 37.33% could be achieved while simultaneously meeting the mechanical requirements, thus significantly reducing both production costs and time. The manufacture of multi-material components is a further potential application of the material extrusion processes. In this work, 316L and 17-4 PH components were additionally manufactured in a single process, resulting in components produced without significant deformation and with a high density that meets the required material quality standards. These application potentials and optimization approaches contribute to the further development of the material extrusion technology and enhance the acceptance in the industrial environment.

Table of contents

List of figures

List of tables

List of abbreviations

Abbreviation	Description
3D	3 Dimensional
ANN	Artificial Neural Networks
AI	Artificial Intelligence
AM	Additive Manufacturing
ANOVA	Analysis of Variance
BAG	Bagging Regression
BJT	Binderjetting
BPP	Bipolar Plate
cGAN	Conditional Generative Adversarial Network
CV	Cross-Validation
DfAM	Design for Additive Manufacturing
DfM	Design for Manufacturing
DM	Desktop Metal
DTR	Decision Tree Regression
EBM	Electron Beam Melting
FB	Filament-Based
FDM	Fused Deposition Modeling
FFF	Fused Filament Fabrication
GBR	Gradient Boosting Regression
JPMA	Japan Powder Metallurgy Association Standard
kNN	k-Nearest-Neighbor
LR	Linear Regression
MEX/M	Materialextrusion of Metals
MEX/P	Materialextrusion of Polymer
MIM	Metal Injection Molding
ML	Machine Learning
MLP	Multilayer Perceptron
MPIF	Metal Powder Industries Federation Standard
NN	Neural Network
PBF-LB/M	Laser Powder Bed Fusion of Metals
CAD	Computer Aided Design
HIP	Hot Isostatic Pressing
ORz	Surface Roughness of the Upskin Area
PEMEC	Proton Exchange Membrane Electrolyzer
PEMFC	Proton Exchange Membrane Fuel Cell
ReLU	Rectified Linear Unit
RF	Random Forest
RFR	Random Forest Regression
RMSE	Root Mean Square Error
SB	Screw-Based
SEM	Scanning Electron Microscope
STL	Standard Tessellation Language
SVM	Support Vector Machines
TGA	Thermogravimetric Analysis
TPMS	Triple Periodic Minimal Surface

Abbreviation	Description
UF	Ultrafuse Filament
URz	Surface Roughness of the Downskin Area
XGBoost	Extreme Gradient Boosting

List of symbols

Symbol	Meaning	Unit
A_M	Acquisition costs	€
B	Binder density	$\frac{g}{cm^3}$
D	Target solution for the regression problem	–
H	Cross-sectional thickness	mm
K_M	Machine costs	€
$\overrightarrow{L_o}$	Shrunk length	mm
$\overrightarrow{\Delta L}$	Initial length	mm
M	Matrix of the input variables	–
Q	Activation energy	$\frac{J}{mol}$
R^2	Coefficient of determination	–
R_z	Line surface roughness	μm
S_a	Area surface roughness	μm
$\vec{S}$	Dimensional shrinkage	–
T	Temperature	$°C$
V	Weight function of the output layer	–
V_B	Binder content	–
W	Weight function of the hidden layer	–
W_B	Removed binder content	–
X	Element of the domain set	–
X_{Charge}	Batch size	–
Y	Element of the target set	–
b	Bias term	–
b_B	Width of a rectangular block	mm
b_{Br}	Width of the installation space	mm
c	Learning rate	–
d_k	Desired neuron	–
f_{SVM}	Decision function of the support vector machine	–
i	Input element	–
j	Hidden neuron	–
k	Output neuron	–
k_B	Boltzmann constant	$\frac{J}{K}$
k_K	Kernel function	–
k_M	Machine hourly rate	$\frac{€}{h}$
l_B	Length of a rectangular block	mm
l_{Br}	Length of the installation space	mm
n	Integer variable	–
n_B	Number of units	–
n_{Charge}	Number of batches	–

Symbol	Meaning	Unit
o_k	Output value of the neuron	–
p	Iteration variable	–
s	Iteration variable	–
t	Time of the binder removal	h
t_2	Debinding process time	h
t_d	Depreciation period	h
t_P	Process step time	h
t_R	Setup time for each machine	h
u	Input vector	–
u^i	Support vector	–
v_{kj}	Weight from the hidden neuron	–
w_{ji}	Weight from the iteration s	–
x^n	Input feature data point vector	–
x_B	Distance between the components	mm
x_i	Input vector of the input element	–
y^n	Label value vector for the n_{th} input	–
$\vec{y}$	Target vector	–
y_i	Output from neuron *j*	–
ϕ_B	Particle fraction	–
$\phi(u)$	Mapping function	–
$\phi(u^{(i)})$	Kernel vector	–
α	Overhang angle (Chapter 5.1 and 6)	°
α	Infill structure rotation around x-axis (Chapter 5.3)	°
α_i	Lagrange multipliers	–
β	Infill structure rotation around y-axis	°
$\overrightarrow{\beta_C}$	Coefficient vector	–
γ	Structure rotation around the longitudinal axis	°
δ_i	Percentage annual utilization	%
$\vec{\epsilon}$	Residual	–
λ	Sigmoid function gradient factor	–
ρ_B	Density after solvent debinding	$\frac{g}{cm^3}$
ρ_S	Sintered density	$\frac{g}{cm^3}$
σ	Target value	–
$\hat{\sigma}$	Predicted value	–
τ	Solvent binder dissolvability factor	–
φ	Infill volume	%
ω_i	Actual value of the dependent variable	–
$\widetilde{\omega}_i$	Predicted value of the dependent variable	–

1. Introduction

In an era of globalization and diminishing resources, technological innovations are playing an increasingly crucial role in optimizing resource use and reducing material waste. From a global economic perspective, selecting the right supply chain is a prerequisite for the success of a product in terms of costs, time, and quality. The decision regarding production location and transportation routes is often made empirically and depends on factors such as technical expertise, the quantity of parts required, national regulations, and available suppliers [1,2]. Manufacturing a product entirely at a different location also carries the risk of unintentionally disclosing technical knowledge [3]. The market launch for innovation of functional products requires nowadays faster development speeds and shorter times [4,5]. Optimizing product life cycles requires the integration of new methods and approaches to meet specific product requirements. One solution is the use of artificial intelligence (AI) from the design phase of prototypes through to testing and quality assurance of ready to market products. By integrating AI and machine learning (ML) algorithms, the iterative process of prototype development — including design, testing, and refinement — as well as the overall costs can be shortened [6]. Utilizing real data can further enhance the efficiency of the design process [7].

The adoption of AI approaches and ML algorithms in additive manufacturing (AM) [8] has increased significantly in recent years and supports the optimization of processes [9–12], the design [12–14], and the simulation [16,17]. AM, in particular, is a key technology as it eliminates challenges related to production location and transportation in the value chain, while also aligning with the trend of integrating new optimization approaches in product development. The underlying factors include the significant design flexibility for components unachievable through conventional manufacturing methods, the automated nature of the process that obviates the need for tooling, and the decentralized, on-demand production enabled by the use of digital 3D models [18–20]. The increasing interest in this manufacturing technology is particularly evident in the metal AM market, which reached a size of EUR 3.47 billion in 2023, reflecting a 14.6% growth compared to the previous year [21]. Among the various AM technologies, powder bed fusion of metals using a laser beam (PBF-LB/M) [22] holds the largest share, representing 71.8% of all installed systems. In contrast, sinter-based AM processes, such as binder jetting (BJT) [23] and material extrusion of metals (MEX/M) [23], account for only 12.5% of the installed machines. This can be attributed to the limited commercial success of material extrusion of metals (MEX/M) in particular, as well as the absence of compelling business use cases [21]. Additional challenges for the process include the reliable prediction of shrinkage, the limited understanding of optimal sintering pathways, and high material costs, which hinder its further development from an industrial perspective [21].

The MEX/M process is a multi-stage method consisting among others of shaping, debinding, and sintering [24]. The process is similar to conventional metal injection molding (MIM), with the primary difference being in the shaping of the feedstock material. The material utilized is metal powder with a high filler content embedded in a

K.Asami, *Analysis of Sinter-Based Additive Manufacturing for Functional Products*, Light Engineering für die Praxis, https://doi.org/10.1007/978-3-032-28433-4_1

polymer matrix (binder) [25]. This process is well-suited for low to medium production volumes and components with at most moderate geometric complexity [5,26,27]. It is being explored for applications in the automotive, aerospace, and healthcare industries [25,28–31]. However, to date, only individual components, replacement parts, or prototypes have been produced that meet the mechanical and material requirements, benchmarks have been already largely achieved by other AM processes [5,21,31,32].

The primary advantage of the MEX/M process lies in leveraging the existing knowledge from the well-established material extrusion of polymers (MEX/P), also known as fused deposition modeling (FDM) or fused filament fabrication (FFF), and the metal injection molding (MIM) process to efficiently and easily manufacture components. Additional advantages include the elimination of the need for molds in the shaping process, thereby reducing lead times. Compared to other AM technologies, another advantage is that no powder handling is required, unlike BJT or PBF-LB/M. Furthermore, no laser is used as an exposure source, and a depowdering step is also unnecessary. Moreover, fully enclosed porous structures can be produced without the challenge of powder removal.

For the process to succeed despite the described challenges and stagnations, it must be systematically analyzed in detail, and potential applications must be identified. To contribute to its further industrial maturation, design aspects with material properties comparable to those of other processes must be thoroughly investigated. In particular, the properties have to align with those of the MIM process to ensure broader application possibilities and establish the process as a complementary solution.

2. State of the art

This chapter describes the basics of additive manufacturing, with a particular focus on the filament-based MEX/M process, which was selected as the process for further investigation.

2.1 Additive manufacturing (AM)

According to DIN EN ISO/ASTM 52900 [23], additive manufacturing processes can be categorized into single-stage and multi-stage methods. In single-stage processes, components with the desired material properties are produced in a single process step [5]. In multi-stage processes, the desired characteristics are only achieved after additional steps, such as the sintering process in BJT and MEX/M. In addition to the stage categories, additive manufacturing can also be classified according to DIN EN ISO 17296-2 based on factors such as geometry, material type, application, energy source, fusion state, process category, and fundamental principles [20].

The AM process differs significantly from subtractive or conventional manufacturing methods. Unlike conventional methods such as casting and molding, AM primarily builds the component by applying material layer by layer [20].

2.1.1 Material extrusion of metals (MEX/M)

The MEX/M process, also known as metal FFF or metal FDM [33], is a five-step AM process that involves the design of a geometry in a computer-aided design (CAD) model and the data preparation via a slicing software, the shaping of the green part using a 3D printer, the solvent and thermal debinding of the brown part, the sintering process to achieve the final material properties and a post-processing step. The geometry is created using commercial CAD software and saved as an STL (Standard Tessellation Language) file. This format allows the design to be prepared for manufacturing with a slicing software. During preparation, process parameters such as temperature, layer height, print velocity, infill, and retraction are defined, and any necessary support structures are generated. Slicing generates a G-code file that, in addition to storing parameters for the geometry structure, also includes the machine path direction and the orientation of the component. The specific type of material extrusion used in the shaping process is not detailed, allowing for various implementation possibilities. The starting material, or feedstock, consists of metal powder embedded in a plastic matrix, also known as a binder. Depending on the feedstock, three types of extrusion can be distinguished: Figure 2.1 shows the filament-based extruder (a), the screw-based extruder (b), and the plunger-based extruder (c). The filament is wound onto a spool and fed into the melting unit using transport wheels (fig. 2.1 (a)). A distinction is made between direct extruders and Bowden extruders based on the position of the transport wheels. If the wheels are located directly

K.Asami, *Analysis of Sinter-Based Additive Manufacturing for Functional Products*,
Light Engineering für die Praxis, https://doi.org/10.1007/978-3-032-28433-4_2

behind the melting unit, the extruder is known as a direct extruder. Conversely, if the wheels are mounted separately from the melting unit within the build chamber, it is referred to as a Bowden extruder [5,34]. The binder in the filament is melted and extruded through a nozzle onto the print bed, which can be heated, layer by layer. The most commonly used filament diameters are 1.75 mm and 2.85 mm [5,35]. In contrast to filament-based material extrusion, screw-based material extrusion utilizes yarn as the starting material (fig. 2.1 (b)). With a screw extruder, the starting material is fed into the heating element. The main advantage is that higher layer-by-layer build-up rates can be achieved compared to filament-based extrusion [5]. The subsequent process step of filament production and preparation is also eliminated [31]. However, attention must be paid to the size of the starting material, which could be in granulate form. Typically, granulate sizes of 3 to 5 mm allow a continuous material feed [36,37]. Selecting the correct granulate size is crucial, as granulate that is too large cannot be mixed evenly in the feed mechanism, and granulate that is too small can block the hopper during filling [31,38]. Unlike the other methods, plunger-based extrusion is a discontinuous process constrained by the stroke volume of the plunger in the extruder (fig. 2.1. (c)) [5]. The primary advantage is the high degree of material flexibility in the selection of the starting material [31]. However, the disadvantages of the extrusion process outweigh the advantages. In addition to the obviously limited amount of material due to the piston volume, where the amount of material must be known for each printing process, air inclusions in the polymer melt can lead to an interruption of the extrusion process [5,39].

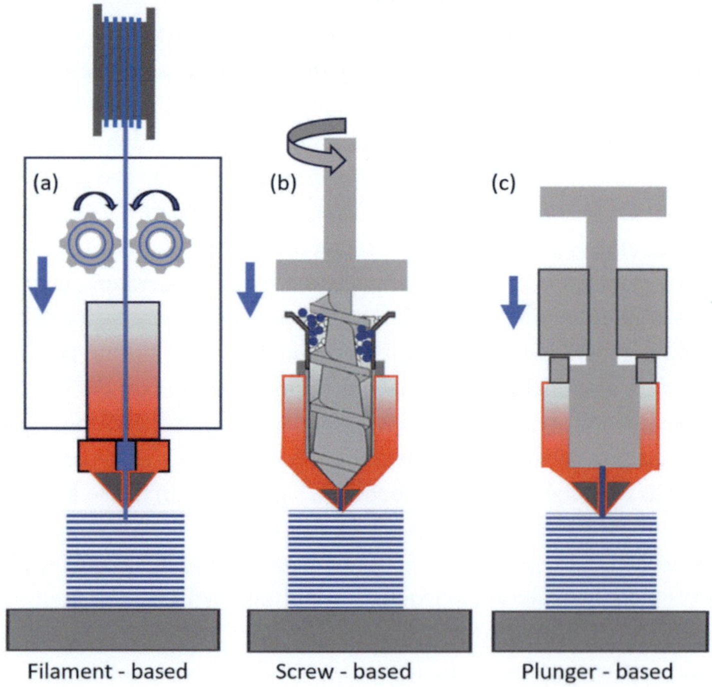

Figure 2.1 Schematic of the common material extrusion implementations. Reproduced from [40] (licensed under CC BY 4.0).

The aim of the debinding step is to remove the polymer content of the extruded component. In practice, debinding is carried out using solvents, or through catalytic and thermal methods [5,31]. The sintering process serves as a heat treatment to bind the metal powder together and compact the previously debinded component. Material properties are achieved after this step, depending on the specific material used [41]. The final process step is post-processing, where any support structures not part of the final component are removed [42]. Depending on the requirements, this step involves optimizing the component surface, enhancing mechanical properties (e.g., through hot isostatic pressing (HIP)), and improving dimensional accuracy [5,43–45].

2.1.1.1 Filament-based shaping

The most widely used MEX/M systems for the shaping step are the filament-based printers [31]. Figure 2.2 illustrates the filament-based shaping process using a printer. The filament is conveyed into the melting zone of the heated nozzle via transport wheels or rollers [40,46–48]. The melted material from the liquefier is pressed through the nozzle layer by layer, first onto the build plate and then onto the previously extruded layer. Depending on the printer configuration, the extrusion head and/or the build plate can move in the x, y, and z directions. During the extrusion process, the extruder is typically moved via stepper motors in the x and y axes according to the defined path of the G-code. After extrusion has been completed in one layer, either the build plate is lowered along the z-axis or the extruder is raised by exactly one layer height and the extrusion starts again until the component is finally printed.

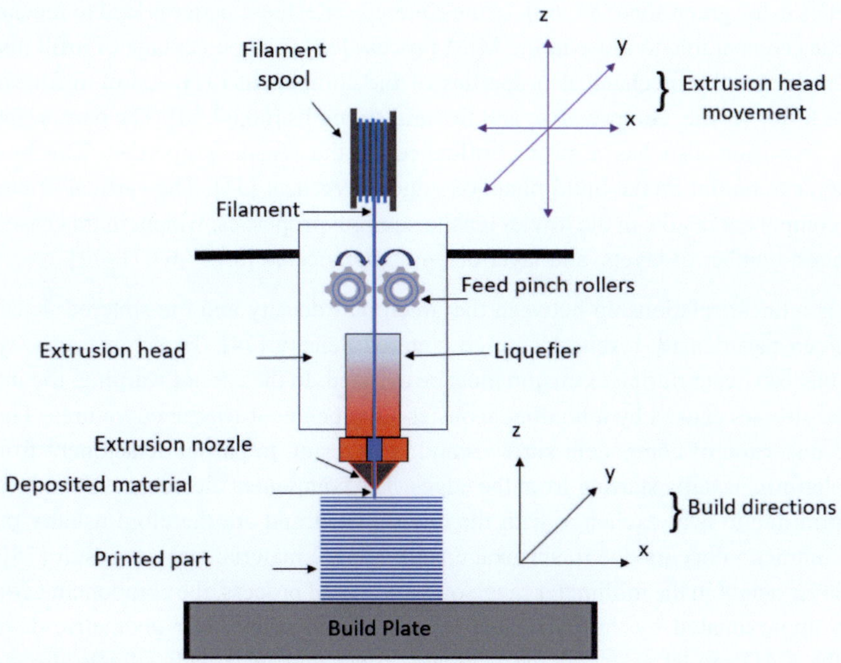

Figure 2.2 Schematic of the filament-based material extrusion with a commercial 3D printer. Reproduced from [40] (licensed under CC BY 4.0).

The control of the selected process parameters is a prerequisite for error-free and defect-free printing, which has a direct influence on the sintered material properties of the part [31,49–53]. Depending on the material properties and the process parameters, satisfactory sintered MEX/M components can be attributed to sufficient interlayer and intralayer bonding [40,54,55]. Numerous process parameters have a significant influence on the layer bond. E.g., the temperature during extrusion and on the build plate, the layer height, the printing speed, the infill pattern, the infill density, the extrusion multiplier, which controls the amount of material for the deposition, and the number of perimeters all play an immense role in achieving the desired result [31,40,56–59]. A reduced layer height, printing speed and increased extrusion multiplier result in a higher density of the green part and minimize voids during the extrusion [36,53,57,60,61]. In contrast to the density, an increase in layer height and printing speed causes an increase in surface roughness [36]. Furthermore, even a slight change in the extrusion temperature can have an influence on the geometric deviation and the sintered material properties [40,62]. The extrusion multiplier can reduce imperfections during layer application. An increase in the extrusion multiplier leads to a higher green part tensile strength [49]. However, a value that is too high can lead to irregular surfaces and must be selected accordingly [63]. The fan speed at which the layers are cooled during extrusion has a significant influence on the quality of the green parts. If the fan speed is too high, it leads to nozzle blockages and increased surface roughness [30]. There are different infill patterns for the printing process. Most infill patterns are rectilinear, honeycomb, and concentric [31]. In contrast to other infill patterns, the rectilinear patterns lead to uniform patterns and enhanced mechanical properties in the green state [64–68]. In the sintered state, these patterns lead to mechanical properties comparable to those in the MIM process [69]. The percentage of infill also has an influence on the mechanical properties of the component [31]. A low infill volume leads to lower tensile, compressive, and flexural strengths [66,67,70]. The part orientation for the extrusion also has a strong influence on the tensile properties. The build-up direction can be flat on the build plate, on edge, or vertical [31]. The vertical orientation of the component results in the lowest tensile strength properties, which, in this case, have the highest number of layers, and layer delamination occurs [53,57,67,71–73].

There is a linear relationship between the green part density and the sintered density. A high green part density results in a high sintered density [74]. To achieve this, typical errors that can occur during extrusion must be avoided. In the case of warping, the internal material stresses caused by inhomogeneous cooling behavior trigger curvatures. This can lead to distortion of component surfaces and, as a result, to partial detachment from the build platform, usually starting from the edges. The component elements at the edges tend to deform due to heat exchange with the environment and are therefore usually printed first. Contraction occurs due to shrinkage, and internal material stresses result [75]. Due to the layer height in the millimeter range of the MEX/M process, the component geometry is only approximated by the individual layers. The layer-by-layer geometric deviation from the CAD model is called the staircase effect and negatively affects the surface roughness [76]. Depending on the component orientation, the number of layers—and thus the staircase effect—can be reduced, as well as the post-processing effort [20].

2.1.1.2 Feedstock

A variety of materials that can be sintered, such as those used in the MIM process, can be used as feedstock in the MEX/M process. Figure 2.3 shows the alloys used in the MEX/M process. It can be seen that the most commonly used material, regardless of the manufacturer, is 316L stainless steel, which accounts for 41.57 %.

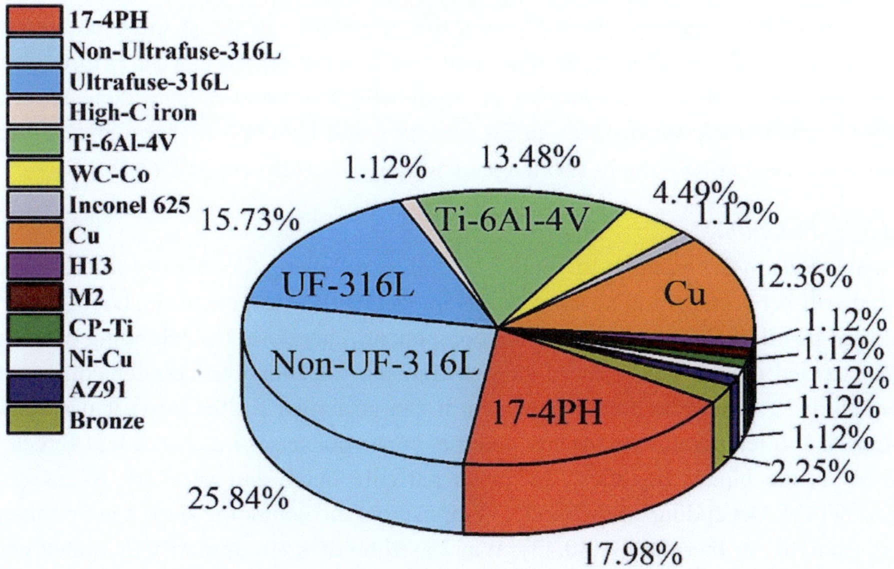

Figure 2.3 Materials used for the material extrusion process. (For the alloy 316L, a distinction is made between BASF SE Ultrafuse 316L (UF-316L) and other providers (Non-UF-316L)). Reproduced from [31] (licensed under CC BY 4.0).

Pre-alloyed metal powders are used as the starting material for the feedstock [31]. The powder preferably has a fine, spherical particle shape with a particle size distribution comparable to that used in the MIM process [77,78]. Typical particle size distributions (D_{50}) in MIM range from 10 to 20 μm [31,38,79–83], which is in a similar range as the particle size of the MEX/M process. The solid loading content of the MEX/M filament depends on the particle size, ranging from 55 to 65 vol.% [30,78,84]. The use of smaller particles with a size of 10 μm leads to a higher filling level of the solid loading at 60 vol.% [85].

The binder consists of several components made up of low and high molecular weight polymers and additives [40,86]. The low molecular weight polymer is called a plasticizer and has a binder content of 59 to 90 vol.% [31,40]. The primary function of the plasticizer is to increase the flow behavior of the feedstock and paraffin or carnauba wax is typically used [31]. The high molecular weight polymer has the task of keeping the geometry of the part during pre-sintering and debinding, also called the backbone [31,40]. EVA [87,88],

PP [81], PE [83], and PA [80] can be used as backbone materials [31]. Additives are added to the binder to ensure fluidity in the liquid state [40,89,90]. The proportion is usually up to 10 vol.%, and the stearic acid is most commonly used [31]. When manufacturing the binder for filaments, thermoplastic elastomers are typically added to ensure strand flexibility [91]. For the MEX/M process, approaches of single-component binder systems exist, which have been successfully produced and used for the production of 316L filament [92].

The appropriate metal binder composite must be selected as the base material to meet a wide range of requirements. In addition to the printability in the shaping process and sinterability of the formulation, requirements, such as consistent extrusion through the nozzle, appropriate rheological properties, rapid solidification after layer deposition, the ability to bridge a certain distance under pressure, and stability in shape and structure during repeated heating without shrinkage or warpage must be met [40,93–96].

2.1.1.3 Debinding

The main target of debinding is to remove the binder following the shaping process. There is a trade-off between debinding time and its impact on the component. It should occur as quickly as possible without damaging the component. Therefore, the debinding process is usually carried out in two or more steps to remove the individual binder components separately [43]. When the binder is heated, it can no longer fix the powder particles in position due to temperature gradients, internal vapor pockets, or gravitational forces. In contrast, as the binder liquefies, the pores partially open, and capillary forces resist distortion [43]. Depending on the binder composition, the debinding steps can be solvent-based, catalytic, or thermal [31,40,43]. Wax-based binders are primarily debinded using solvents [31,77]. In solvent-based debinding, the green part is immersed in the solvent during thermal heating. The time t required to remove the binder content depends on several influencing factors, mainly the temperature T and the cross-sectional thickness H of the green part, as described in Equation 2.1.

$$t = \frac{H^2}{\tau} \ln\left(\frac{V_B}{1-\phi_B}\right)^{\frac{Q}{k_B T}} \tag{2.1}$$

The factor τ depends on the dissolvability of the binder in the solvent, and V_B is the binder content to be removed. ϕ_B is the particle fraction, the difference of which represents the initial binder fraction. Q is the activation energy, and k_B is the Boltzmann constant. Equation 2.1 shows that the debinding rates depend on the molecular mobility, which increases with smaller cross-sectional areas and higher temperatures [43].

Typical faults in solvent debinding are bulges or cracks in the component caused by excessively high or low temperatures during the process. Higher temperatures increase the debinding rate and thus reduce the overall debinding time. However, the components swell and tend to deform at excessively high temperatures. Conversely, cracks are more likely to occur at temperatures that are lower, so temperature control is necessary to ensure the process step is successful [43].

Thermal debinding is the final process step aimed at completely removing the backbone and other binder components from the part [31,40,43]. The main factors influencing thermal debinding are the temperature, heating rate, atmosphere, and holding time to remove the binder without defects. Typical heating rates are 1 to 5 $°C/min$, and several holding times at different temperatures can be introduced, which depend on the backbone binder formulation [31,43]. The selection of the temperatures can be carried out by means of thermogravimetric analysis (TGA) of the starting material or the binder, which is usually completely debinded between 450°C and 600°C [31,77]. The atmosphere has a significant influence on the oxidation state of the components, which is associated with high debinding rates and can negatively impact sintering, as well as the temperature control [43]. In general, argon, nitrogen, hydrogen, and vacuum are used as the atmosphere for stainless steel [71].

There is a relationship among the weight of the removed binder content (W_B), the porosity, the density of the binder (B), and the time of the process (t_2), as described by Equation 2.2, assuming a homogeneous binder distribution in the component [43].

$$W_B{}^2 = B\, t_2 \tag{2.2}$$

Rearranging the Equation 2.2 leads to the differential Equation 2.3, which specifies any point in time of the debinding rate [43].

$$dW_B = \frac{B}{2\, W_B}\, dt_2 \tag{2.3}$$

There is a trade-off between achieving a good, sinter-ready brown component and rapid debinding. High temperatures can cause cracks or blisters in components due to thermal stress gradients, and when the binder softens due to viscous flow under gravitational stress, this may further lead to distortion [43]. The evaporation of the binder leads to internal compressive stresses which, at high temperatures, can exceed the component's strength, forcing the component to deform and causing cracks or distortion [43]. Warping also occurs, caused by anisotropic binder discharge or volume changes due to local oxidation [43].

2.1.1.4 Sintering

In the sintering process, the spherical particles are joined together and densified in a heat treatment [31,41,97]. The driving forces for sintering are the surface stresses on the curved surface of the individual particles and capillary forces [43,97]. The transport mechanisms include surface and volume diffusion, as well as mass transport, through evaporation and condensation, and constitute the initial stage of sintering. Due to the increase in temperature, the atoms migrate from the particle surfaces to the contact points of the adjacent particles, strengthening the bond by forming sinter necks [5]. When the necks are formed between adjacent particles, the strongest mechanisms are plastic and viscous flow, as well as volume diffusion and grain boundary diffusion [31,43,77,97]. Depending on the powder type, size, distribution, material, solid loading, print quality, sintering temperature, and sintering time, the shrinkage at this stage is in the range of 12 % to 20 % [31,77,98].

If the starting material consists of a uniform powder with a high packing density, warpage is reduced because more particles are in direct contact and the bonding process is intensified [43]. The dimensional shrinkage $\vec{S}$ is defined as:

$$\vec{S} = \frac{\overrightarrow{\Delta L}}{\overrightarrow{L_o}} \tag{2.4}$$

The initial length depending on the spatial direction is $\overrightarrow{L_o}$ and the shrunk length is $\overrightarrow{\Delta L}$.

Furthermore, the sintered density is defined as follows:

$$\rho_S = \frac{\rho_B}{(1-|\vec{S}|)^3} \tag{2.5}$$

The sintered density ρ_S results from the brown density ρ_B after debinding divided by the difference between 1 and the dimensional shrinkage $\vec{S}$. Equation 2.5 shows that shrinkage and sintered density are dependent on one another. The greater the shrinkage, the greater the achieved density, whereby a density of 99 % can be realized [31,41,43]. The movement of atoms to create a bond is the trigger for solid-state sintering [43]. The grain boundary diffusion is, in this case, the most useful densification process, where the atoms move to the grain boundaries and create a flow of mass into the pores [43]. The main factors influencing the achievement of a high density are maximum temperature, heating rate, atmosphere, and temperature holding times. Table 2.1 shows the influencing factors utilized experimentally in the MEX/M process for the sintering runs of the material 316L. It can be recognized that hydrogen and argon were mainly used as atmospheres. The sintering temperatures, on the other hand, vary in the range of 1250 °C and 1400 °C. A heating rate of 5 °C/min and a sintering time of 3 hours were used most frequently. Not all information has been entered for some references; these are marked with "-".

Table 2.1 MEX/M sintering parameters for stainless steel 316L adapted and extended from [31] (licensed under CC BY 4.0).

Sintering Temperature (°C)	Sintering Time (h)	Heating Rate (°C/min)	Sintering Athmosphere	Reference
1350	1	2.17	Ar	[57]
1250	-	-	H_2	[99]
1280	2	-	Ar	[53]
1250	1.5	20	Vacuum	[100]
1100	-	5	Ar	[101]
1330 - 1360	2	0.2	Vacuum	[30]
1380	3	5	H_2	[67,102,103]
1380	2	5	Ar	[68]
1380	3	5	Ar	[104]
1380	3	5	H_2	[66]
1050 (pre sintering), 1380	0.75 / 3	5	H_2	[105]
1320	2	10	Ar	[73]
1310 - 1400	1, 6, 12	3	Ar	[106]
1380	3	5	H_2	[92]
1250	-	10	H_2	[107]
1350	4	5	H_2	[85]
1360	3	5	H_2	[108]
1286 - 1320	2	2.5	Ar	[109]

The essential requirements of AM parts depend on the geometric accuracy and the mechanical properties, which are influenced by the factors mentioned. There is currently no standard for MEX/M components. Therefore, the standard for MIM parts is used, such as the Metal Powder Industries Federation (MPIF) Standard 35 for Metal Injection Molding Materials for stainless steel [110] and the Japan Powder Metallurgy Association (JPMA) Standard [31].

Figure 2.4 shows the dimensional shrinkage versus the solid loading content of material 316L in different investigations. For the MIM process, the typical shrinkage is between 12 % and 20 %, which is also marked in Figure 2.4.

Overall, it can be recognized that the shrinkage is largely within the typical shrinkage range for MIM. There is a tendency towards a dependency between the solid loading content and the shrinkage. The shrinkage decreases with higher solid loading content [31]. With a solid loading content of 55 % by volume, the shrinkage is in the range of 14 % to 20 %, and with a solid loading content of 62 % by volume, it is in the range of 13 % to 25 %. Furthermore, it can be observed that the shrinkage in the x- and y-directions is lower than in the z-direction [31,60,67], which is due to higher temperature gradients in this direction, the friction-free z direction, the gravitational force, and thus also the layer height and orientation in the green part's print [31,60,111].

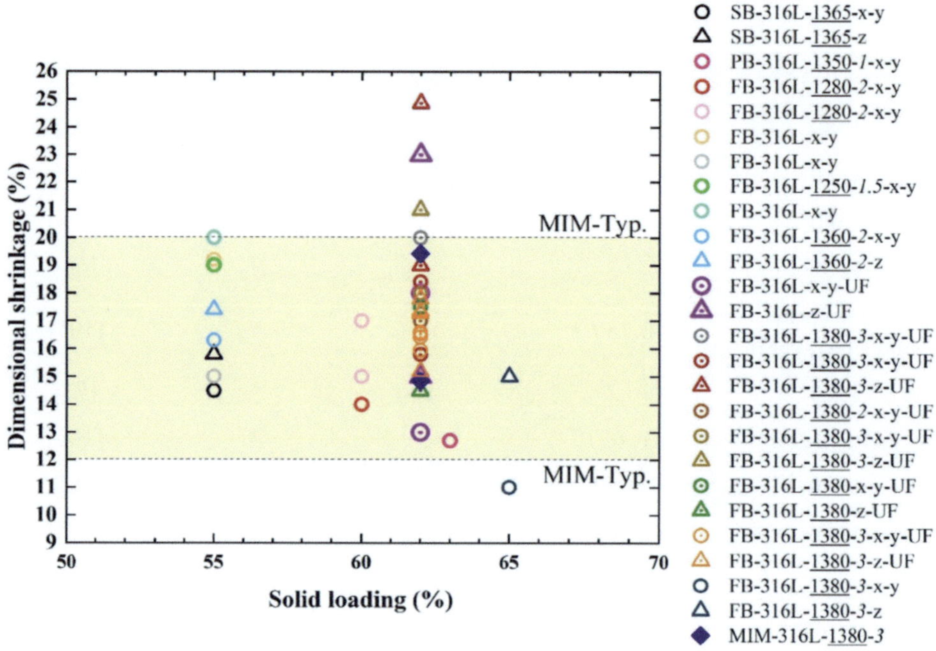

Figure 2.4 Dependency of the dimensional shrinkage on the solid loading content for the alloy 316L in the MEX/M process. SB stands for screw-based, and FB for filament-based. The underlined values are the maximum sintering temperatures (°C), and the italicized numbers represent the sintering time (h), and UF is Ultrafuse filament. The terms x, y, and z indicate the direction of the measured shrinkage. Reproduced from [31] (licensed under CC BY 4.0).

Figure 2.5 shows the relative density of the final components as a function of the sintering temperature. The MPIF Minimum Standard 35 for MIM material in relative density is 96.6 % and is marked as the green area. It can be seen that a large number of data points do not reach the MPIF minimum standard and have a density of less than 96 %. The range of determined relative densities is 72 % to 99 %, whereby the density increases with higher sintering temperature [31]. It can also be observed that, in addition to the plunger-based Desktop Metal system, the Ultrafuse filament in the MEX/M process in particular achieves the MPIF minimum standard. Additionally, it can be recognized that the only other filament-based process to achieve the highest density was with a sintering temperature of

1320 °C and subsequent hot isostatic processing, achieving 99.7 % [73]. When infill structures are used, the relative density decreases significantly and is below the MIM MPIF minimum standard, without effectively worsening the mechanical performance [66]. The tensile sample with an infill of 50 % has a higher density (approximately 79 %) than the one with an infill of 75 % (approximately 75 %). A comparison with the bending specimens used does not show this inconsistency, as the sample with 50 % infill has a relative density of less than 80 %, while the sample with 75 % infill has a density of approximately 83 % [66]. This suggests that geometry plays a role in density. Furthermore, the vertical build orientation leads to a lower relative sintered density than the flat and on-edge build directions [31].

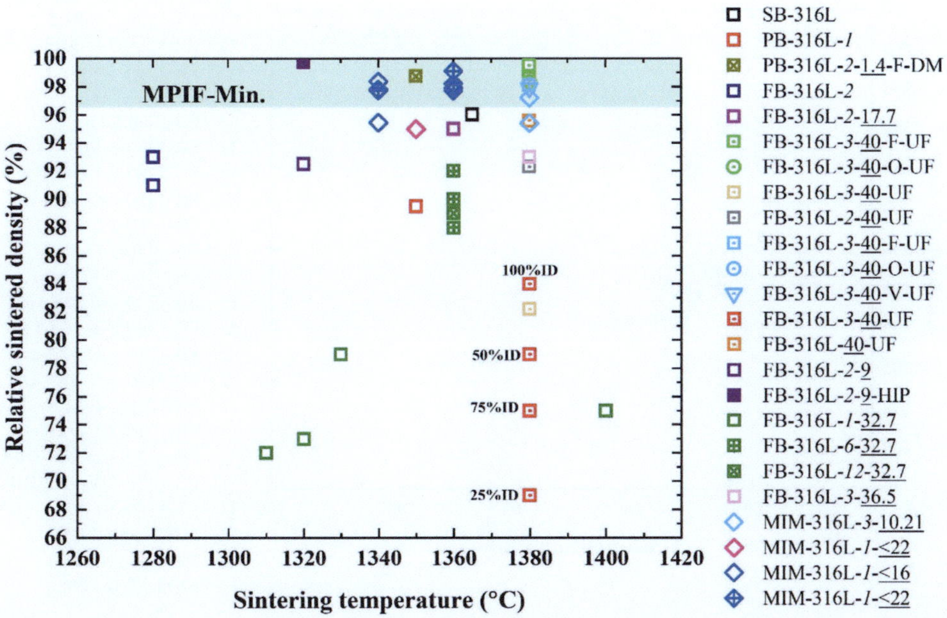

Figure 2.5 Relative sintered density as a function of the sintering temperature for 316L in the MEX/M and MIM process. SB stands for screw-based, PB for plunger-based, and FB for filament-based. The sintering time (h) is illustrated in italic numbers, and the powder size (μm) is underlined. The letters F, O, and V stand for Flat, On-edge, and Vertical printed orientations, DM stands for Desktop Metal, UF for Ultrafuse, HIP for hot isostatic pressing, and %ID for infill density. Reproduced from [31] (licensed under CC BY 4.0).

Figure 2.6 shows the tensile strength versus elongation at break for the MEX/M and MIM methods. The MPIF standard is marked in light blue. It can be observed that the majority of MEX/M materials cannot achieve the MPIF minimum standard in either tensile strength or elongation at break. For the most filament-based MEX/M samples, the tensile strength is in the range of 100 MPa to 540 MPa, and the elongation at break is in the range of 3 % to 50 %. In general, the flat and on-edge print directions achieve higher tensile strengths than the vertical build direction. This can be demonstrated by the same filament, the Ultrafuse, in Figure 2.6 [31]. The different printing and sintering parameters are significant factors influencing the observed variation of the tensile strengths [31]. It can also be seen that the tensile specimens with honeycomb infill structure [66] exhibit lower tensile strength due to the lower amount of material used, but no lower elongation at break compared to specimens sintered at 1380 °C. In the filament-based MEX/M process, only the Ultrafuse filament and an additional HIP process were able to achieve the MIM standard in terms of both tensile strength and elongation at break. The main causes were identified as errors in the shaping process, indicating that an optimized process route with high green part densities can produce good mechanical properties in the sintered state and achieve MIM standards [31].

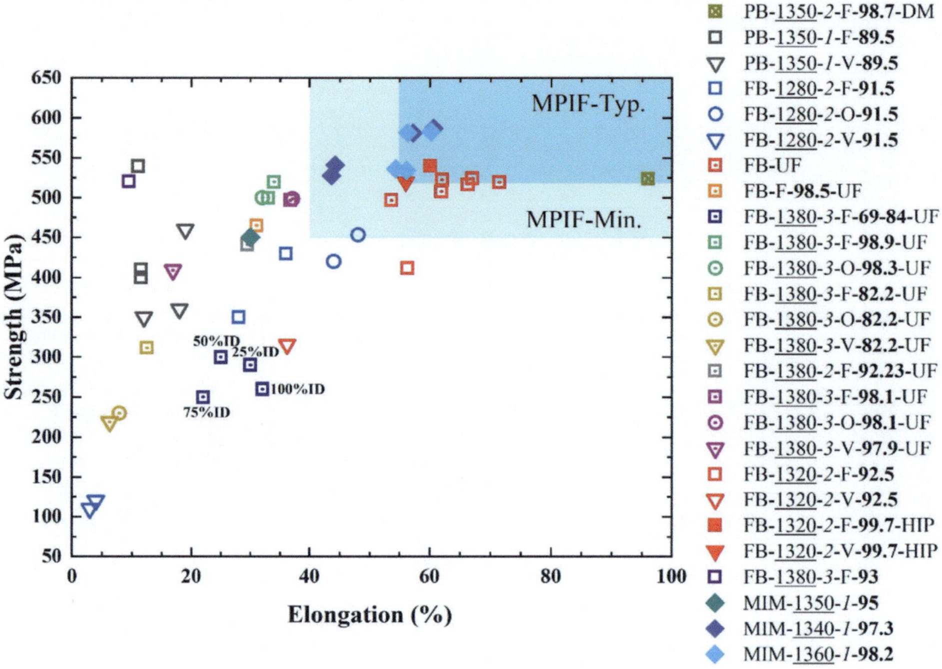

Figure 2.6 Tensile strength as a function of elongation at break for the MEX/M and MIM process with 316L. The sintering temperature (°C) is underlined, followed by the sintering time (h) in italics, and the number in bold stands for the density achieved. The letters F, O, and V stand for Flat, On-edge and Vertical printed orientations, DM stands for Desktop Metal, UF for Ultrafuse, HIP for hot isostatic pressing, and %ID for infill density. Reproduced from [31] (licensed under CC BY 4.0).

2.1.2 Design for additive manufacturing

Design for Manufacturing (DfM) strives to reduce manufacturing costs through constructive measures and to meet the requirements for quality characteristics [112]. AM has the potential to produce components that cannot be produced conventionally due to the high degree of design freedom [113]. However, there are still design limitations in AM [114] that need to be identified in a holistic approach for a specific material, as in the Design for Additive Manufacturing (DfAM), in order to realize the full potential of the manufacturing process and reduce production waste as well as costs [115]. By developing design guidelines, the evaluation parameters for the pure function fulfillment of the component can be optimized, and the parameters of cost and time can be reduced through minimizing design iterations [112]. In different guidelines [114–117], basic geometries such as walls, overhangs, bridges, bars, and boreholes were examined. These geometries were varied in their dimensions, and their dimensional accuracy after production was evaluated depending on the geometric fidelity.

Furthermore, DIN EN ISO/ASTM 52910:2020-02 provides design recommendations for the quantitative and qualitative assessment of the geometric performance of AM components using benchmark test specimens [118].

Design guidelines exist for various AM processes such as PBF-LB/M, MEX/P, Electron Beam Powder Bed Fusion (PBF-EB/M), and BJT, which lead to increased adaptation of the technology [112,115,117,119–121].

The examples given show the need for a guideline in the MEX/M process. As described in Chapters 2.1.1.1 and 2.1.1.4, the main focus is on the optimum layer bond of the individual layers of the green parts, which depend on the process parameters. Although there are design recommendations from the system manufacturers, they are not understandable without documentation of the deviations [122].

There exist qualitative guidelines for the MIM process that provide general recommendations, such as avoiding large volume changes or material accumulation [77,117]. However, these are not directly related to the component geometries and must therefore always be taken into account. A direct transfer of the guidelines to the MEX/M process is not possible and has to be further investigated.

The guidelines illustrate the production limits of the geometries for a specific system and material, for an optimized set of process parameters, which can then be transferred to the functional products with the given requirements. Furthermore, by using these geometries and their variations with the same material and process parameters, a comparison can be made among different systems.

2.2 Machine learning

This chapter describes the basics of ML and focuses mainly on the models and algorithms used in this thesis. In addition, the current state of the art in the context of AM is analyzed.

2.2.1 Fundamentals of machine learning (ML)

The basic definition by Tom Mitchell states that a program learns from experience E with a task T and a performance measure P: its performance on T, as measured by P, improves with increasing experience E [123].

Machine learning is defined as a subset of AI that uses statistical methods to enable computers to learn from data and make decisions independently without explicit programming. It is based on the approach that computers can extract knowledge from data, recognize patterns, and draw conclusions with minimal human intervention [124–126]. ML can be divided into three main areas: supervised, unsupervised, and reinforcement learning [124,127]. The goal of each of these categories is to correctly map the mathematical function in Equation 2.6, where each element of the domain X is assigned exactly one element of the target set Y.

$$f: X \rightarrow Y \tag{2.6}$$

Due to the databases, several elements Y are assigned to an element X because of statistical effects or measurement errors. This leads to inaccuracy and ambiguity in the assignment and must be taken into account in the algorithms used [124].

In the case of unsupervised learning, the models' patterns must be found in unlabeled data (e.g., clustering). In reinforcement learning, the models receive rewards or penalties, based on the actions performed, which are rooted in trial and error procedure [127].

Supervised learning is a key machine learning technique in which models are trained on labeled data sets. In this approach, the algorithm learns to match the input data to the correct output by using examples that contain both the input features and the corresponding labels [128–130]. Basically, supervised learning is divided into two categories: classification and regression, depending on the output type. In the classification approach, the target quantity is determined discretely through labels or categories [124,127,131]. For example, if the labels consist of 0 and 1, only these discrete values are the output, not values in between such as 0.5 [124]. In the regression approach, analogous to the classification approach, a task is solved in which the target variable assumes values from a continuous subset. Equation 2.7 describes the regression problem with the continuous target set Y and X, the space of feature vectors from the initial set [124,132].

$$D = \{(x^1, y^1), (x^2, y^2), \dots, (x^n, y^n)\} \subset X \times Y; Y \subseteq \mathbb{R}^n \tag{2.7}$$

In the simplest case of linear regression, the equation to be solved is represented as Equation 2.8:

$$\vec{y} = M\overrightarrow{\beta_C} + \vec{\epsilon} \tag{2.8}$$

Where $\vec{y}$ is the vector of the target variables, M is the matrix of the input variables of the initial quantity, $\overrightarrow{\beta_C}$ is the coefficient vector, and $\vec{\epsilon}$ is the residual. The aim of linear regression is to estimate the coefficients $\overrightarrow{\beta_C}$ in such a way that the sum of the squared errors (residuals) is minimized to find the local minimum. This is known as the least squares method and described by Equation 2.9 [124]:

$$\min_{\beta_C} \|y - M\beta_C\|^2 \tag{2.9}$$

If there are non-linear relationships between the individual variables, non-linear functions can be used to solve Equation 2.9. Polynomial regression models extend linear regression and use polynomial approaches to capture the non-linear relationships in the data [133]. For overly complex models, regularization techniques such as ridge regression or lasso regression can be applied to avoid overfitting [127].

The main task of ML is for the algorithms to predict unknown inputs accurately, which is referred to as generalization when predictions are accurate with the unseen data [132]. In training ML models, a distinction is made between training error and generalization error, also known as test error [132]. To develop a well-adapted ML algorithm, the aim is to minimize the training error and keep the deviation between the training and test error as small as possible [132]. If the model is unable to achieve a satisfactorily low error in the training data set, this is referred to as underfitting [132]. Furthermore, if the deviation between training and test errors is too large, it is referred to as overfitting [132]. When a data set is too small, the even smaller test data set may fall in the range of average error, making it difficult to assess whether a model is performing well or poorly. Cross-validation (CV) techniques are used to address this issue by repeatedly mixing and dividing the data into different subsets for training and testing [132]. This process artificially expands the existing data set, allowing the test error to be iteratively evaluated within each respective subset.

Other supervised models include support vector machines, which only output a class identity. The solution uses kernels k_K to predict according to Equation 2.10:

$$f_{SVM}(u) = b + \sum_i \alpha_i k_K(u, u^{(i)}) \ with \ k_K(u, u^{(i)}) = \phi(u)^\top \phi(u^{(i)}) \tag{2.10}$$

Multi-Layer Perceptrons (MLPs) are a type of artificial neural network (NN) distinguished by their tiered design, which includes an input layer, one or more intermediate layers, also called hidden layers, and an output layer. They are mainly employed for activities such as classification and regression. MLPs are made up of interconnected nodes, or neurons, arranged in layers. Each neuron in one layer is linked to every neuron in the next layer. To introduce non-linearity, each neuron applies an activation function, such as sigmoid or Rectified Linear Unit (ReLU), to its input. For training, MLPs utilize backpropagation to adjust weights based on the difference between the output error and the expected result [131,134,135]. The Equation 2.11 and 2.12 describe the output layer weight V and hidden layer weight W with backpropagation to localize the minimum of the misclassification error in MLPs [131].

$$V_{kj}(s+1) = v_{kj}(s) + c\lambda(d_k - o_k)o_k(1 - o_k)y_j(s) \tag{2.11}$$

$$W_{ji}(s+1) = w_{ji}(s) + c\lambda^{2y_j}(1 - y_j)x_i(s)\left(\sum_{p=1}^{P}(d_k - o_k)o_k(1 - o_k)\, v_{kj}(s)\right) \tag{2.12}$$

Through the hidden layer of neurons, the input vector x_i is connected to the output layer [131]. Here, i represents the input element, j denotes the hidden neuron, k is the output neuron, λ is a factor for the gradient of the sigmoid function, d_k is the desired neuron, and o_k is the actual output [131].

Decision tree regression (DTR) is a technique utilized in statistics and machine learning to forecast continuous outcomes. This method functions by dividing the data set into subsets according to feature values, thereby forming a tree-like model of decisions. Each internal node signifies a decision based on a feature, and each leaf node corresponds to a predicted outcome [131]. DTR is able to capture non-linear relationships between features and the target variable [136]. Furthermore, DTR is prone to overfitting the training data, particularly with complex data sets, which can result in poor generalization to unseen data [137]. In comparative studies, DTR frequently underperforms when compared to ensemble methods such as Random Forests (RF), which enhance accuracy and robustness by combining multiple trees [138,139].

The RF algorithm is a machine learning technique for classification and regression tasks. It functions by building multiple decision trees during training and then outputs the mode of their predictions for classification tasks or the mean prediction for regression tasks [140]. It combines the predictions of several trees to improve accuracy and control overfitting, which is a common issue with single decision trees [140]. Each tree is trained on a random subset of the data, a process that aids in reducing variance and enhancing model robustness [141]. Random Forest can offer insights into the significance of various features in making predictions, which aids in feature selection [142]. Another regression-based algorithm is gradient boosting, which adds predictors to the residual error in each tree [133].

K-nearest neighbors (kNN) works on the basic principle of finding the k nearest data points (neighbors) to a given input and making a prediction using weights or labels [143]. kNN uses a distance metric (e.g., Euclidean, Manhattan) to determine the proximity of data points [143]. The algorithm works on the principle that the distances between the sample to be measured and the known class samples are calculated first. The samples are then sorted according to the size of these distances. In the third step, the k points with the smallest distances to the point to be examined are selected. The frequency of the categories of the k closest points is then counted. Finally, the category with the highest frequency among the k points is returned as the predicted classification of the point to be examined [143].

When selecting from the multitude of ML models, the general approach is to use the model with the lowest complexity that has very good generalization properties, depending on the unknown data [144]. The ML models are evaluated differently depending on the ML type. Mean Squared Error (MSE), Root Mean Squared Error (RMSE), and the Coefficient of Determination (R^2) are used as error measures for the regression models [144–146]. Equations 2.13 and 2.14 define the MSE and RMSE.

$$MSE(\sigma,\hat{\sigma}) = \frac{1}{n}\sum_{i=0}^{n-1}(\sigma_i - \hat{\sigma}_i)^2 \tag{2.13}$$

$$RMSE(\sigma,\hat{\sigma}) = \sqrt{MSE(\sigma,\hat{\sigma})} \tag{2.14}$$

R^2 is a statistical measure that indicates the proportion of the variance in the dependent variable that can be explained by the independent variables in a regression model [145–147]. R^2 is defined in Equation 2.15 as follows:

$$R^2 = 1 - \frac{\sum_{i=1}^{n}(\omega_i - \tilde{\omega}_i)^2}{\sum_{i=1}^{n}(\omega_i - \bar{\omega}_i)^2} \tag{2.15}$$

2.2.2 ML in AM

ML in AM is being researched for various reasons. Figure 2.7 shows the areas of application of ML for various objectives. Basically, the research opportunities can be divided into five major areas: defect detection and monitoring, design optimization, sustainability, material analysis, and process parameter optimization. ML can support design optimization and take into account constraints such as component properties, manufacturability, and material usage [148]. Furthermore, ML can optimize sustainability aspects by detecting and predicting material waste and energy inefficiencies [148]. In material analysis, it is used to predict material properties and thus can identify the most suitable combination of materials [148]. With suitable training, ML models can optimize process parameters with a focus on component properties [148].

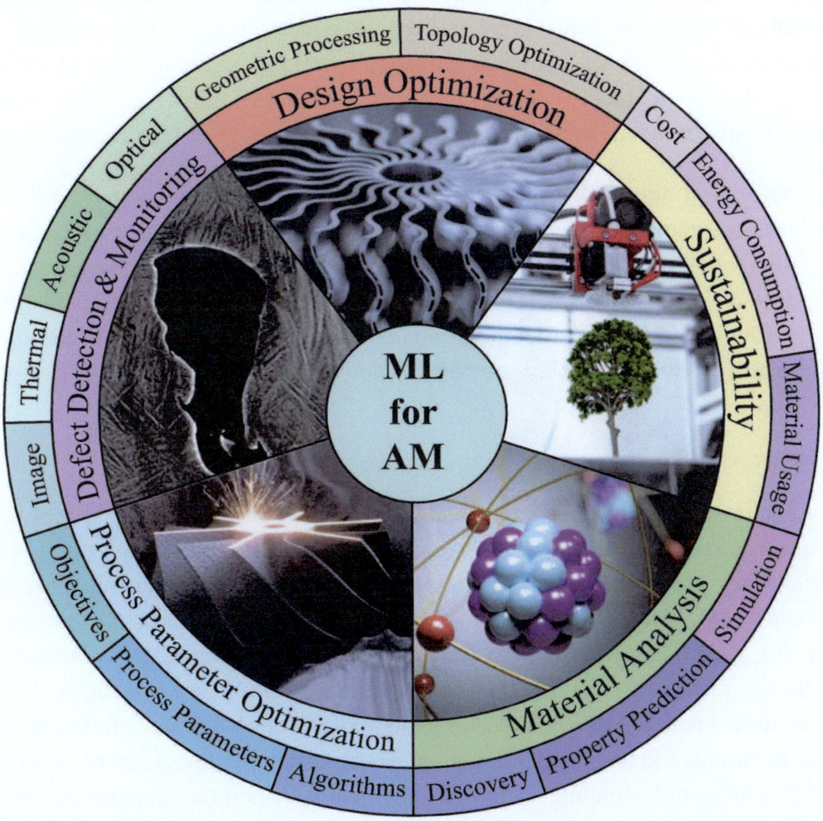

Figure 2.7 Machine learning investigation in different research fields: design optimization, sustainability, material analysis, process parameter optimization, defect detection, and monitoring. Reproduced from [148] (licensed under CC BY 4.0).

Furthermore, ML in AM can support the manufacturing process by detecting anomalies via sensors. In this way, defects can be prevented during production. In each of these sub-areas, the most widely used AM technology, PBF-LB/M, is also the one in which the most studies are conducted [148–150]. In design optimization, for example, ML models such as support vector machines (SVM), DTR, or Conditional Generative Adversarial Network (cGAN) are used to optimize the shape from the image-based or CAD-based input [6,151,152].

ML with different types of models in the MEX/P process is used in the parameter optimization with the aim of improving component quality and properties [153–158]. The used models are LR, SVM, deep learning algorithms, Gaussian process regression, and genetic algorithms. The optimization of the design by ML has been carried out using models of deep learning algorithms, Gaussian process regression, and Bayesian networks [148,159–163]. The main used data types were CAD models, 3D coordinates, images, and construction orientation. One approach uses artificial neural networks (ANN) to generate rectangular grid infill structures. Here, the data is generated by a slicer as input. It was shown that ANN enables infill savings of 45.9% compared to the benchmark infill from the slicer with comparable component properties [164]. However, this was only investigated using polymer as the material and simple cubes as test geometries.

Finally, there is an approach to use ML algorithms for the prediction of shrinkage in MEX/M, where, among others, LR and neural networks have been used as models, with the result that the deviation in the prediction of the shrinkage with neural networks is less than 1 mm in width and 0.153 mm in height [165]. By using these models, a realistically accurate prediction of shrinkage could be made for simple geometries. Additionally, by using more hidden layers, complex geometries should also be better predicted in the future.

2.3 Lattice structure

Lattice structures consist of repeating patterns of unit cells with open spaces between them, with the aim of reducing weight without changing the contour structure [166,167]. In addition, they have been tested in various applications due to their high strength-to-weight ratio [166]. There are different ways of subdividing the structures according to type, pattern, functional, and spatial extension [166]. For example, a distinction can be made between different structures such as foam, honeycomb, and two-dimensional unit cells [168,169]. Foam structures consist of randomly shaped unit cells in which the cell walls can be oriented in any spatial direction. These structures can occur in wood or cork [166,170]. Honeycomb structures consist of identically shaped two-dimensional unit cells that are lined up to form a honeycomb and extruded in the orthogonal direction [166,171–173]. Triply periodic minimal surface cells (TPMS) are structures that have zero mean curvature at each point of the cell [166,174–176]. Mathematically, these structures do not describe their own overlapping surfaces, which form a periodic 3D pattern with a minimal surface [177].

Various evaluation approaches exist for lattice structures, such as relative density, unit cell parameters, and volume fraction [166]. Bending-dominated structures perform better than stress-dominated structures in reducing deformation dissipation in terms of relative density [166,178]. In general, higher density leads to enhanced mechanical properties of the manufactured lattice structure in terms of strength and stiffness improvement, which has a nonlinear relationship [166,179–181].

In addition, there are graded structures that can also be realized using topology optimization approaches, which can improve mechanical properties and optimize stress distribution in torsion and shear [166,182–184].

The orientation of the lattice structures also plays an important role and has a significant influence on the mechanical properties, where a specific stiffness and energy absorption of 187 % can be realized with a simultaneous weight reduction of 20 % with optimal orientation [185].

3. Research hypotheses, methodology, and investigation approach

Sinter-based AM technologies benefit from the research results of MIM technologies in the debinding and sintering process steps. The MEX/M process, in particular, further benefits from the MEX/P process, which has been intensively investigated with regard to potential errors, influencing factors of process parameters, and the use of lattice structures. In Chapter 2.1, it was shown that in some cases the MPIF MIM standard 35 [110] can be achieved in terms of material properties and is potentially suitable for various applications, but is still too unexplored from an industrial perspective and has therefore not been adopted yet. Furthermore, there exists a lack of scientifically validated design guidelines for the MEX/M process in order to achieve greater acceptance from an economic perspective and identify its full potential, as is the case with other AM technologies discussed in Chapter 2.1.2.

The use of ML for the prediction of design parameters has shown, as discussed in Chapter 2.2.2, that it can be effective for different AM processes, and thereby the experimental effort can be reduced. ML models for predicting shrinkage exist in the MEX/M process, but these only focus on simple geometry approaches.

Lattice structures are used in various AM processes to reduce weight while simultaneously meeting mechanical property requirements, enabling component optimization.

The current state of the art poses a number of challenges in the manufacture of functional products using the MEX/M process. To date, the focus has been on the development of material mixtures for filament production and the optimization of process parameters for the stabilization and repeatability of the process chain. The consideration of design aspects and the use of lattice structures in combination with ML models have not yet been explored.

3.1 Research hypotheses

The following research hypotheses have been formulated based on the current state of the art.

H1: Primary research hypothesis

Functional products can be manufactured through the systematic development of process parameters and design guidelines in the MEX/M process.

H2: Secondary research hypothesis

By investigating lattice structures and ML approaches, functional MEX/M products can be optimized with minimal effort.

K.Asami, *Analysis of Sinter-Based Additive Manufacturing for Functional Products*, Light Engineering für die Praxis, https://doi.org/10.1007/978-3-032-28433-4_3

<u>H3: Tertiary research hypothesis</u>

The prediction of target parameters such as surface roughness and density is possible using ML in the MEX/M method with minimal experimental data.

In order to identify the process-related limitations of MEX/M technology, design guidelines are developed in this thesis. These guidelines consider features that can occur in functional products. The main advantage of the MEX/M process chain is that it does not use loose powder and can produce outwardly closed porous structures. The investigation of lattice structures and ML models enables the optimal selection of parametric structures to achieve the desired mechanical properties. Furthermore, ML models are used to predict surface roughness and density to investigate optimization possibilities and, consequently, quality parameters in the multi-stage process chain.

3.2 Methodology

Figure 3.1 illustrates the methodological approach followed in this thesis. Design limits are systematically mapped along the MEX/M process chain, and a functional components are manufactured to validate the optimization approaches.

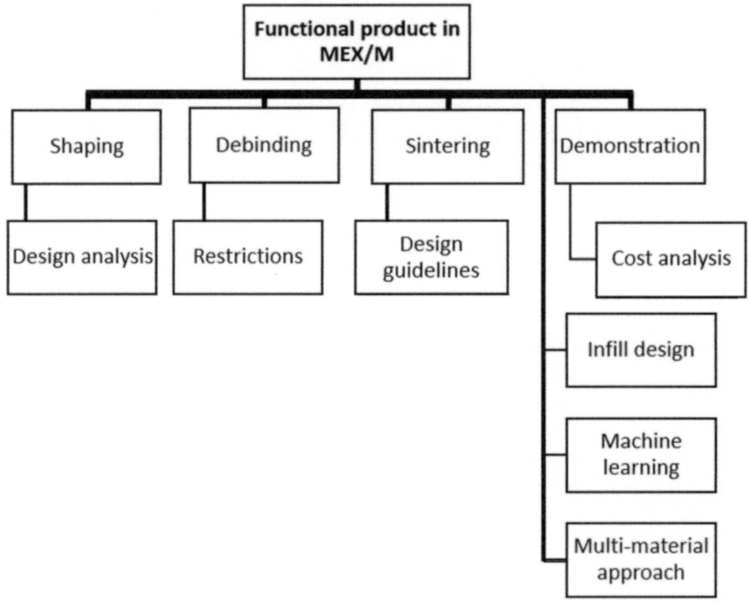

Figure 3.1 The methodological approach to the aspects examined in this thesis.

Chapter 4 starts by identifying process parameter limits. The selected parameters are then used for the entire thesis. Chapter 5.1 examines design restrictions for green part production. The resulting design guidelines for the production of green parts are then developed. In Chapter 5.2, restrictions for the debinding and sintering process are identified and design guidelines are further developed. In Chapter 5.3, porous structures are simulated and feasibility constraints are identified. This data basis is used for the development and optimization of ML models in Chapter 5.4. In Chapter 6, quality parameters are investigated using ML models and optimization approaches are developed. In Chapter 7, the identified design restrictions and optimization approaches from Chapters 5 and 6 are applied and validated on different functional products. In Chapter 7.1, different designs of bipolar plates are manufactured using the MEX/M and PBF-LB/M processes and analyzed geometrically. Chapter 7.2 illustrates the production of an injection mold with a parametric infill design for a bipolar plate. Chapter 7.3 then uses a cost estimation tool to evaluate the production costs of both the bipolar plate and the injection mold. Finally, a multi-material approach for the MEX/M process is presented in Chapter 8. Chapter 9 discusses and contextualizes all the results in relation to the state of the art and provides an outlook on future research directions.

3.3 Investigation approach

The test samples were produced using a Renkforce RF2000 (Conrad Electronic SE, Hirschau, Germany) 3D printer for the production of green parts. The CAD models of the components were created using Inventor 2021 (Autodesk Inc., San Francisco, USA), SolidWorks (Dassault Systèmes, Vélizy-Villacoublay, France), and Synera (Synera GmBH, Bremen, Germany). The Simplify3D (Cincinnati, USA) slicer was used to prepare the build jobs. The used material is AISI 316L (1.4404) filament from PT+A GmbH (Dresden, Germany) with a diameter of 2.85 mm. The debinding process was carried out with thermally heated acetone as a solvent. The thermal debinding and sintering were conducted in an EXSO90 (NEW AIM3D GmbH, Rostock, Germany) sintering and debinding furnace in an argon atmosphere. The experimental measurements and evaluations of the samples were carried out using a feeler gauge, a caliper gauge, a Keyence (Keyence Corporation, Osaka, Japan) VK-X150, a VHX-S600E reflected light microscope (SEM), a Zeiss (Carl Zeiss AG, Oberkochen, Germany) Supra 55 VP scanning electron microscope, a Keyence VL-570 optical 3D coordinate measuring machine, and the Q500 temperature calibration unit (UB TA Instruments, New Castle, United Kingdom) for thermogravimetric analysis.

The statistical analysis was carried out using Statgraphics 18 (Statgraphics Technologies Inc., The Plains, Virginia, USA).

The simulative study was carried out using the software nTopology (nTopology, New York City, USA), Inventor 2022 (Autodesk, San Francisco, USA), and Hyperworks 2021 (Altair Engineering Inc., Troy, Michigan, USA).

The ML models were developed and implemented in Python (v3.8.19) with the Scikit-learn (v1.3.0) library

4. Material and process parameter analysis

This chapter examines the used AISI 316L material and includes preliminary investigations with the aim of identifying the optimum process parameters, which are the basic prerequisite for the subsequent process steps. Some of the findings have already been presented in a publication [109] and are partially adopted.

4.1 Feedstock material

The filament AISI 316L (1.4404) is specified by the manufacturer with the following starting powder composition, which is within the range according to DIN 17440 and is displayed in Table 4.1.

Table 4.1 MEX/M starting powder composition for stainless steel 316L.

Amount	Cr	Ni	Mo	Mn	Si	P	C	Fe
Weight [%]	16.8	10.1	2.2	1.38	0.74	0.02	0.018	bal.

The filament manufacturer has indicated in the particle size distribution that fine powder was used. An analysis of the debinded filament, shown in Figure 4.1 as an SEM image, reveals that the powder size ranges from 2 μm to 15 μm and is predominantly spherical [109]. The comparison with Chapter 2.1.1.2 shows that the particle size is within the range of typical MIM and MEX/M powders, and no anomaly in the feedstock can be determined.

Figure 4.1 SEM visualization of the debinded AISI 316L filament. Reproduced and adapted from [109] (licensed under CC BY 4.0).

K.Asami, *Analysis of Sinter-Based Additive Manufacturing for Functional Products*, Light Engineering für die Praxis, https://doi.org/10.1007/978-3-032-28433-4_4

An examination of the filament width, conducted using the VHX-S600E microscope on four different sections each approximately 600 mm in length, showed that the filament diameters range from 2.826 mm to 2.857 mm. It can be observed that the feedstock material shows a deviation of up to 0.024 mm in filament diameter.

4.2 Process parameter identification

The aim of this subchapter is to identify suitable process parameter combinations, enabling component manufacturing based on target parameters such as dimensional accuracy, density, and roughness.

4.2.1 Shaping

According to the manufacturer, a nozzle with a diameter of at least 0.4 mm should be used for green part printing. They recommend an extrusion temperature of 120 °C to 150 °C and a printing speed of 30 mm/s to 80 mm/s.

75 cubes with a side length of 10 mm were produced with different parameter combinations to investigate the influence of printing speed, layer height, printing temperature, and extrusion multiplier on the target parameters green part density, geometric accuracy, and surface roughness (see Appendix A.1, Table A.1.1). The density was determined using Archimedes' principle in ethanol on a precision scale. The roughness S_a was determined with the VHX-S600E. The side surfaces of the cubes were measured and averaged. The geometry was measured with a caliper on all edge surfaces and in the center and also averaged. It can be observed from the measurement data that the geometric deviation increases with increasing layer height and extrusion multiplier. Similarly, the density and surface roughness also tend to increase. A trade-off between dimensional accuracy, roughness, and density must therefore be selected.

For the statistical analysis of the measurement data (Table A.1.1.), a full factorial design of the influencing factors and target variables was carried out as in [186]. Analysis of variance (ANOVA) and regression analysis were used to determine the influence of the individual parameters on the target values. For the target criterion density, there are strong dependencies on the parameters extrusion multiplier, printing speed, and temperature. All four parameters have a major influence on surface roughness. The geometric deviation is mainly influenced by the temperature, the printing speed, and the extrusion multiplier. The analysis shows that the optimum process parameters are 120 °C for the temperature, 70 mm/s for the printing speed, 0.2 mm for the layer height, and an extrusion multiplier of 1.2. The statistical evaluation does not take into account the infill structure and possible defects, which have an influence on the strand bonding. As can be seen from Chapter 2.1.1.1, a reduction in printing speed and layer thickness leads to an increase in density and reduces the occurrence of voids. A high printing speed also leads to an increase in surface roughness. The glass transition temperature, which is responsible for the deformation of the shape, was empirically determined to be between 50 °C and 60 °C.

Therefore, a lower temperature of 40 °C is used for the building platform in order to avoid incorrect prints. The following optimum parameters were therefore selected empirically:

Table 4.2 Selected process parameters for the production of green parts in the shaping step.

Parameter	Unit	Value
Extrusion temperature	$°C$	135
Heat bed temperature	$°C$	40
Printing speed	mm/s	35
Extrusion multiplier	-	1.35
Layer height	mm	0.2
Skirt layer	-	2
Retraction distance	mm	4.5
Infill pattern	-	rectlinear
Nozzle Diameter	mm	0.4

4.2.2 Debinding

The debinding step is divided into solvent and thermal debinding, which are described in the following sections.

4.2.2.1 Solvent debinding

The exact binder composition is proprietary and not provided by the manufacturer. Depending on the wall thickness, a solvent-based debinding process in acetone at 42–45 °C for approximately 12 hours is recommended. The resulting mass loss in terms of weight should then be in the range of 4.65 % to 5.35 %.

Different cubes with an edge length of 4 mm to 12 mm were used for the experimental investigation of solvent-based debinding. These were placed in 400 ml of acetone and heated with a hotplate, without distillate recirculation. Initial preliminary tests showed that a defect-free debinding is guaranteed with an acetone ratio of 8:1 concerning the component volume. Similar observations were made for solvent-based debinding [187]. The samples were then heated to temperatures of 30 °C, 40 °C, 42 °C, and 50 °C and weighed after 3, 6, and 9 hours to determine the mass loss in %. Figure 4.2 illustrates the mass loss over time at different temperatures. It can be observed that the minimum debinding quantity specified by the manufacturer was only reached by the cube with a 4 mm edge length after 9 hours at the temperature of 30 °C. At a heating temperature of 40 °C, samples with an edge length of up to 7 mm were able to exceed this threshold at the same time. At a temperature of 42 °C, the cubes with a edge length of 10 mm showed a mass loss of approximately 4.65 % after 9 hours. At a temperature of 50 °C, all samples exceeded this threshold value of mass loss after just 6 hours. Note that at a temperature of

42 °C, cracks were already visible in the samples with a 6 mm edge length after 3 hours, which indicates an anisotropic temperature distribution and thus thermal stress in the solvent [43]. To successfully prevent this error, the temperature was set to 40 °C, and the debinding time was extended to 12 hours. All samples exceeded the threshold value after 12 hours at this temperature, indicating that a temperature of 40 °C and a debinding time of 12 hours with an acetone ratio of 8:1 were selected as optimal process parameters.

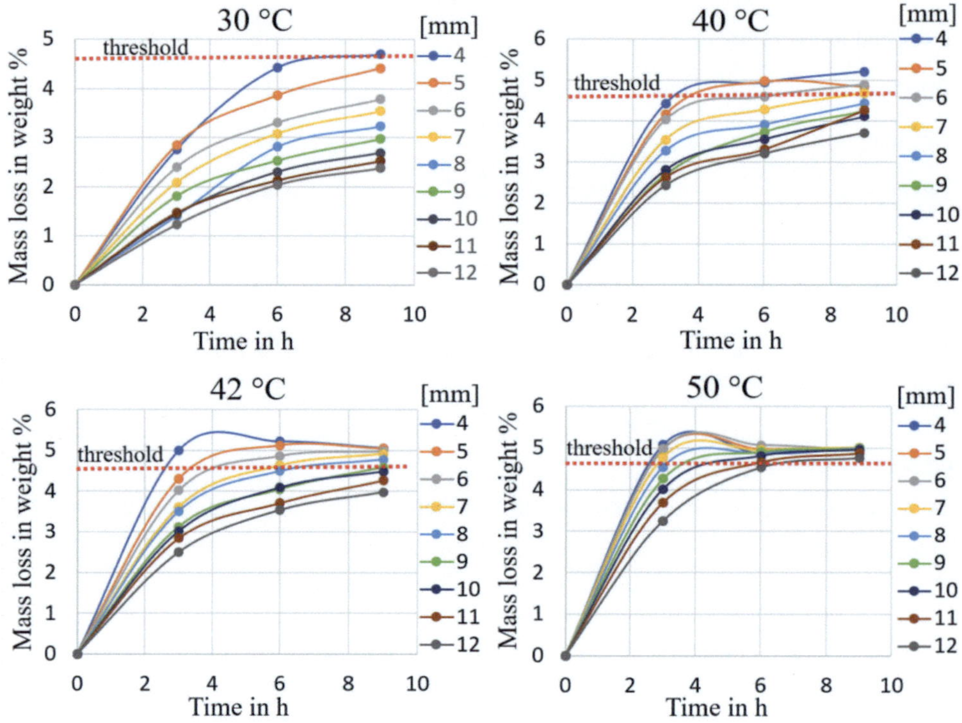

Figure 4.2 Percentage weight loss as a function of time for the 4-12 mm cubes at temperatures of 30 °C, 40 °C, 42 °C, and 50 °C. The red dashed line represents the threshold value of 4.65 % mass loss.

4.2.2.2 Thermal debinding

The manufacturer recommends carrying out thermal debinding in a hydrogen atmosphere with two holding times: 350 °C for 150 minutes and 550 °C for 150 minutes.

Chapter 2.1.1.3 shows that TGA analyses are carried out to assess the thermal debinding process and determine the binder reduction. A TGA analysis is therefore carried out to determine the holding temperatures. Two 50 mm long pieces of filament were used as samples, one of which was debinded in the solvent and the other not. Figure 4.3 illustrates the percent weight loss as a function of the temperature of the two filament variants. It can be observed that the binder starts to evaporate at a temperature of approximately 200 °C for both variants. The last binder component is removed at approximately 470 °C for the non-debinded filament and at approximately 440 °C for the debinded filament. The mass loss at this stage is approximately 7.5 (weight) % and approximately 3 (weight) %, respectively. It can be seen that the thermal debinding process is already complete due to the previous solvent-based debinding at lower temperatures, which shortens the debinding time. However, as the debinding process depends on the wall thickness of the components, it is expected that the binder is completely burned out at a temperature of 550 °C. The manufacturer's recommendations are therefore used to select the debinding temperature and the holding time.

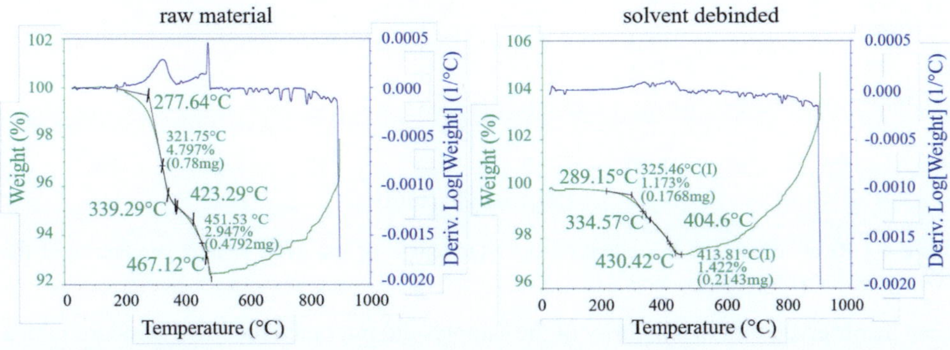

Figure 4.3 TGA analysis of the non-debinded 316L filament (left) and the solvent-based debinded filament (right).

4.2.3 Sintering

The filament manufacturer recommends sintering at 1380 °C for 90 minutes with a hydrogen atmosphere. The temperature increase should be 6 °C per minute.

Four different temperatures (1250 °C, 1286 °C, 1315 °C, and 1380 °C) were investigated to determine the optimum sintering temperature. The temperature increase is based on the recommendation of the filament manufacturer. The investigated mechanical properties include shrinkage, microstructure, density, and tensile strength.

4.2.3.1 Density

For the density tests, cubes with an edge length of 3 mm to 12 mm were sintered at the four temperatures. While density is primarily a material property, different edge lengths were examined to assess whether the printed samples exhibited any density variation using the selected process parameters in the shaping process. Figure 4.4 illustrates the measured densities, determined using Archimedes' principle in ethanol, as a function of the edge length.

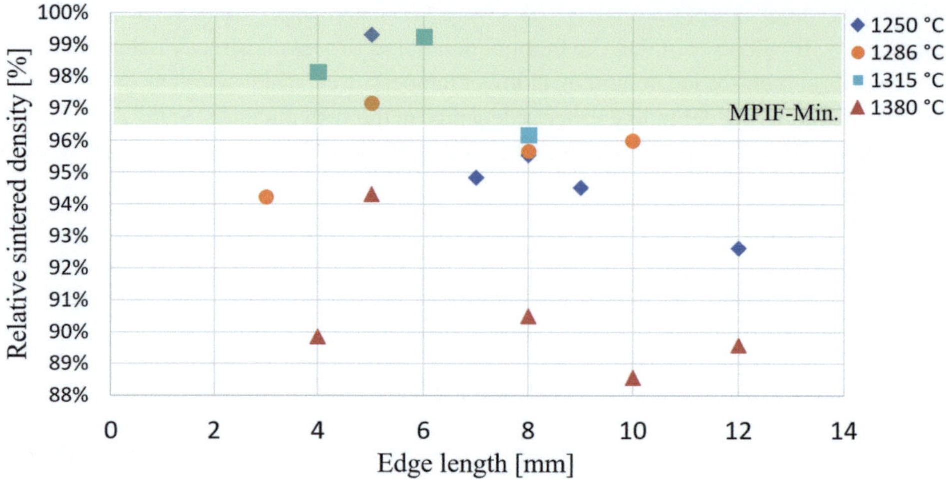

Figure 4.4 Relative density dependent on the edge length of the 316L sample geometries at the investigated sintering temperatures.

It can be observed that the majority of the samples did not reach the MIM standard with a density of 96.6 %. This observation was already made in Chapter 2.1.1.4. Only samples with an edge length from 4 mm to 6 mm and sintering temperatures of 1286 °C to 1315 °C achieved a density of over 96 %. Furthermore, it can be recognized that the 1380 °C sintering temperature results in the lowest densities regardless of the edge length. These samples swelled during the sintering process and exhibited severe deformation. One possible cause of the deformation is that the inert gas was trapped in the pores due to the excessively high temperature, and the thermal gradient within the sample contributed to the deformation [188,189]. The resulting pressure difference favors the expansion of the gas, causing the sintered particles to be torn apart, thereby lowering the density.

4.2.3.2 Microstructure

To examine the microstructure, the cubes were metallographically prepared and then analyzed using an incident light microscope. Figure 4.5 shows an exemplary microstructure of the ground samples perpendicular to the build-up direction (Figure 4.5a and 4.5b) and parallel to the build-up direction (Figure 4.5c) at a sintering temperature of 1250 °C. It can be seen that the grain sizes are similar irrespective of the build-up direction and that there are no precipitates. The only observation is that with larger geometries, more pores appear in the grains and at the grain boundaries compared to smaller geometries.

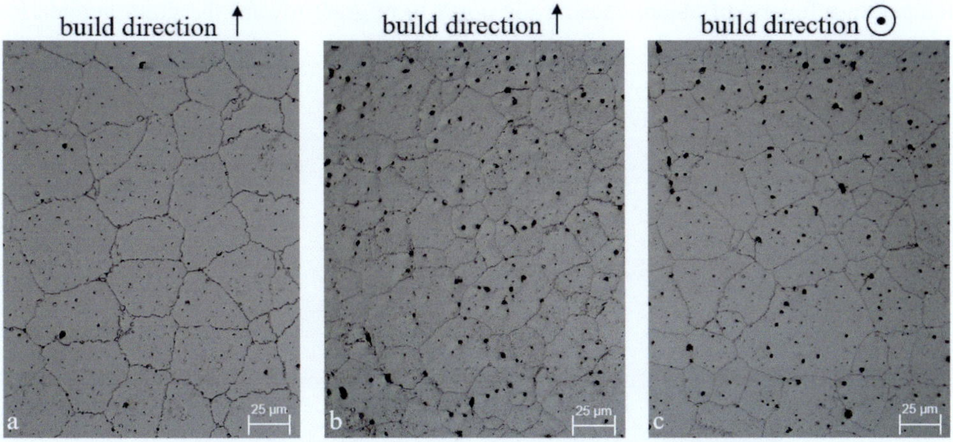

Figure 4.5 Microstructure at the sintering temperature of 1250 °C for the polished samples: edge length a) 7 mm, b) 12 mm, c) 9 mm.

The microstructure at a sintering temperature of 1315 °C is generally more inhomogeneous than at a sintering temperature of 1250 °C, as different grain sizes and orientations can be observed to a greater extent. This can be observed regardless of the build-up direction (compare fig. 4.6a and fig. 4.6c). This is an indication of an uneven temperature distribution within the sintering chamber, which creates a thermal gradient and can lead to component deformation. Furthermore, it can be observed that precipitates have formed along the grain boundaries. The phases from the solid solution result in an equilibrium state at this temperature, leading to precipitation.

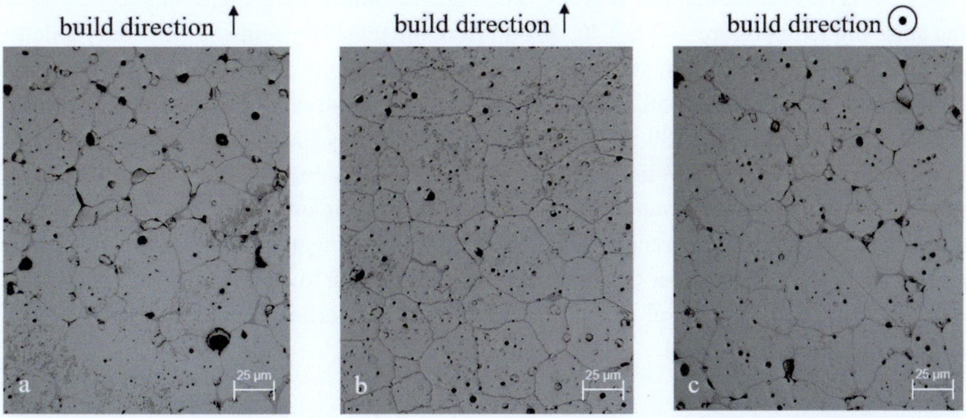

Figure 4.6 Microstructure at the sintering temperature of 1315 °C for the polished samples: edge length a) 4 mm, b) 8 mm, c) 6 mm.

Figure 4.7 illustrates the microstructure at a sintering temperature of 1380 °C. The microstructure is similar in both build-up directions and shows clear grain and pore growth. This growth contributes to the deformation of the samples and their low density. At this temperature, the pores are increasingly found at the grain boundaries. The high temperature favors diffusion, causing the pores to migrate toward the grain boundaries, where they become trapped.

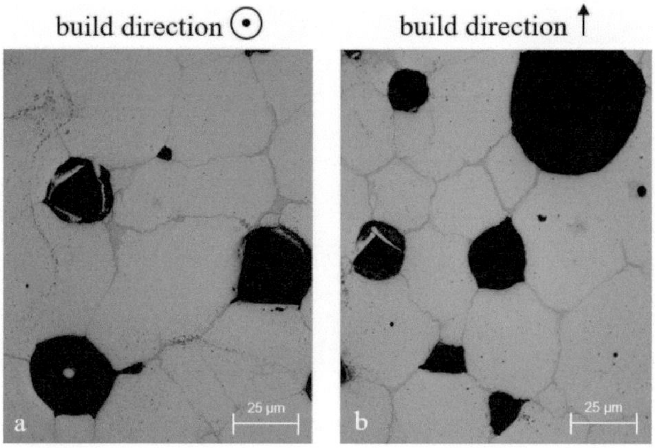

Figure 4.7 Microstructure at the sintering temperature of 1380 °C for the polished samples: edge length a) 12 mm, b) 10 mm.

4.2.3.3 Tensile strength

The tensile specimens were manufactured and tested in accordance with ISO 6892-1:2019(en) (see Appendix A.2, Figure A.1). The tensile tests were carried out on an MTS 810 hydropulser test rig (MTS Systems GmbH, Berlin, Germany). Figure 4.8 shows the tensile strength and elongation at break determined for the tested sintering temperatures. It can be seen that no tensile specimen was able to achieve the minimum requirement of MPIF standards for elongation at break. The highest value here is approximately 22 % at the sintering temperature of 1250 °C. In addition, the tensile specimens were only able to achieve the MPIF minimum standard in terms of ultimate tensile strength at a sintering temperature of 1250 °C at 582 MPa. The lowest tensile strengths were determined at a sintering temperature of 1380 °C. This can be explained by the increase in grain size and the larger pore formation. With finer and more uniform microstructures, the dislocations migrate faster to the grain boundaries and prevent the sliding processes, resulting in higher strength [190]. An inverse trend in tensile strength and elongation at break can be seen in this series of tests. The higher the sintering temperature, the lower the tensile strength and elongation at break. A general increase in mechanical properties can be achieved by using a nitrogen atmosphere [191].

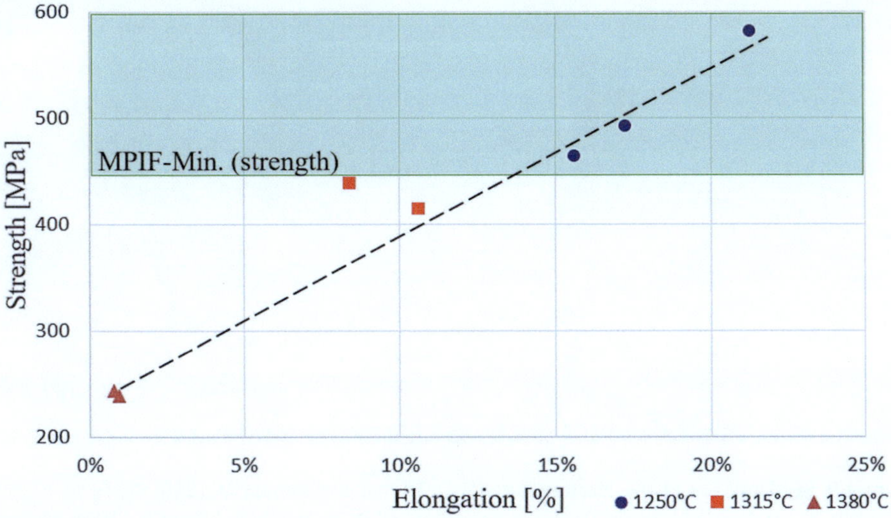

Figure 4.8 Tensile stress versus elongation diagram of the investigated tensile specimens at the different sintering temperatures.

SEM images were also taken of the fracture surfaces of the tensile specimens, which are displayed in Figure 4.9. Figure 4.9a illustrates the fracture surface of the tensile specimen at 1250 °C, and Figure 4.9d depicts a detailed view of the surface. Figures 4.9b and 4.9e represent the fracture surface and its detailed view at 1315 °C. Finally, Figures 4.9c and 4.9f show the fracture surface and its detailed view at 1380 °C. Generally, it is even clearer that the grain size and pore size increase as the temperature rises. The detailed views reveal that the number of pores decreases with increasing temperature. For instance, Figure 4.9d illustrates a large number of micropores, which only appear sporadically in Figure 4.9e. The examination of the fracture surfaces in Figures 4.9e and 4.9f shows that these are purely intercrystalline cleavage fractures. In this case, the components break at the grain boundaries, which are contaminated by dirt or the larger appearance of precipitates. These phenomena are also called embrittled grain boundaries [192]. In addition to the intercrystalline cleavage fracture, the fracture surface at the sintering temperature of 1250 °C also exhibits regions where honeycomb fractures occur. The material breaks directly within the grains, and honeycombs are formed on the fracture surface after the sample is destroyed. These honeycomb fractures (also known as sliding fractures) occur in tough materials and at low loading speeds, which contribute to better material properties in terms of ultimate tensile strength [192].

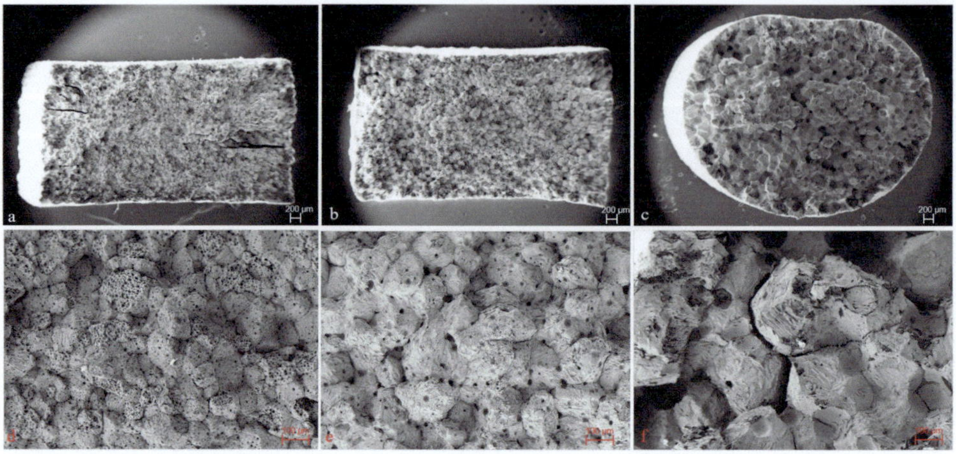

Figure 4.9 SEM images of the fracture surfaces: a) tensile specimen at 1250 °C, b) at 1315 °C, c) at 1380 °C, d) detailed view of the fracture surface of a), e) detailed view of the fracture surface of b), and f) detailed view of the fracture surface of c).

4.2.3.4 *Shrinkage*

The filament manufacturer specifies a scaling factor of 1.1705 to 1.1825 for the production of components in the MEX/M process. This corresponds to a shrinkage of 17.05 % to 18.25 %, which would be in the range of typical shrinkage for MIM components. The measurements of the cube components with a caliper at the edges and in the center showed that the shrinkage does not increase as the edge length increases, but fluctuates around a value. The average shrinkage at 1250 °C is 13.21 % in the x-direction (parallel to the build-up direction), 12.69 % in the y-direction (parallel to the build-up direction) and 14.07 % in the z-direction (perpendicular to the build-up direction). At the same sintering temperature, the mean shrinkage of the tensile specimens is 13.97 % in the x-direction, 14.27 % in the y-direction and 14.63 % in the z-direction. It can be recognized that the shrinkage perpendicular to the build-up direction is higher, on average, than in the other dimensions, which is typical for the process chain. The average shrinkage at 1315 °C is 13.54 % for the cubes in the x-direction, 12.69 % in the y-direction, and 14.64 % in the z-direction. The mean shrinkage of the tensile specimens is 14.62 % in the x-direction, 14.71 % in the y-direction and 14.98 % in the z-direction. The same behavior of the anisotropic dimensional shrinkage can be observed in the z-direction. It can clearly be determined that the shrinkage is higher in all directions than at a sintering temperature of 1250 °C. However, all measured data is within the typical MIM shrinkage and below the manufacturer's specifications. Due to the swelling and deformation of the samples at a maximum sintering temperature of 1380 °C, no shrinkage could be measured.

In summary, the MIM properties could be achieved, in some cases, for lower sintering temperatures. Extending the sintering time can have a positive effect on the material properties. Therefore, the sintering temperature of 1286 °C with a sintering time of 180 minutes is selected for the subsequent investigations.

5. Design analysis

This chapter examines the design for all subprocess steps in the MEX/M process and investigates features that are likely to occur in functional components.

5.1 Green part feature manufacturing

The following chapter is mainly based on the results from source [193], which is available as open access at [194] (see Appendix B.1). Figure 5.1 illustrates the investigated design features with their respective variations, which are commonly used in functional products. The features were measured at the edges and in the middle of the samples using a caliper and a feeler gauge. Microscope measurements of the roughness values were taken from the downskin surface.

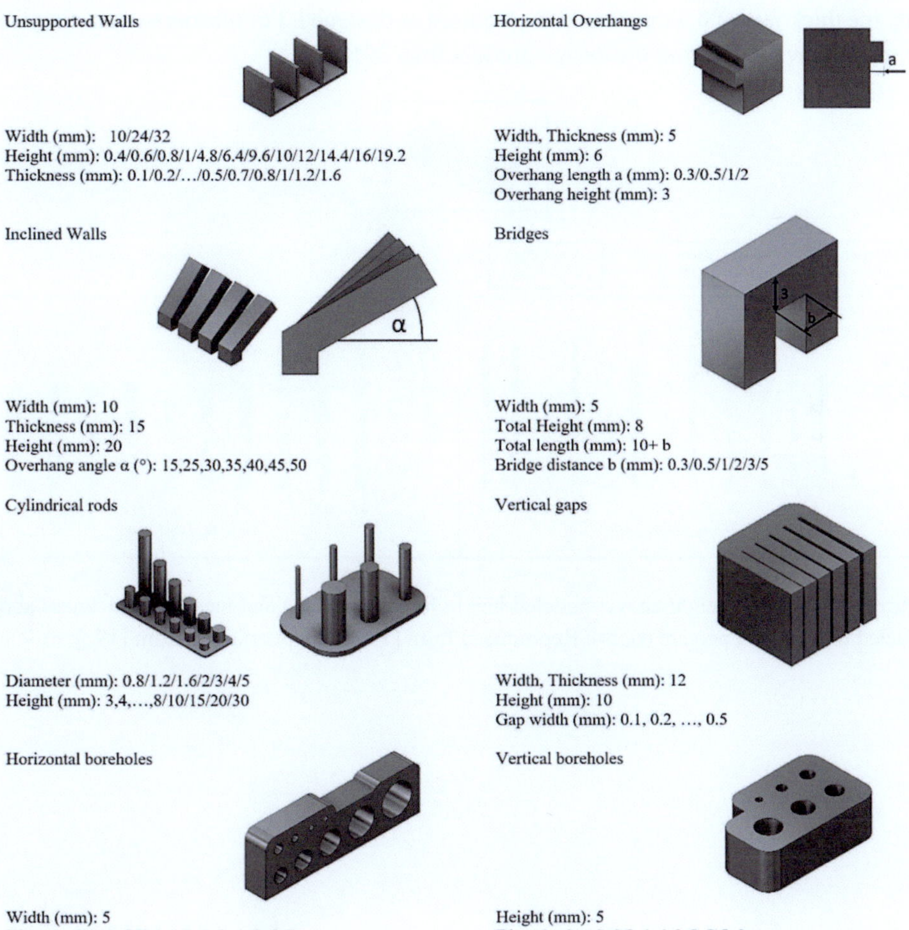

K.Asami, *Analysis of Sinter-Based Additive Manufacturing for Functional Products*, Light Engineering für die Praxis, https://doi.org/10.1007/978-3-032-28433-4_5

Figure 5.1 The design features investigated were manufactured using the MEX/M process with the material 316L. Reprinted from [194] with permission from [193].

5.1.1 Unsupported walls

Unsupported walls were manufactured with widths ranging from 1.2 mm to 10 mm and lengths from 10 mm to 32 mm, with different heights, as depicted in Figure 5.2 (left). In addition, a variation of thicknesses was measured for walls with a width and length of 10 mm (fig. 5.2 right). Observations of the investigated heights lead to the conclusion that all unsupported walls could be built, and from a height of 0.6 mm, the deviation in height is less than 5 %. This finding is independent of the wall's length and width. It is noticeable that walls with a width of 1.2 mm have a reduced height starting at 12 mm. In this case, the height-to-width ratio of 8:1 is exceeded. Considering the thickness, it can be stated that, except for the 0.7 mm sample, all samples were manufactured thicker than designed. Only the thickness of 0.7 mm was manufactured as designed. For thicknesses starting from 1 mm, the deviations from the design are less than 20%.

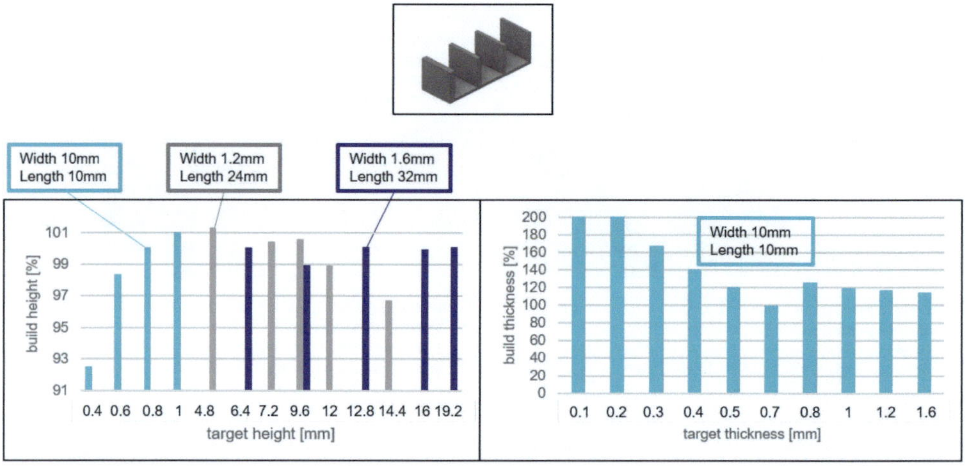

Figure 5.2 Measured height above modeled height in percent (left) and measured thickness above modeled thickness in percent (right). Reproduced from [194] with permission from [193].

5.1.2 Bridges, overhangs, and inclined walls

All variations of the distances between the bridges could be produced (see fig. 5.3). Deformations in the overhang area are evident starting from a bridge length of 2 mm. A closer examination of an overhang length of 1 mm showed a tendency toward deformation.

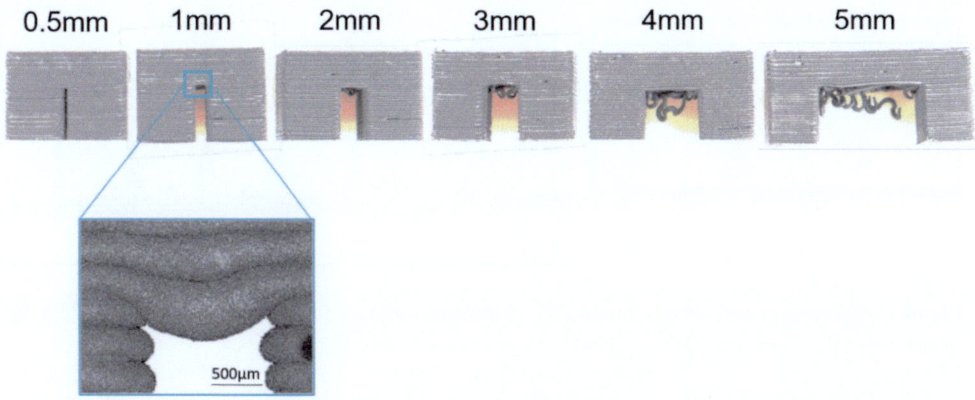

Figure 5.3 Manufactured bridge distance sizes. Reproduced from [194] with permission from [193].

Overhang samples supported on one side have been produced with overhang lengths ranging from 0.5 mm to 2 mm and are illustrated in Figure 5.4. It should be noted that the filament layer sinks when the overhang length is 0.5 mm. The approach of using different infill patterns to prevent this deformation has failed.

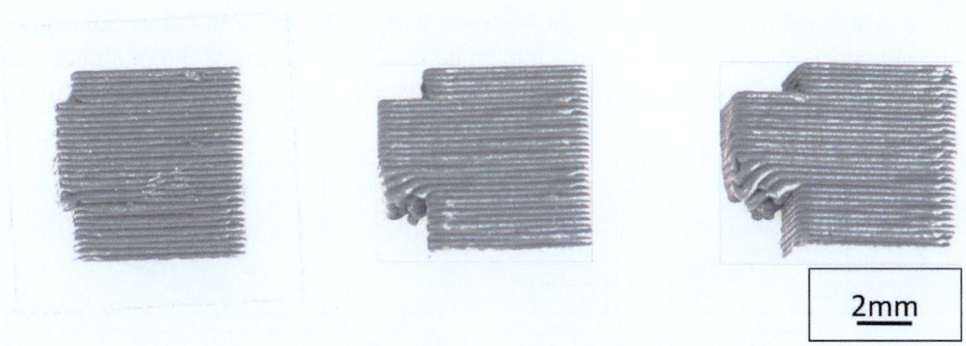

Figure 5.4 Manufactured horizontal overhangs supported on one side. Overhang distances are 0.5 mm (left), 1 mm (center), and 2 mm (right). Reproduced from [194] with permission from [193].

The inclined walls have been manufactured from 0 to 50°, measured at the downskin surface, and the measured average surface roughnesses (R_z) are presented in Figure 5.5. With the exception of an overhang angle of 0°, all test geometries could be produced from an overhang angle of 15°. The average surface roughness at this angle is 376.6 μm. The roughness is significantly lower at 25° with 227.7 μm, compared to 30°, which has an average of 423.6 μm. As the surface roughness tends to decrease with increasing angle from 35°, it can be assumed that the roughness measured at 25° must be an outlier. The measured roughness for 35° is 179.94 μm, and for 50°, it is 110.45 μm.

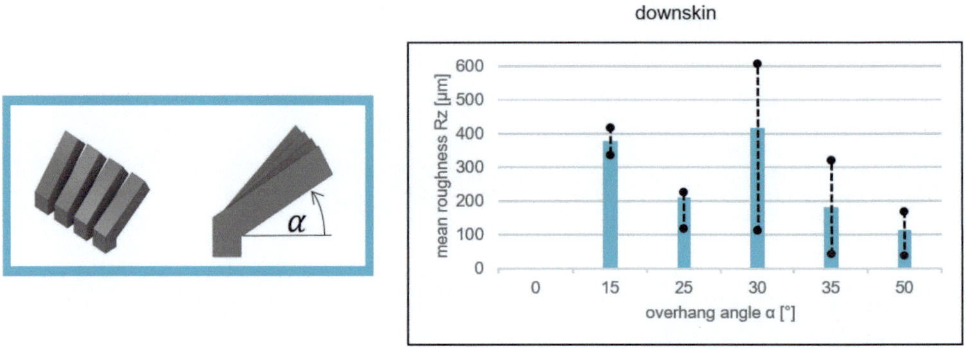

Figure 5.5 Average surface roughness (R_Z, downskin area) of the inclined wall from 0° to 50° overhang angle. Reproduced from [194] with permission from [193].

5.1.3 Cylindrical rods

Different height segments with varying thicknesses were produced from the cylindrical features and are illustrated in Figure 5.6 (left). The small cylindrical structures were designed up to 8 mm in height, and the large ones up to 30 mm. It should be noted that all sample geometries could be produced. At lower heights, it can be seen that the cylinders are larger than designed. With the larger cylinder samples, the deviation decreases significantly, so that there is a maximum deviation of 0.5 %. Furthermore, the thickness of the cylindrical rods at a fixed height was examined, as illustrated in Figure 5.6 (right). The rod thicknesses from 0.8 mm to 2 mm could not be manufactured. From a thickness of 3 mm, the thicknesses are consistently 3 % smaller than the designed cylindrical rods.

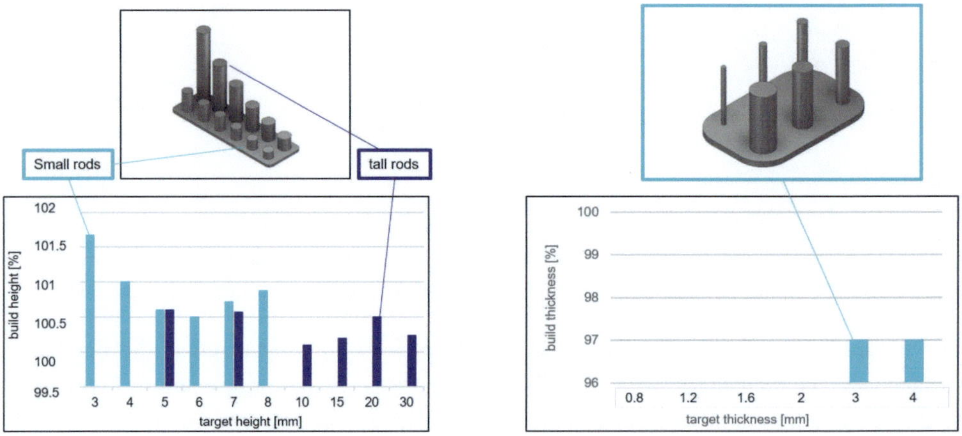

Figure 5.6 Measured height as a percentage of the designed height (left) and measured thickness as a percentage of the designed thickness (right) of the cylindrical rods. Reproduced from [194] with permission from [193].

5.1.4 Vertical gaps

The vertical gaps were modeled with widths ranging from 0.1 mm to 0.5 mm. It should be noted that gaps up to 0.2 mm width could not be produced. An examination of these areas showed that the individual layers were extruded together. Starting from a vertical gap width of 0.3 mm, the deviation of the gaps decreases steadily with increasing width. With a vertical gap of 0.4 mm, the deviation is approximately 60 %. One explanation for this is the extrusion multiplier, which was set at 35 % over-extrusion. If the over-extrusion occurs on both edges of the vertical gaps, the reduction in the actual width is theoretically 70 %.

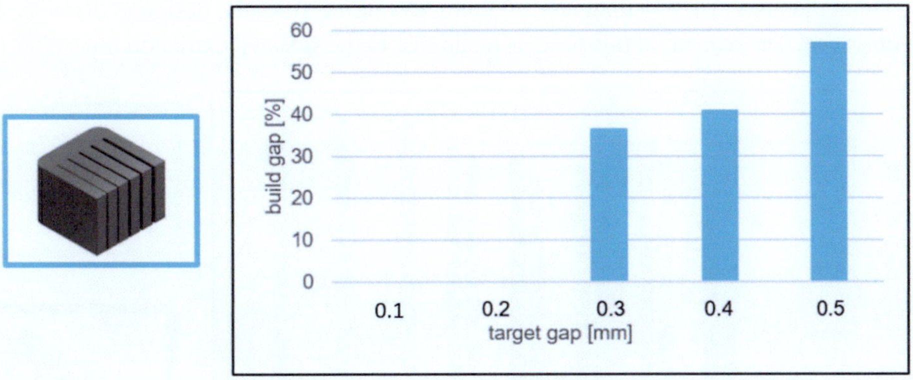

Figure 5.7 Actual vertical gap width as a percentage of the designed vertical gap width. Reproduced from [194] with permission from [193].

5.1.5 Boreholes

Boreholes were manufactured in different diameters in the vertical and horizontal directions to the building platform, and are shown in Figure 5.8. Boreholes that were produced perpendicular to the build direction were only possible in the range from 1 mm to 4 mm. Overall, the measured diameters were smaller than the modeled diameters. This deviation decreases with increasing diameter. From a diameter of 4 mm, the material starts to collapse like it does with the bridges, and the use of support structures is necessary. The use of a teardrop shape is also conceivable, as was investigated in the case of [115]. The boreholes parallel to the build direction have all been manufactured except for those with a 0.8 mm diameter. Up to a diameter of 5 mm, deviations from the designed diameter can be observed. The reason, in this case, is again due to the selected extrusion multiplier.

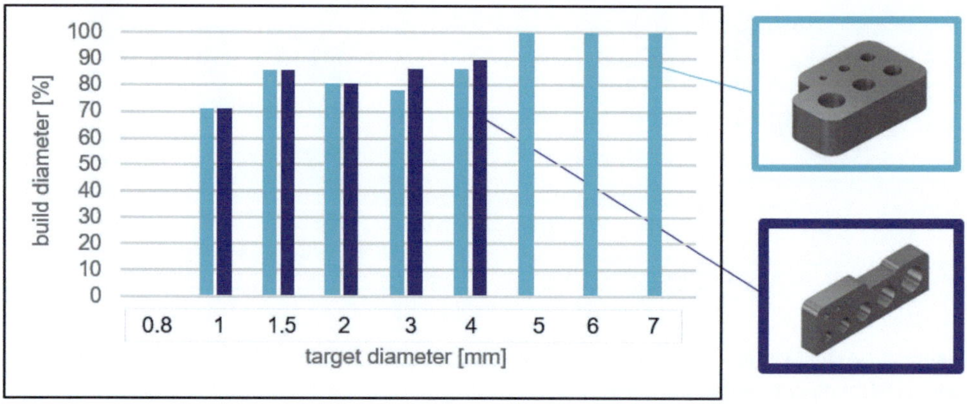

Figure 5.8 Produced diameter as a percentage of the designed diameter of the boreholes, both parallel and perpendicular to the build direction. Reproduced from [194] with permission from [193].

The identified boundary conditions from the examined design features serve as the basis for the development of the design guidelines which were published in [193] and [194].

5.2 Debinded and sintered feature manufacturing

This chapter is based on the publication [195] (see Appendix B.2). The identified boundary conditions for green part production were taken into account, and new design features were produced that consider the entire process chain. Only the vertical gaps and the boreholes parallel to the build direction were omitted, as these can be manufactured true to size using a scaling factor. All test results, illustrations, measurements, and design features have been made available in an open-source file [196]. Figure 5.9 shows the investigated design features and variations.

An investigation of the shrinkage behavior using a caliper at the edges and in the middle of all samples after the shaping, debinding, and sintering process confirmed that the shrinkage mainly takes place after sintering. After solvent-based debinding, the mean shrinkage is 1.8 % with a standard deviation of 0.2 % along the x- and y-direction. In the z-direction, the mean shrinkage is 0.8 % with a standard deviation of 0.3 %. In this case, the smaller shrinkage in the z-direction can be explained by the fact that fewer binders were ejected in this axial direction. The shrinkage after sintering has increased as expected and is 15.5 % with a standard deviation of 0.4 % in the x- and y-directions and 14.9 % in the z-direction with a standard deviation of 0.2 %. Contrary to expectations, the shrinkage in the z-direction is smaller than in the other dimensions. One possible explanation could be that the interlayer connection showed more voids, which were positioned against the shrinkage in the z-direction.

The features have been examined with a 3D scan. For the qualitative and quantitative investigation, the sample geometries were dimensioned with a scaling factor of 1.183 in the x- and y-directions and 1.176 in the z-direction. The scaling factors are within the range specified by the manufacturer as outlined in Chapter 4.2.3.4.

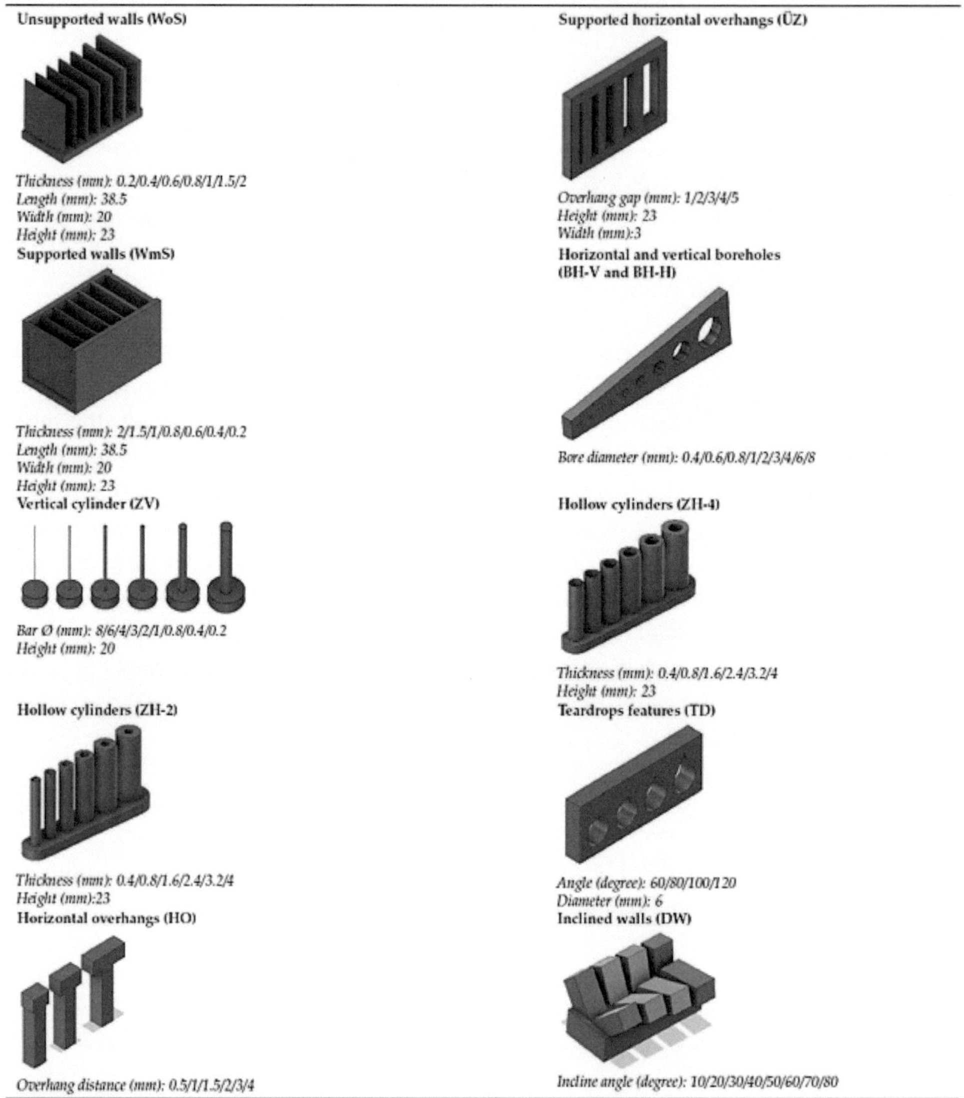

Figure 5.9 Selected design features for the development of design guidelines in the MEX/M process. Reproduced from [195] (licensed under CC BY 4.0).

5.2.1 Unsupported and supported walls

Unsupported walls with a wall thickness of 0.6 mm could be built, as illustrated in Figure 5.10. A deviation of 0.2 mm is visible at the upper end of the wall, as previously recognized in Chapter 5.1.1. After the solvent-based debinding, all walls were deformed, making it impossible to measure the deviation. The deformation was due to an uneven temperature distribution, which led to an anisotropic diffusion process. In principle, walls with a thickness of 0.6 mm can be manufactured; however, during a controlled debinding and sintering process, dimensional shrinkage affects the walls.

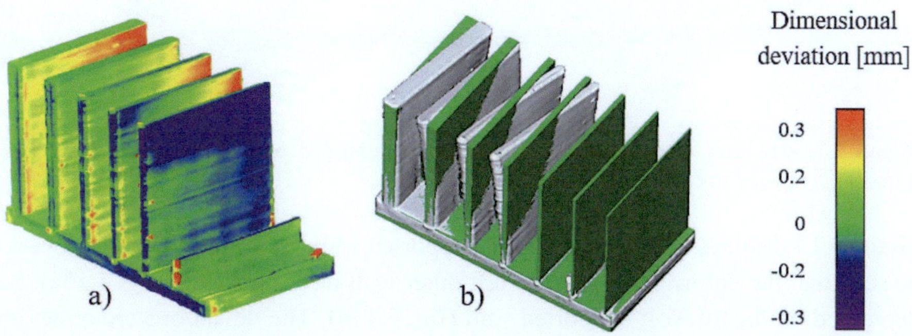

Figure 5.10 3D scan of the unsupported walls: a) Green part, b) Debinded part. Reproduced from [195] (licensed under CC BY 4.0).

The supported walls could be built up from 0.2 mm and the green part is illustrated in Figure 5.11. The green part of the supported walls shows that the deviations from the modeled component are mostly below 0.2 mm. The deviations on the top right-hand side are in the range of 0.2 mm to a maximum of 0.3 mm.

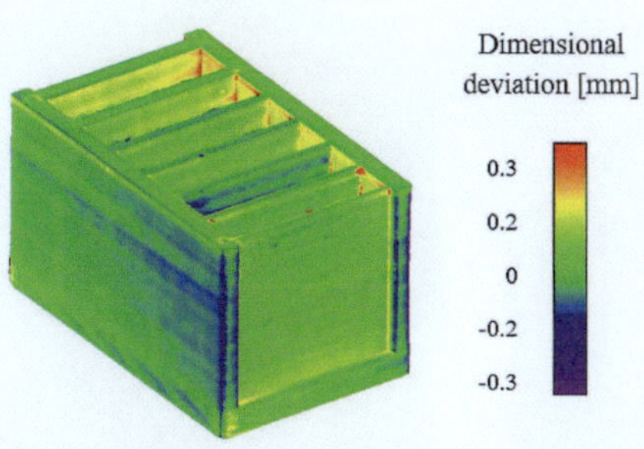

Figure 5.11 3D scan of the supported walls as a green part. Reproduced from [195] (licensed under CC BY 4.0).

After solvent-based debinding, a deviation of the rear and front walls is more noticeable, as shown in Figure 5.12. The dimensional deviation is 0.3 mm for both the outer walls, the thickest 2 mm walls, and the thinnest 0.2 mm supported walls. The mid walls exhibit a maximum deviation of 0.2 mm.

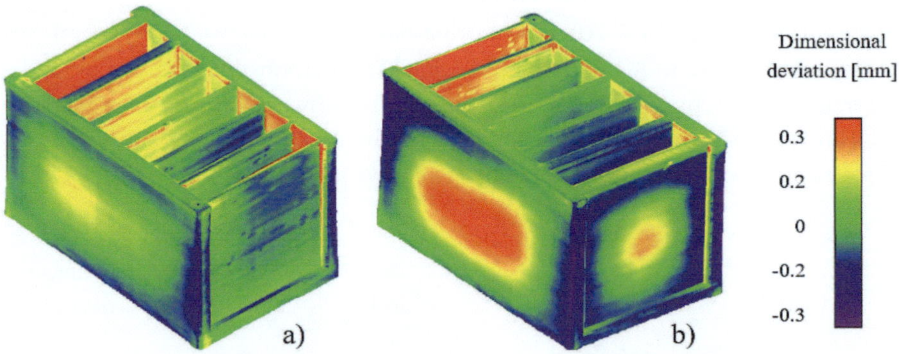

Figure 5.12 3D scan of the supported walls as a debinded part: a) Front view b) Rear view. Reproduced from [195] (licensed under CC BY 4.0).

Figure 5.13 displays the 3D scan of the supported walls after the sintering process. It can be seen that the deformations on the left outer wall have increased (fig. 5.13a), but have decreased on the thickest supported wall (fig. 5.13b). The deviations are a maximum of 0.4 mm for the walls with a thickness of 0.2 mm. From a thickness of 0.6 mm, the deviation is reduced to a maximum of 0.2 mm. The walls with a thickness of 0.8 mm show deviations of up to a maximum of 0.2 mm. At a wall thickness of 1 mm, the deviation reaches 0.2 mm again and is more noticeable at 1.5 mm. The 2 mm wall thicknesses again show a deviation of 0.3 mm. Although these walls could be produced with a thickness of 0.2 mm, the deviation is large at 0.4 mm, which corresponds approximately to a deviation of one nozzle width.

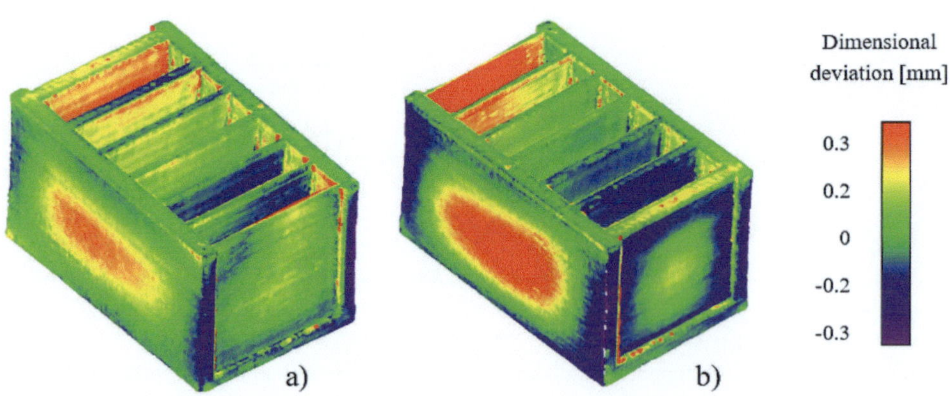

Figure 5.13 3D scan of the supported walls as a sintered part: a) Front view b) Rear view. Reproduced from [195] (licensed under CC BY 4.0).

5.2.2 Cylindrical structures

The production of the cylindrical structures was successful from a diameter of 0.6 mm, as shown in Figure 5.14. After printing, deviations of more than 0.3 mm in the cylindrical rods, ranging from 0.2 mm to 0.6 mm, are recognizable (fig. 5.14a). At 0.8 mm, the deviation is only 0.3 mm at the top area. A 0.2 mm cylinder distortion can be seen during debinding, so no measurement could be made (fig. 5.14b). After sintering, the deviation decreases noticeably for larger diameters. For a diameter of 0.8 mm, an average deviation of 0.2 mm is recorded in the z-direction due to shrinkage (Fig. 5.14c).

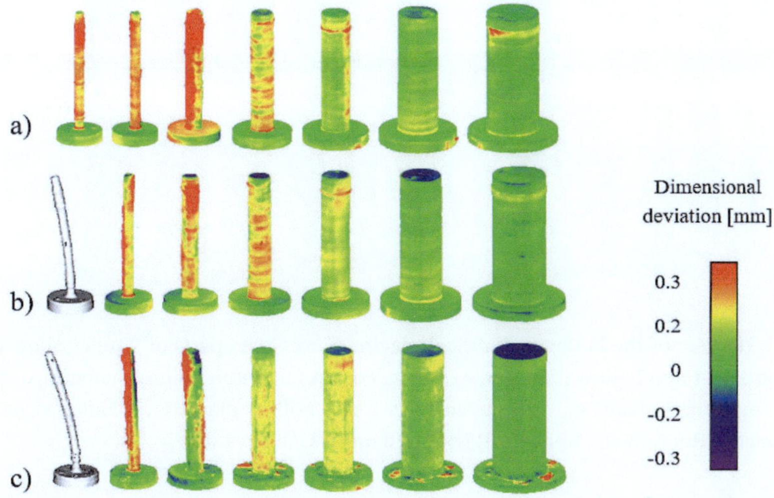

Figure 5.14 3D scan of the cylindrical specimen: a) Green parts b) Debinded parts c) Sintered parts. Reproduced from [195] (licensed under CC BY 4.0).

5.2.3 Hollow cylinders

The hollow cylinders were manufactured with a hollow diameter of 2 mm and 4 mm, which are illustrated in Figure 5.15. In the green state, the outer hollow cylinders showed a slight inward inclination towards the center axis of the component, and this deformation was more pronounced in the larger samples (fig. 5.15a-b). These deviations of 0.3 mm at the upper side edges occur primarily with the smallest and largest wall thicknesses. The deviation becomes even clearer during debinding, primarily affecting the lateral surface (fig. 5.15c-d). After sintering, the deformations of the 2 mm hollow diameters can be recognized at the wall thicknesses of 0.4 mm and 0.8 mm, with a deviation of more than 0.3 mm. These structures are bent due to the height-to-width ratio during the debinding stage and cannot be produced defect-free at this height. Similar deformation can also be observed with the 4 mm wall thickness. A lower debinding temperature, which inevitably extends the debinding duration, can prevent this deformation. With a 4 mm hollow diameter, bends can also be detected at the edges of the samples. However, they are less noticeable than with the 2 mm diameter.

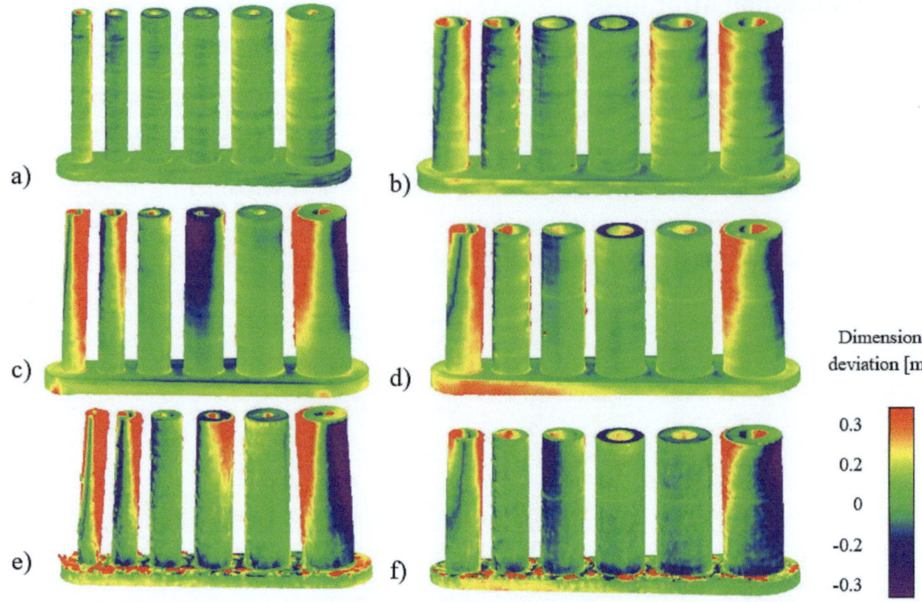

Figure 5.15 3D scan of the hollow cylindrical specimen: a) Green parts of 2 mm hollow diameter, b) Green parts of 4 mm hollow diameter, c) Debinded parts of 2 mm hollow diameter, d) Debinded parts of 4 mm hollow diameter, e) Sintered parts 2 mm hollow diameter, f) Sintered parts 4 mm hollow diameter. Reproduced from [195] (licensed under CC BY 4.0).

5.2.4 Horizontal overhangs

Horizontal overhangs supported on one side already exhibit significant deformations from an overhang length of 0.5 mm, as depicted in Figure 5.16. An overhang length of 1 mm indicates considerable sagging of the material.

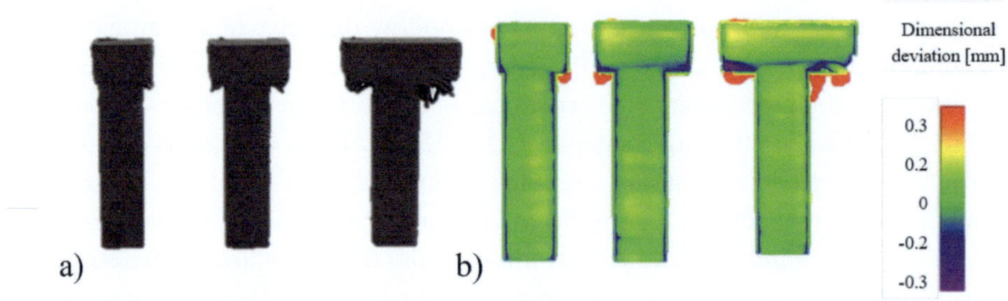

Figure 5.16 One side supported horizontal overhang: a) Green parts as build b) 3D scan of the green parts. Reproduced and adapted from [195] (licensed under CC BY 4.0).

The components could be debinded and sintered without any significant changes to the components being observed. Figure 5.17 illustrates the sintered state of the 0.5 mm overhang.

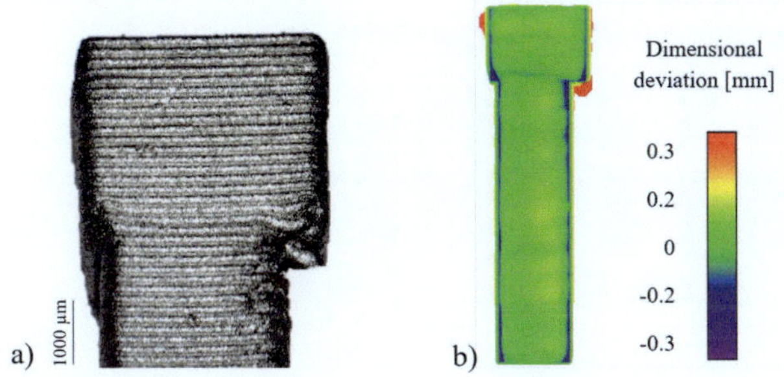

Figure 5.17 One side supported horizontal overhang: a) Sintered part as built, b) 3D scan of the sintered part. Reproduced and adapted from [195] (licensed under CC BY 4.0).

It can be observed from Figure 5.17 that the deformations that occurred in the green part remained after sintering. The primary restriction in this case is the production of the green part.

Two-stage supported overhangs could be produced as green parts without defects for overhang lengths up to 2 mm. At an overhang length of 2 mm, material sinking is observed, as shown in Figure 5.18.

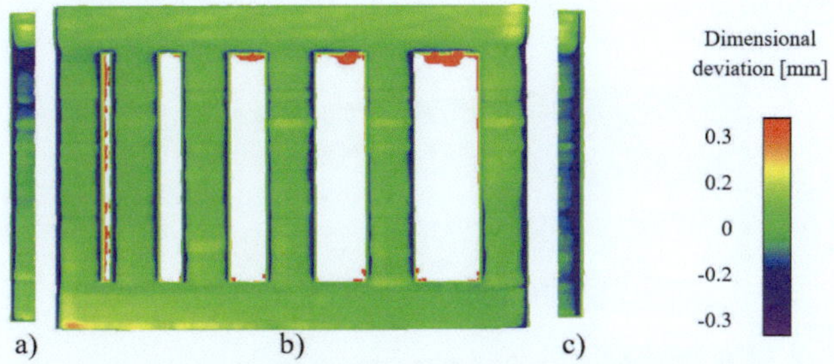

Figure 5.18 3D scan of two sided supported horizontal overhang as a green part: a) Right view, b) Front view, c) Left view. Reproduced and adapted from [195] (licensed under CC BY 4.0).

The subsequent solvent-based debinding and the sintering process do not reduce the deformation. On the contrary, a deviation in the overhangs measuring 1 mm is also observed after the sintering process (see fig. 5.19b). After the debinding process, a deformation of the samples can be seen at the corners, caused by thermal stress (see fig. 5.19a). In each case, two corners exhibit positive bending and the other two corners exhibit negative bending. A comparison of the sintering process of the samples shows that the deformations could be reduced after sintering, though they are still 0.4 mm in the positive direction and 0.37 mm in the negative direction.

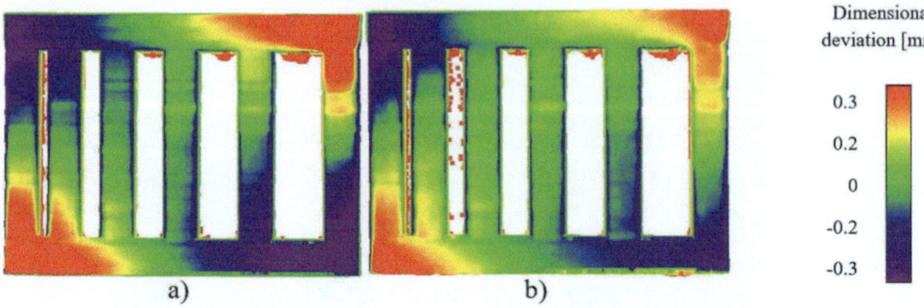

Figure 5.19 3D scan of two sided supported horizontal overhang: a) Debinded part, b) Sintered part. Reproduced from [195] (licensed under CC BY 4.0).

5.2.5 Boreholes and teardrop shapes

Circular boreholes could be successfully produced in the vertical and horizontal directions. Figure 5.20 shows the samples in the respective orientations after each process step of the MEX/M process chain. As already observed in Chapter 5.15, the boreholes exhibit under-extrusion up to a diameter of 3 mm, resulting in an average size approximately 30% smaller than the modeled dimensions, regardless of the orientation (fig. 5.20a,b). From a diameter of 3 mm, the free overhang length decreases after debinding and sintering (fig. 5.20c,e). In the horizontal direction, there is no deviation from the designed diameter of 4 mm (fig. 5.20d,f).

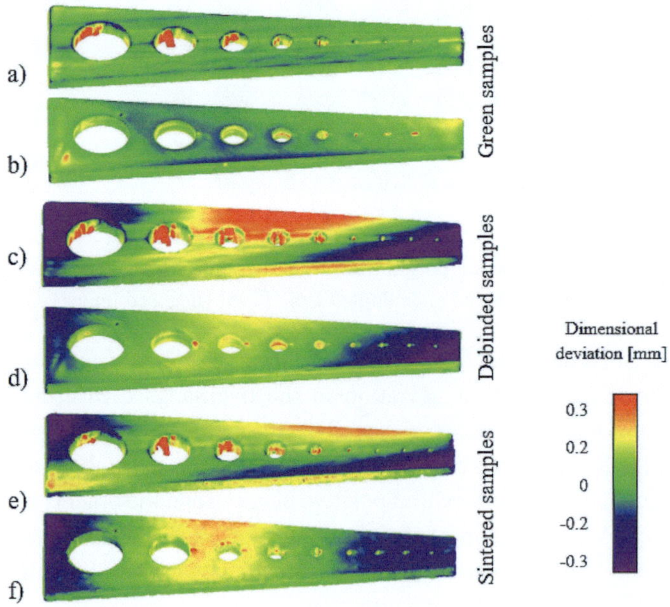

Figure 5.20 3D scan of boreholes manufactured in horizontal and vertical direction: a) Green part vertical direction, b) Green part horizontal direction, c) Debinded part vertical direction, d) Debinded part horizontal direction, e) Sintered part vertical direction, f) Sintered part horizontal direction. Reproduced from [195] (licensed under CC BY 4.0).

The production of teardrop shape boreholes in the vertical direction is feasible without material fall-off for diameters of 6 mm and 14 mm, which is shown in Figure 5.21a and 5.21b. The stair-case effect can be observed in all inclined planes of the teardrop structures, causing a deviation of up to 0.3 mm in these areas.

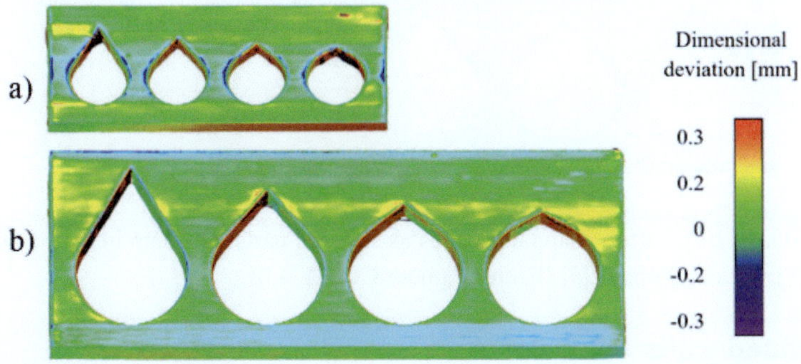

Figure 5.21 3D scan of teardrop shape boreholes as green parts: a) 6 mm in diameter, b) 14 mm in diameter. Reproduced from [195] (licensed under CC BY 4.0).

The debinding of the teardrop shape samples has remained consistently and shows a deviation of up to 0.26 mm at the corners of the sample for the 6 mm diameter (fig. 5.22a). For the 14 mm diameter, deviations of 0.19 mm to 0.35 mm can be observed (fig. 5.22b).

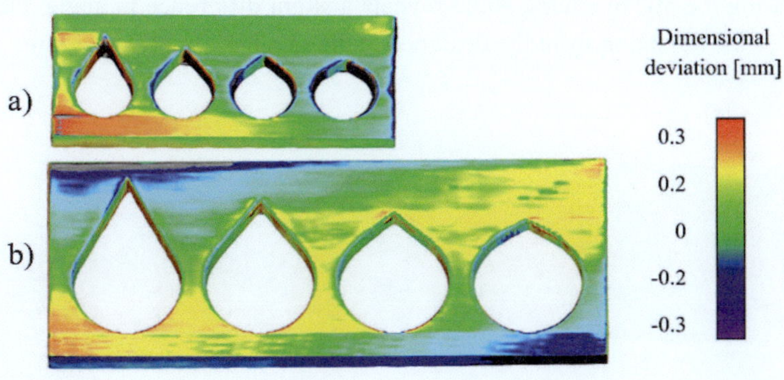

Figure 5.22 3D scan of teardrop shape boreholes as debinded parts: a) 6 mm in diameter, b) 14 mm in diameter. Reproduced from [195] (licensed under CC BY 4.0).

After sintering, the deviations increase, and the teardrop shapes with angles of 120° begin to round off (fig. 5.23a,b). However, the deviation depends on the angle: the steeper the angle of the free overhang, the smaller the deviation.

Figure 5.23 3D scan of teardrop shape boreholes as sintered parts: a) 6 mm in diameter, b) 14 mm in diameter. Reproduced from [195] (licensed under CC BY 4.0).

5.2.6 Inclined walls

Inclined walls have been manufactured at different angles from 10° to 80° in 10° increments and are shown in Figure 5.24. The overhang angles of 10° showed sinking in the individual layers and could therefore not be successfully produced in the green part (fig. 5.24a-b). Angles from 30° to 50° clearly show the staircase effect of the individual layers, resulting in deviations of up to 0.4 mm at 50°. At an overhang angle of 80°, this phenomenon is reduced, and the deviation reaches a maximum of only 0.25 mm. After sintering, shrinkage of the entire sample with the overhang angles is observed (fig. 5.24c-d). Comparing the 50° overhang angle reveals a slight difference in angle. Densification also reduces the deviation from the designed angle to approximately 0.4 mm.

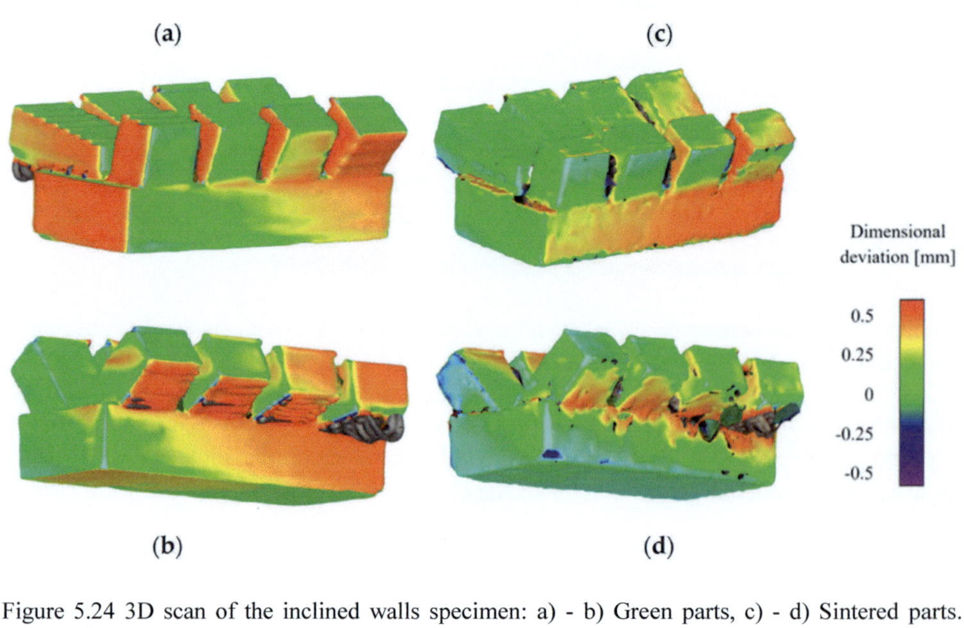

Figure 5.24 3D scan of the inclined walls specimen: a) - b) Green parts, c) - d) Sintered parts. Reproduced from [195] (licensed under CC BY 4.0).

The findings and limitations of the entire process chain have been taken into account for the development of design guidelines, which are published in [195].The validation was successfully carried out on an impeller demonstrator, considering the restrictions related to wall thickness, hollow cylinders, and enclosed walls (see Appendix B.2).

5.3 Infill simulation

The use of infill structures consisting of lattice structures is one way of optimizing the weight of a component without changing the external geometry of the component. Chapter 5.2 has already shown that fine and complex structures are difficult to manufacture in the MEX/M process. Therefore, the investigation is limited to simpler geometries. Chapter 2.3 demonstrates that honeycomb structures in particular have essential advantages in the application. The current chapter is mainly based on publication [167] (see Appendix B.3).

Compression and 3-point bending simulations, according to ASTM E9 [197] and ISO 7438:2020(en) [198], were carried out for the simulative investigation of the infill structures. The test specimens for the simulations were varied based on the infill volume. The dimensions of the contour surface were kept constant for the compression specimens, having a diameter and height of 40 mm. The bending specimens are rectangular, measuring 80 mm in length, with a cross-sectional edge length of 20 mm. The parameterization of the infill structures was carried out using identical unit cells that represent the infill structure and the respective pores. The parametric models are designed in Autodesk Inventor 2021 and are made available in [199]. In addition to the infill volume, the infill structures were also varied in terms of dimensional orientation and wall thickness, as shown in Figure 5.25. The change in angle α rotates the structure around the x-axis, β around the y-axis, and γ rotates the structures around the longitudinal axis of the specimen. The compression specimens were not varied around the y-axis because of their rotational symmetry.

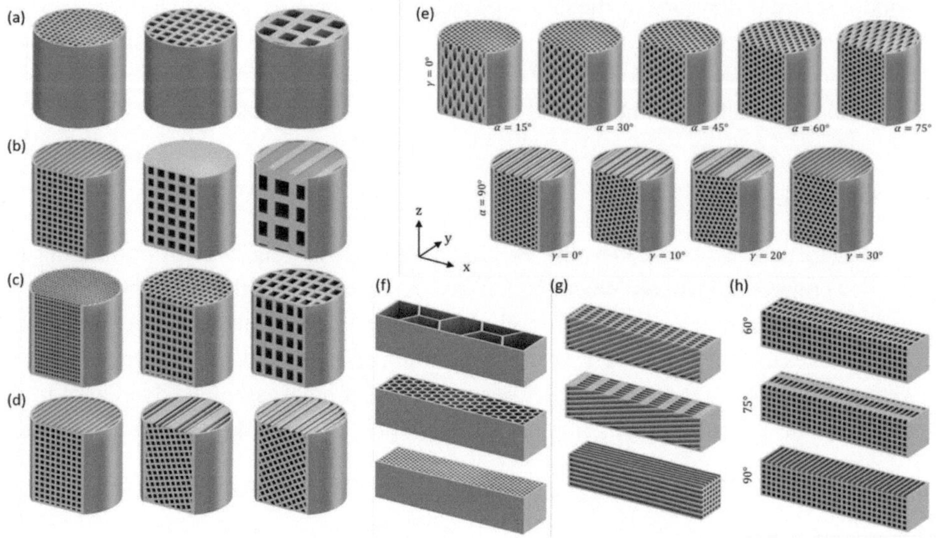

Figure 5.25 Variation of the infill structures in terms of dimensional orientation, infill type (honeycomb and grid), and wall thickness for the simulation tests: (a)-(e) investigated compression samples, (f)-(g) examined 3-point bending specimens. Reproduced from [167] (licensed under CC BY 4.0).

The simulations of the bending and compression tests were carried out using Hyperworks 2021. The element size for the simulation has been set to 2 mm, with a minimum element size of 0.1 mm. Table 5.1 shows the simulation parameters used. Tetra elements were used for the cylindrical test geometries. Depending on the variation, the bending specimens consist of pyramid and tetrahedral elements to ensure the most accurate approximation, respectively. A convergence analysis was carried out to validate the mesh. Ninety-six bending simulations and eighty-four variations of the cylindrical compression specimen were simulated. The loads selected for the compression simulations were 50 kN, 100 kN, and 200 kN. 20 kN were selected for the loads of the bending simulations. AISI 316L stainless steel with the material properties in Table 5.1 and a linear elastic model were selected for the simulations.

Table 5.1 Material characteristics and simulation parameters.

Element type	Tetra, tetrahedra, pyramid
Element size	0.01 – 2 [mm]
Young's modulus	200 [Gpa]
Density	7.9 [g/cm^3]
Poisson's ratio	0.27

The compression simulation for honeycomb and grid infill structures was performed with a 1 mm wall thickness at a load of 50 kN and the maximum stress over the infill content is illustrated in Figure 5.26. The figure shows that the global and local stresses decrease with increasing volume infill content. Furthermore, it can be observed that both honeycomb and grid structures have a similar level of maximum stress with the same vertical orientation and the same volume fraction.

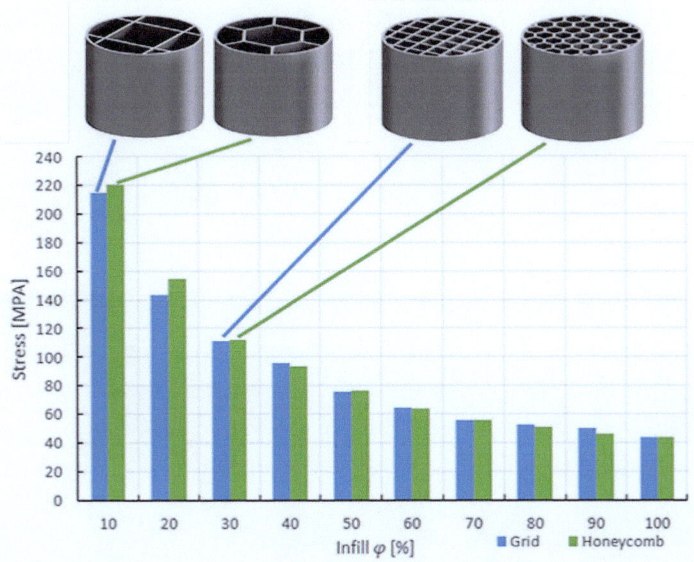

Figure 5.26 Maximum stress over the infill content in the compression samples is shown for a 50 kN load and 1 mm wall thicknesses. Reproduced from [167] (licensed under CC BY 4.0).

Besides, it can be observed that the maximum stress decreases asymptotically with increasing infill content. From an infill proportion of 60 %, the percentage decrease in maximum stress is less than 15 % with increasing infill volume, regardless of which structure is used. Therefore, an infill percentage of 60 % can be identified as the threshold value. Increasing the load to 100 kN leads to a maximum stress of 126 MPa with 60 % infill in the grid structure and to a maximum stress of 252.3 MPa with the same infill but a load of 200 kN. Both stress values are below the maximum tensile strength of 316L and are sufficient to withstand the applied load.

The orientations of the infill structures at 90° (transverse to the load) and at 0° (in the load direction) are shown in Figure 5.27 as a function of the maximum stress and the infill volume. It should be emphasized that the maximum stress generally decreases with increasing volume. Additionally, it can be observed that the 90° orientation exhibits higher maximum stresses than the 0° orientation. The highest maximum stress for all volume fractions occurs in the honeycomb structures at a 90° orientation. At an infill volume of 60 %, the maximum stress of the lattice structures is almost isotropic, with the 90° and 0° orientations having similar maximum stresses. Finally, the force distribution can be seen in Figure 5.27. The distribution occurs not only within the infill structures but also on the contour surfaces of the sample geometries.

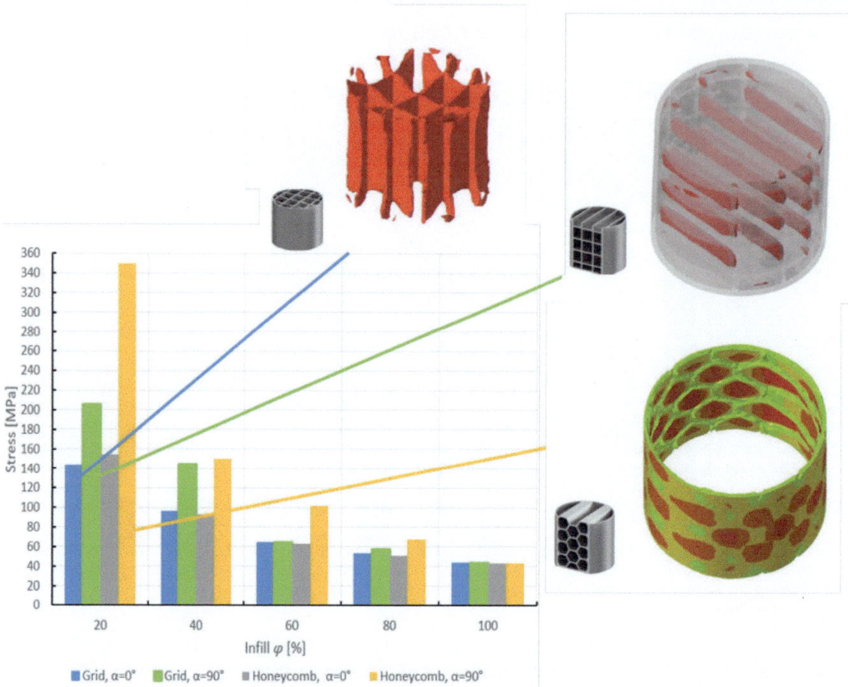

Figure 5.27 Maximum stress over infill percentage for honeycomb and lattice structures at 0° and 90° orientations (1 mm wall thickness, 50 kN load). Reproduced and adapted from [167] (licensed under CC BY 4.0).

A closer look at a 60% infill with different orientations of angle α is presented in Figure 5.28. The maximum stresses increase with increasing angles and reach their peak at an orientation of 45°. Thereafter, the maximum stresses decrease, with those of the honeycomb structures being higher than those of the lattice structures between 15° and

75°. The illustrations of the force distribution also demonstrate that the stress occurs continuously in both the infill structures and the contour surface of the sample geometries.

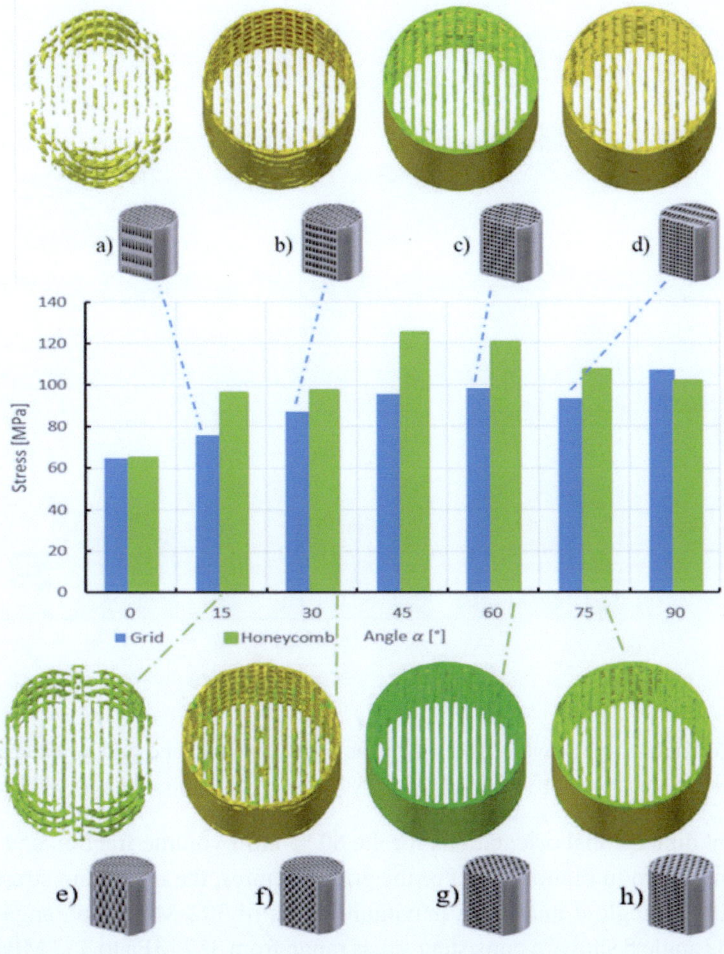

Figure 5.28 Maximum stress over the angle α at 60 % infill for grid (a-d) and honeycomb (e-h) structures: 1 mm wall thickness, 50 kN load. Reproduced from [167] (licensed under CC BY 4.0).

Varying the wall thicknesses from 1 mm to 10 mm showed no significant change in the maximum stress for the honeycomb and grid structures at the same 0° orientation. However, maximum stress tended to increase with greater wall thicknesses at the 45° and 90° orientations.

The 3-point bending simulations were carried out at a load of 20 kN and a wall thickness of 1 mm, with increasing infill content, and the results are shown in Figure 5.29. A similar trend to the compression simulations can be observed, although the maximum stresses are consistently higher in this case. The maximum stresses for the honeycomb structures are higher than those for the grid structures, regardless of the volume fraction. From an infill content of 60 %, the maximum stress is 400 MPa for grid structures and 466 MPa for honeycomb structures, which is below the maximum tensile strength of the 316L material. Therefore, as with the compression simulations, the same threshold can be identified in this case.

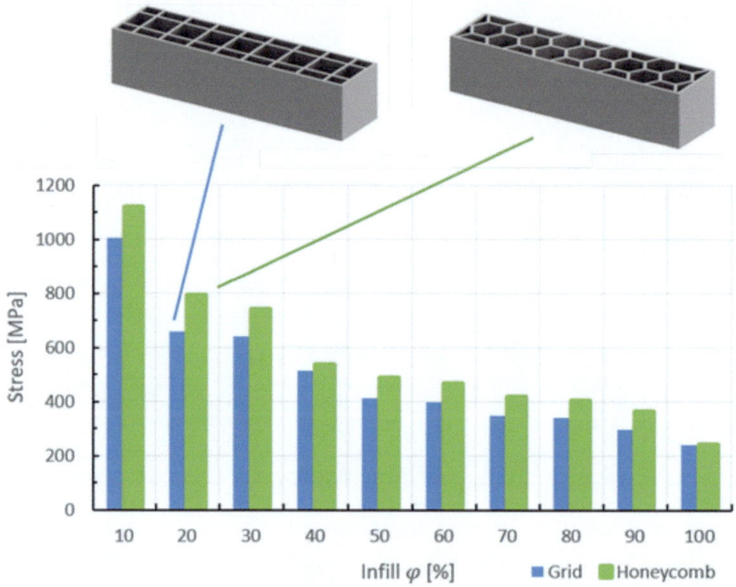

Figure 5.29 Maximum stress over infill percentage for honeycomb and grid structures with a wall thickness of 1 mm and a load of 20 kN. Reproduced from [193] (licensed under CC BY 4.0).

The different dimensional orientations for the 60 % infill volume fraction and 1 mm wall thickness are shown in Figure 5.30. For the grid structures, the maximum stress decreases steadily with the angle α and has a maximum stress of 302 MPa at an angle of 90°. A change in the angle β shows a consistent stress range from 387 MPa to 357 MPa. At angles greater than 60°, the maximum stress decreases to a value of 340 MPa. With honeycomb structures, the maximum stress increases gradually from 335 MPa to 347 MPa as angle α increases. Changing the angle β reduces the maximum stress from 354 MPa at 15° to 300 MPa at 90°. The maximum stresses are consistently lower compared to grid structures. The highest maximum stresses occur at the angle γ for both infill structures compared with the stresses at the angles α and β. For the grid structures, the maximum stress increases from 15° at 411 MPa to 464 MPa at an angle of 45°. For honeycomb structures, the stress at 15° and 30° is constant at 441 MPa and tends towards a lower maximum stress with increasing angle γ, so that at 45° it is only 402 MPa.

In conclusion, the investigations of the angle variations show that the lowest stresses occur in the grid structures when the angle α changes. If the angle β is varied, honeycomb structures exhibit lower maximum stresses and should therefore be used. If a load case occurs at angle γ, honeycomb structures exhibit lower maximum stresses as the angle gamma increases.

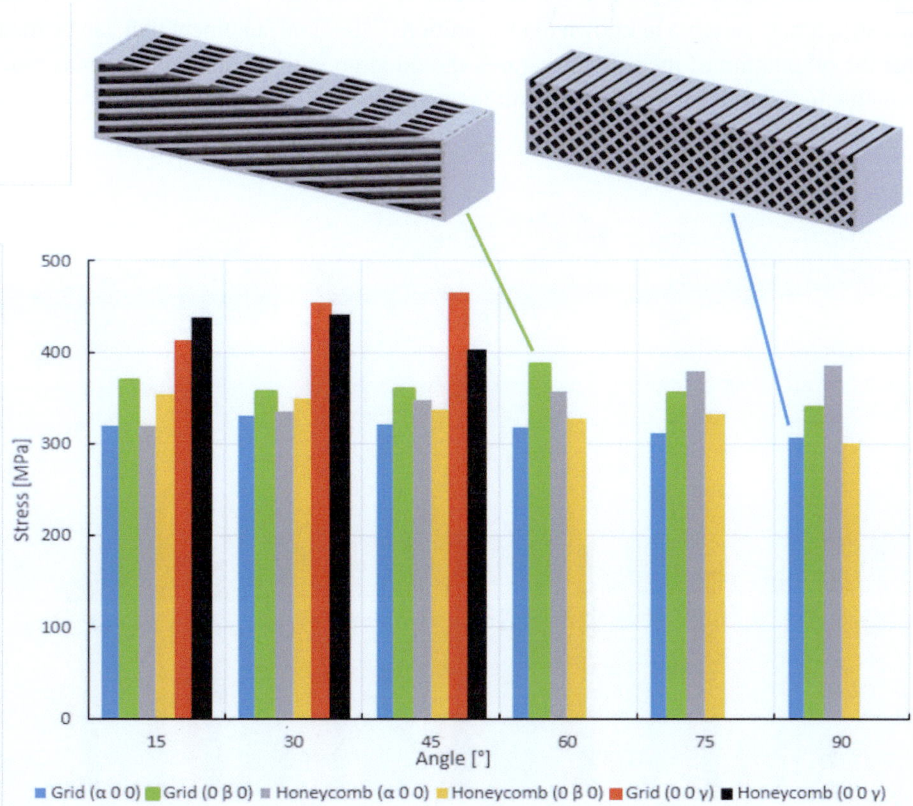

Figure 5.30 Maximum stresses vs. angular changes (α, β, γ) for honeycomb and grid structures with 1 mm wall thickness and 60 % infill at 20 kN load. Reproduced from [193] (licensed under CC BY 4.0).

The investigation of the influence of the wall thickness as a function of different angles is shown in Figure 5.31. The 2 mm wall thickness exhibits the lowest maximum stress in the grid structures with an orientation of 0° for α, β, and γ. The wall thickness of 8 mm in this orientation was not simulated because the predetermined threshold value of the maximum stress was exceeded. The highest maximum stress occurs at a wall thickness of 4 mm with an orientation of α at 90°, while β and γ remain at 0°. The honeycomb structures with 0° orientation for α, β, and γ show an increase in maximum stress with increasing wall thickness. One reason for this is that the stress concentrations occur in the voids of the structures, and the force flow is not directly absorbed by the material. The same observations can also be made for the α 90° orientation and the β 90° orientation, up to a wall thickness of 4 mm. A wall thickness of more than 4 mm leads to a reduction in the maximum stress, as fewer stress peaks are localized in the voids. Generally, there is a tendency for the highest stresses to occur at a wall thickness of 4 mm to 8 mm for both infill structures.

The simulative investigation has revealed that grid and honeycomb structures have advantages and disadvantages in terms of maximum stresses, depending on the parametric design and orientation. No general statement can be made, and it is only possible to decide which structure is suitable in known load conditions. The final statement that can be made is that the orientation of the infill structures should align with the load direction as much as possible in order to minimize the resulting stresses.

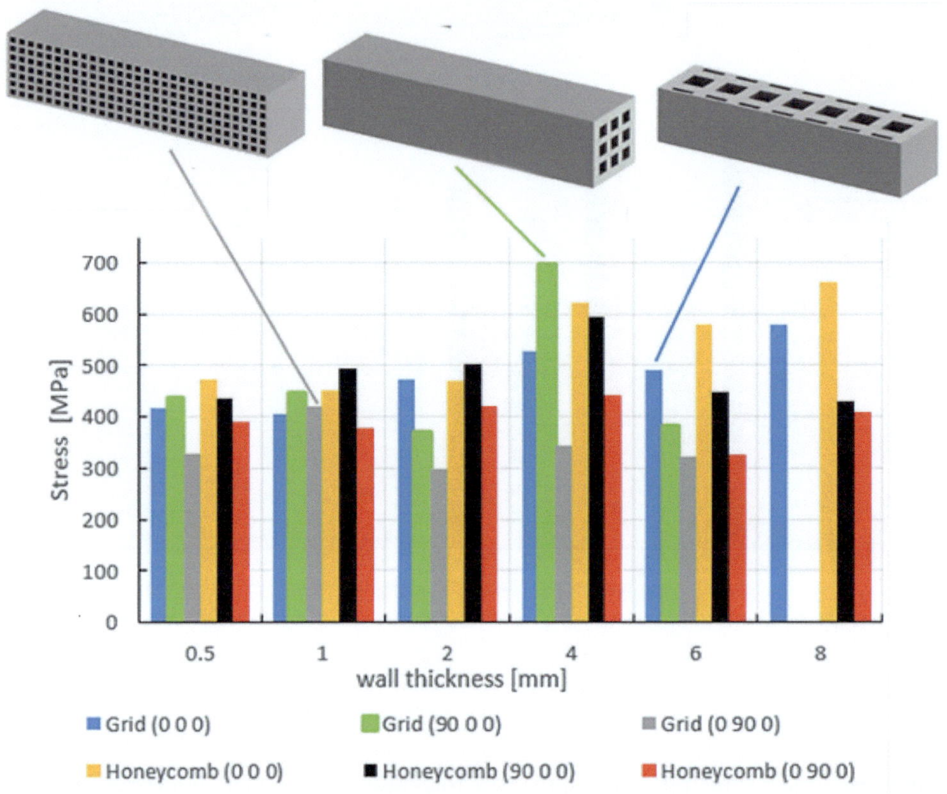

Figure 5.31 Maximum stresses as a function of wall thickness and orientation of honeycomb and grid infill structures at 60 % infill volume. Reproduced from [193] (licensed under CC BY 4.0).

5.4 Machine learning for infill design

The simulation results from the study in the previous chapter have been used as a data set for the development and evaluation of ML models with the aim of predicting stresses in the design process of infill structures without performing a simulation. This approach reduces simulation time to a minimum, providing fast and reliable statements about the stress situation when using an infill structure. Furthermore, it eliminates the need for experienced designers in AM and can help reduce the trial and error process in the development of a component.

This chapter is based on publication [146], and the main findings are described below (see Appendix B.4). The source code for the reproducibility of the results is stored in publication [200]. Python 3.8 and Scikit-learn (v1.3.0) were used to develop the ML models. Four different regression models were examined. The models trained and evaluated were linear regression (LR), decision tree regression (DTR), gradient boosting regression (GBR), and random forest regression (RFR). One hundred sixty-seven data from the compression and bending simulations were divided into 70 % training data and 30 % test data for all models. As evaluation metrics, the R^2 value and the RMSE were calculated for all models and compared with each other.

Figure 5.32 illustrates the methodology used to develop the ML models. The blue parameters were used as training and test data. The final goal is to be able to predict the maximum stress directly by entering the four different parameters via the ML models.

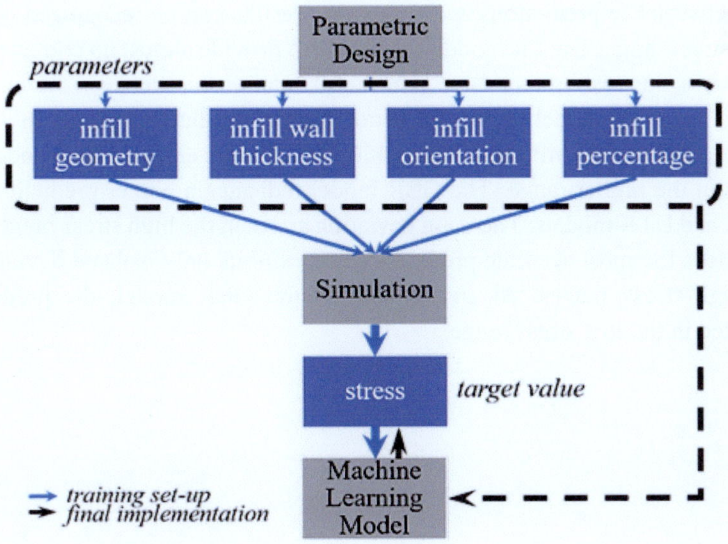

Figure 5.32 Methodology of the ML models for the prediction of the maximum stresses of the infill design. Reproduced from [146] (licensed under CC BY 4.0).

Figure 5.33 shows the stress distribution in the test and training data under load. It can be seen that both test and training data occur for all load cases. Fewer test data points occur only at a load of 200 kN. Given the large bandwidth of the input data and the fact that few data points were simulated, a cross-validation was carried out for the test and training data.

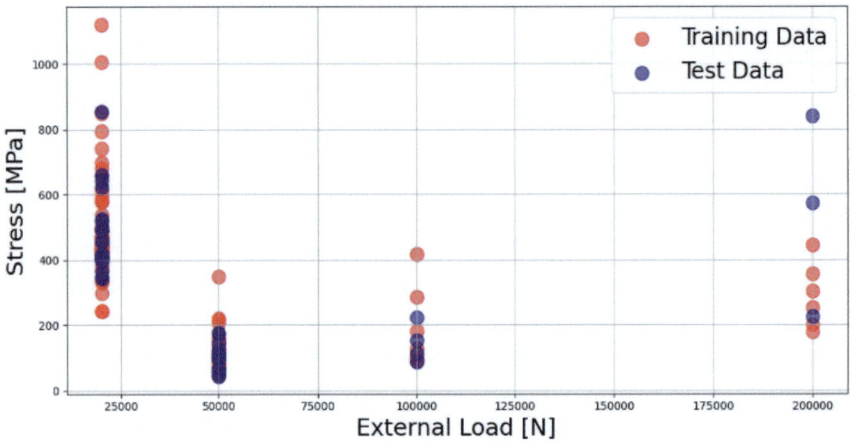

Figure 5.33 Stress distribution in simulation data: analysis of test and training data set splits. Reproduced from [146] (licensed under CC BY 4.0).

Figure 5.34 illustrates the predictions of the simulation results for the four ML investigated models. The red dashed line represents the optimal prediction, which corresponds to the simulation results. Since only 20 kN, 50 kN, and 100 kN loads were collected, a greater deviation between the predictions and simulation results can be recognized in the lower and higher stress range. The LR model was unable to provide almost no correct predictions about the stress, suggesting that there is no linear relationship between the input and the output data. The DTR model performed better and can predict acceptably in the medium stress range of 50 to 150 MPa and 400 to 450 MPa. However, the model underestimates the stresses in the higher range. The GBR model indicates a higher prediction accuracy than the LR and DTR models. The main deviation arises in the high stress range. The RFR model provides the most accurate predictions and exhibits only isolated deviations in the low and high stress ranges. As in the case of the GBR model, the predictions are overestimated in the low stress range.

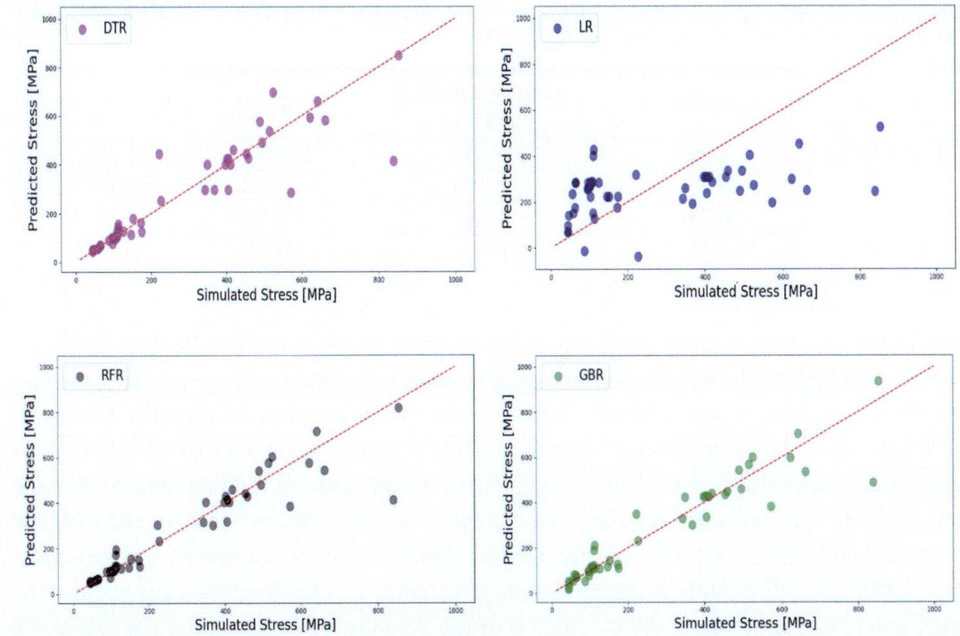

Figure 5.34 Prediction plots of simulated stress results for the ML models: DTR, LR, RFR, and GBR. Reproduced from [146] (licensed under CC BY 4.0).

To examine the influencing factors between the input parameters and the output, the Pearson correlation coefficient was also evaluated and compared, which is shown Table 5.2. As expected, the strongest correlation is found between the stress and the load. The wall thickness and orientation showed the weakest correlation with the stress. Only the infill volume percentage exhibits a stronger correlation with stress than wall thickness or orientation. The infill geometry also demonstrated a minor influence on the stress output, as none of the geometries examined accounted for a significant amount of stress, as already observed in Chapter 5.3.

Table 5.2 Correlation of the maximum stress with the varied input parameters for the GBR, RFR, and DTR models. Reproduced from [146] (licensed under CC BY 4.0).

Model	Load	Infill geometry	Wall thickness	Infill orientation	Infill percentage
GBR	0.723	0.004	0.017	0.015	0.24
RFR	0.719	0.008	0.020	0.014	0.24
DTR	0.716	0.009	0.014	0.018	0.24

Furthermore, the ML models are evaluated in terms of RMSE and R^2 value and are illustrated in Table 5.3. As already observed in Figure 5.34, the LR model has a high deviation in the stress prediction and consequently has a large RMSE. Therefore, the model is unsuitable for making robust predictions. The GBR model has the smallest

Table 5.3 Root mean squared error and R^2 value of the analyzed ML models GBR, RFR, DTR, and LR. Reproduced from [146] (licensed under CC BY 4.0).

Model	Average RMSE [MPa]	R^2 [-]
GBR	67.31	0.91
RFR	77.07	0.88
DTR	86.67	0.84
LR	197.1	0.20

RMSE and can consequently make the most accurate predictions. The RFR model has a slightly higher RMSE, which means that comparably accurate results can be achieved. The DTR model has a higher RMSE, which results in less accurate predictions. However, the RMSE is of the same order of magnitude, and it can be assumed that the predictions are relatively reliable. A modification of the distribution between training data (80 %) and test data (20 %) revealed no significant change in its influence on the RMSE.The assumption that the models are overfitted could, therefore, not be confirmed. The generally large RMSE for all models is due to the large variance of the simulation results and the small number of input data. An evaluation of the R^2 shows that no model has reached a value of 1, which indicates that no overfitting has occurred. The R^2 for the LR model is 0.2, which is very low and confirms that the model is unsuitable for predicting stress. The GBR model has the highest R^2 value, followed by the RFR and DTR models, which are of similar size. All three models appear to be fundamentally suitable for predicting stress. The available data set demonstrated that the GBR model is best suited for predicting the stress and has the smallest RMSE.

The study demonstrated that stress prediction is feasible from the input parameters and that the stress state of the finished component can be predicted from the simulation data. This means that by using the ML models, computer-aided optimization in the design process can be avoided and computational time reduced.The validation was carried out with components manufactured in the PBF-LB/M process. Since no explicit material model for AM was used in the simulation, it can be assumed that comparable results can be achieved in the experimental investigation if the sample geometries are manufactured in the MEX/M process with similar material properties as in the PBF-LB/M process. Therefore, the methodology is generally valid.

6. Machine learning for the prediction of quality parameters

The use of ML has already been demonstrated in the previous chapter, showing that predictions of material characteristics are feasible and can reduce the optimization process. As discussed in Chapter 2.2.2, it is evident that ML is used for a variety of applications in AM, such as process optimization in manufacturing. Therefore, this chapter, which is based on [201], focuses on the application of ML in the prediction of quality criteria in the MEX/M process and the reduction of iteration steps (see Appendix B.5). The models and data sets are provided in source [202].

6.1 Specimen preparation and ML methodology

Different overhang angles from 10° to 80° were produced, and the surface roughness (R_z) of the downskin area (URz) and the upskin area (ORz) was measured. Figure 6.1 shows the different overhang samples and the surface roughness measurements. The respective surface was measured three times using the Keyence VHX-S600E and the average value was calculated.

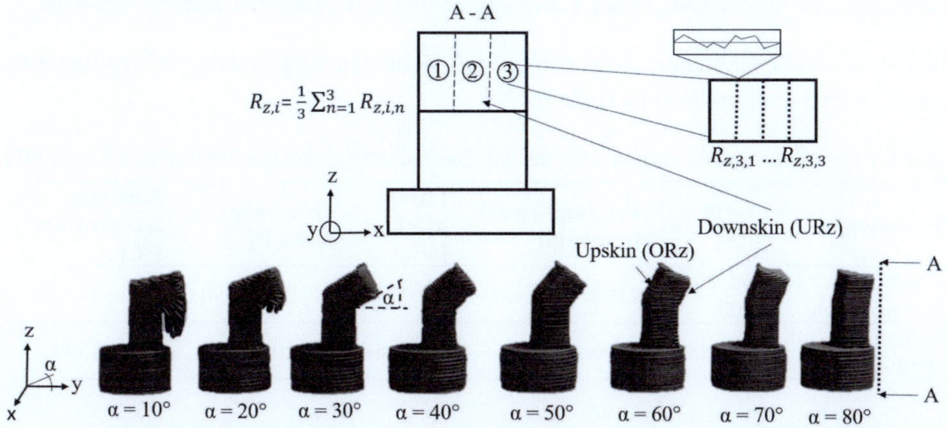

Figure 6.1 Experimental data generation: Overhang specimen manufactured as green parts. Roughness measurement approach. Reproduced from [201] (licensed under CC BY 4.0).

The roughness measurement results are illustrated in Figure 6.2. The respective process parameter variations are used as the numbering. As expected, the roughness measurements show greater values at small overhang angles, and tend to decrease as the angle increases. Some deviations in the measurements of the up-skin area at the low overhang angles are present, as the staircase effect was increasingly recognizable, thus making an accurate measurement difficult.

K.Asami, *Analysis of Sinter-Based Additive Manufacturing for Functional Products*, Light Engineering für die Praxis, https://doi.org/10.1007/978-3-032-28433-4_6

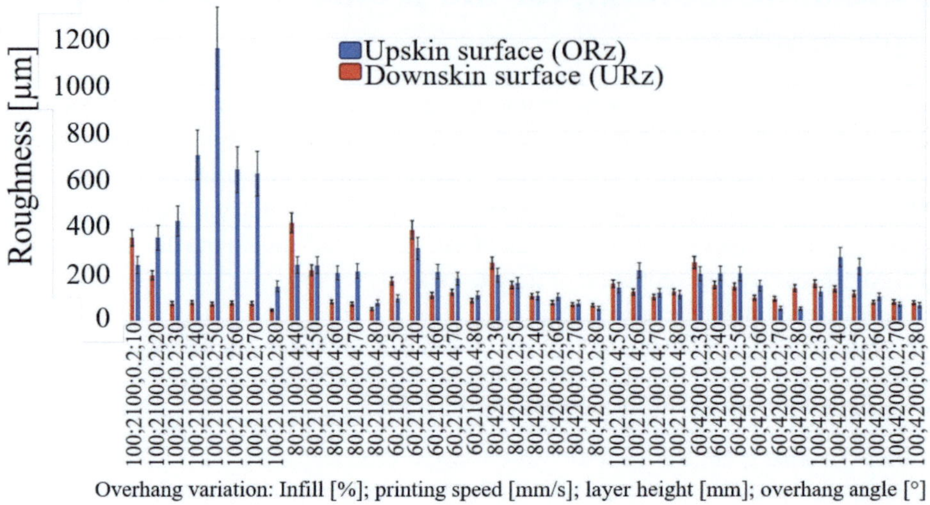

Figure 6.2 Roughness measurements of the upskin and downskin areas for all process parameter variations of the green parts. The first number indicates the percentage infill [%], the second number indicates the printing speed [mm/s], the third number stands for the layer height [mm], and the fourth number indicates the overhang angle [°]. Reproduced from [201] (licensed under CC BY 4.0).

Different process parameters were also printed, debinded, and sintered to investigate the density, which are presented in Table 6.1.

Table 6.1 Process parameter variation for the data base of the ML models. Reproduced from [201].

Variation	Infill [%]	Printing speed [mm/min]	Layer height [mm]	Overhang angle [°]	Sintering temperature [°C]
Green part: 40 combinantions	60, 80, 100	2100, 4200	0.2, 0.4	10, 20, …, 80	-
Sinter part: 48 combinations	50, 100	2100, 4200	0.2, 0.4	10, 20, …, 80	1250, 1300

The methodology for developing the ML models from the experimental data is shown in Figure 6.3. The left part of the methodology illustrates the experimental part, which is then used directly as the basis of data for the prediction of density and surface roughness in the right part.

The input parameters used were the percentage of infill, the printing speed, the layer thickness, the overhang angle, and the sintering temperature. The output parameters are the downskin surface roughness [µm], the upskin surface roughness [µm], and the relative density [%].

The model development, training, and testing of the data is done in Python (v3.8.19) with the use of the scikit-learn (v1.3.0) library.

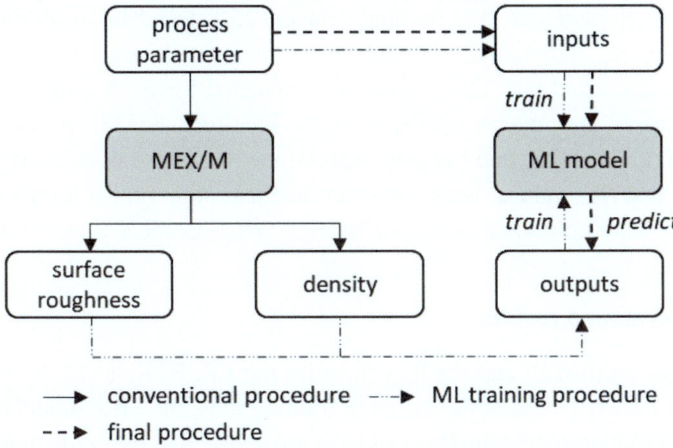

Figure 6.3 Methodical approach for the prediction of quality criteria (density and surface roughness) in the MEX/M process. Reproduced from [201] (licensed under CC BY 4.0).

6.2 Model development

In a preliminary step, a correlation analysis of the input and output data was carried out. The calculation matrix is illustrated in Figure 6.4 and shows the correlation between the input and output parameters.

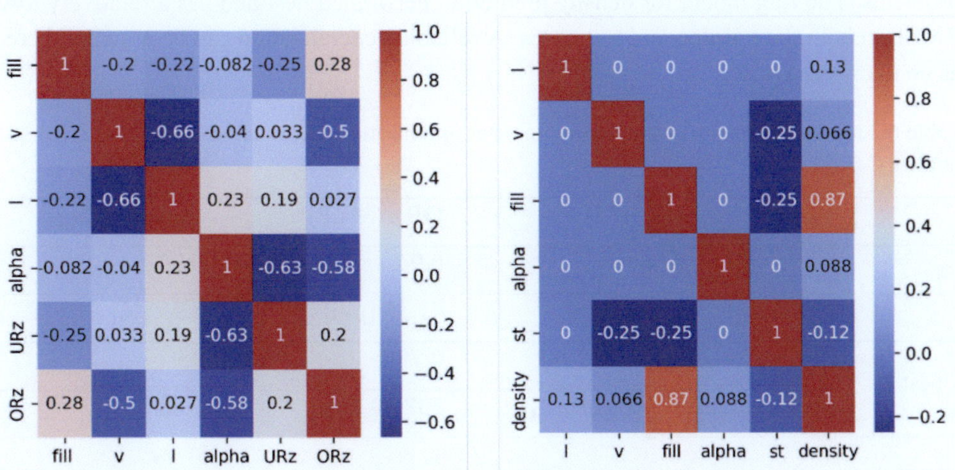

Figure 6.4 Correlation matrices for surface roughness (left) and density (right) from the input and output parameters. Reproduced from [201] (licensed under CC BY 4.0).

The correlation matrices show that the target parameter URz has the strongest negative correlation with the angle α (-0.63). ORz, on the other hand, correlates strong negatively with both the anlge α (-0.58) and the printing speed (-0.5). As can also be seen, density shows a strong positive correlation with the infill percentage (0.87).

The data has been divided into 75 % training data and 25 % test data. In addition, stratified shuffle was applied, polynomial features were calculated, and a feature selection (SelectKBest) was carried out. The models were selected based on criteria of predictive accuracy, robustness, and interpretability from the literature. For this reason, the models LR, RFR, SVM, kNN, MLP, and bagging regression (Bag) were developed, trained, and tested. RMSE and R^2 values were used to evaluate the model performance. The hyperparameters were optimized using grid search (GridSearchCV) and cross-validation.

6.3 ML performance results

The evaluation of the RMSE and the R^2 values for the LR, RFR, SVM, kNN, MLP, and Bag models has shown that, after the different training runs, the MLP model is best suited for roughness prediction, while the Bag model is most effective for predicting density and has achieved the most accurate predictions according to the evaluation criteria.

The RMSE and R^2 values of the selected models were determined using the optimized hyperparameters and the results are presented in Table 6.2. The R^2 value for the prediction of URz with the MLP model is 0.96 in training and 0.87 in testing, indicating that the models are well generalized. The RMSE is 50 % lower in training than in the test case, which may indicate overfitting, as the model performs worse in predicting test data. For ORz, the R^2 value is also higher in the training with 0.94 than in the test with 0.79. The RMSE also rises significantly in the test to 138.99 µm and indicates generalization problems. The Bag model for density prediction performed best and has a relatively low RMSE in both the training and test cases. A reliable generalization can be assumed here, as no overfitting arises.

Table 6.2 Model evaluation of the hyperparameter optimization. Reproduced from [201] (licensed under CC BY 4.0).

	Value	URz (MLP)	ORz (MLP)	Density (Bag)
Train	R^2	0.96	0.94	0.99
	RMSE	17.38 [µm]	41.58 [µm]	0.95 [%]
Test	R^2	0.87	0.79	0.95
	RMSE	32.93 [µm]	138.99 [µm]	1.88 [%]

A further evaluation criterion for the models was developed and is defined as:

$$Average\ accuracy = 1 - \frac{RMSE}{\sum_1^n R_{Zn}} \tag{6.1}$$

The average accuracy of ORz after the calculation of Equation 6.1 is very low at 39.26 %, indicating that no accurate prediction can be made from the existing data set. The reason for this is the wide dispersion of the measurement data and the small number of measurements. A significantly higher accuracy of 75.71 % was calculated for URz. It can be stated that a general prediction is possible with the data set, but it cannot be stated to be highly accurate. However, it is possible to obtain an initial tendency of the achievable roughness to reach the desired process parameter combinations more efficiently through experimental investigation. The calculation of the accuracy for the density with the bag model results in 97.44 % and thus represents the highest accuracy. It can make very accurate predictions about the density of the experimental samples, with a low error of 1.88 % in the relative density, and is robustly suitable for predicting the density. This model can be used to optimize the target density with the variation of the given data set.

7. Manufacturing of the functional products

This chapter is largely based on the results from source [203] and is evaluated in the following sections (see Appendix B.6). The data set of the models and the measurement results are stored in [204].

The growing movement towards renewable energy has made hydrogen a key resource. However, existing methods of hydrogen production are based on fossil fuels and continue the production of greenhouse gases. An alternative approach to hydrogen production can be realized by means of water electrolysis using proton exchange membrane electrolyzers/fuel cells (PEMECs/PEMFCs). They provide a variety of advantages, such as the capability of producing compressed gases for direct use and storage, high-purity hydrogen, and swift response to current fluctuations. Bipolar plates (BPPs) are an essential component of PEMECs/PEMFCs and are responsible for the distribution of reactants and products, as well as for conducting current. BPPs account for approximately 40 % to 51 % of manufacturing costs. Conventional materials used in production are titanium or graphite. AISI 316L, in particular, can be used as an alternative material due to its good mechanical properties and corrosion resistance, thereby reducing costs.

7.1 Design and manufacturing of BPPs

Six different flow fields were designed in Autodesk Inventor 2023 to investigate the BPPs. These include classic serpentine structures, which were further varied with additional design elements to enhance the performance of the BPPs. The BPPs are 3 mm high and have dimensions of 33 mm by 33 mm. For comparison, the different serpentine structures illustrated in Figure 7.1 were additionally manufactured using the PBF-LB/M process. The investigation is limited to the surface of the BPPs, as the MEX/M process obviously produces less accurate results than the PBF-LB/M process due to the higher layer thickness. The machine used and the process parameters are shown in Table 7.1. The material used was AISI 316L, but the particle size ranged from 19 µm to 43 µm and is therefore larger than in the investigation using the MEX/M process. The optical evaluation was carried out with the Keyence 3D scan system, and 20 points measured at the corners, in the channels, and on the bars were systematically carried out for all manufactured samples.

Table 7.1 Machine, process parameter, and material used for the manufacture of the BPPs [203] (licensed under CC BY 4.0).

AM machine	One Click Metal MPrint
Laser power	200 W
Scan speed	800 mm/s
Hatch distance	0.1 mm
Layer thickness	40 µm
Laser diameter	70 µm
Material	316L stainless steel

K.Asami, *Analysis of Sinter-Based Additive Manufacturing for Functional Products*, Light Engineering für die Praxis, https://doi.org/10.1007/978-3-032-28433-4_7

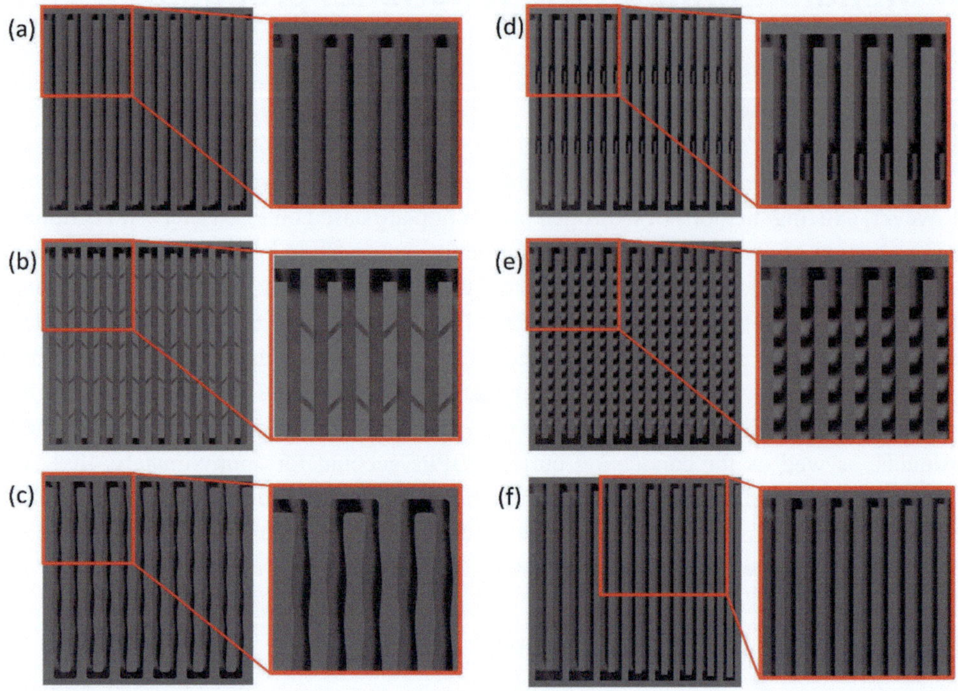

Figure 7.1 Design of the flow fields of the BPPs: a) serpentine, b) ribbed grooves, c) wave structure, d) triangular baffles, e) wavy bottom, and f) narrowing serpentine. Reproduced from [203] (licensed under CC BY 4.0).

The channel width was set to 1 mm for all flow field designs, except for the flow fields in Figure 7.1, c) and f), as these have varying channel widths. When measuring the BPPs manufactured using the MEX/M process, it was found that the channels were smaller and the ribs larger than designed. The restriction is based on the selected process parameters of the extrusion multiplier and shrinkage during sintering.

Figure 7.2 indicates the mean and standard deviation of the measured points of all samples manufactured with both processes. The standard deviation shows that the MEX/M process has a lower deviation at the measuring points for both the serpentine structures and the rib grooves compared to the PBF-LB/M manufactured counterparts. An optical evaluation of the PBF-LB/M samples generally showed warping at the edges due to the energy input into the thin structures. This high energy input is absent in the MEX/M process during layer-by-layer buildup, which is why no warpage was detected. The lowest mean standard deviation of 0.03 mm was measured in the MEX/M process within the uniform serpentine structures. This indicates that the design restrictions determined and applied to the column width, as discussed in Chapter 5.1.4, could be validated by this functional component. Larger mean standard deviations can be observed in the MEX/M for the triangular baffle (0.077 mm) and wavy bottom (0.081 mm) structures. In comparison, the deviations for the PBF-LB/M process are smaller at 0.047 mm and 0.056 mm. In these cases, the smaller design features are better achieved by the PBF-LB/M process in manufacturing.

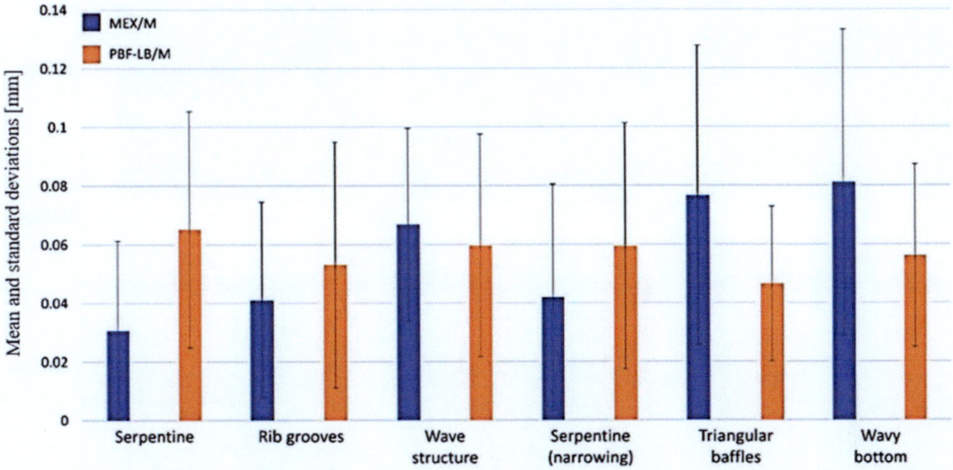

Figure 7.2 Mean standard deviation of the different BPP designs manufactured using the MEX/M and PBF-LB/M processes. Reproduced from [203] (licensed under CC BY 4.0).

Figure 7.3 shows the 3D scan of the different BPPs MEX/M structures, comparing their deviations with the CAD model.

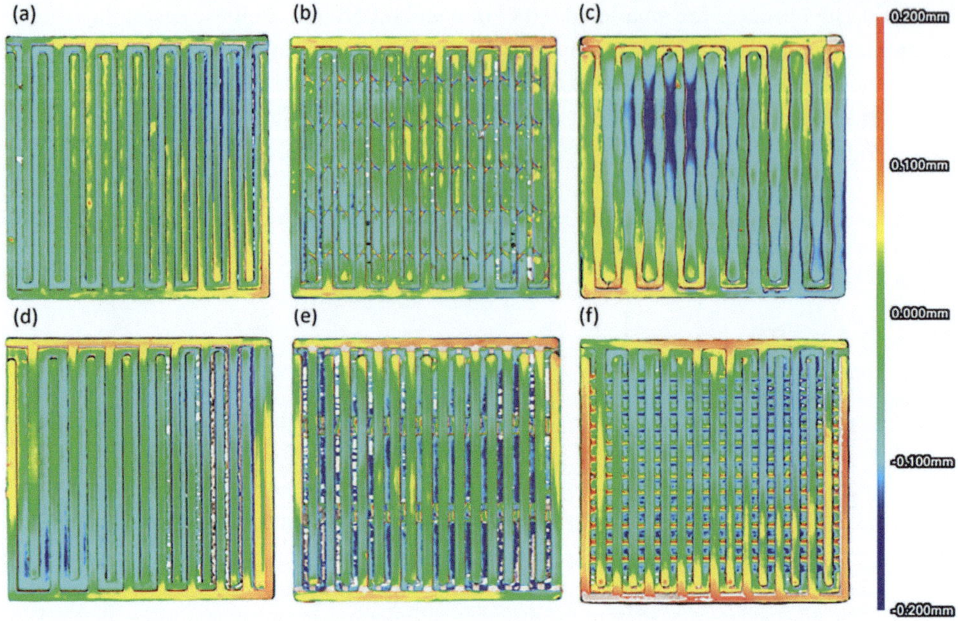

Figure 7.3 3D scan illustration of the BPP structures produced with the MEX/M process, a) serpentine, b) ribbed grooves, c) wave structure, d) narrowing serpentine, e) triangular baffles, and f) wavy bottom. Reproduced from [203] (licensed under CC BY 4.0).

An examination of the 3D scans and the determined standard deviations provides additional information about the overall detected deviations. Generally, it can be observed that the greatest deviations are localized at the edges and the channel floors. In the structures with the largest standard deviations (fig. 7.3 e) and f)), it can be seen that the small features also exhibit the largest deviations in the design.

Through the optical evaluation and comparison with the PBF-LB/M, it can be concluded that functional products in MEX/M can be manufactured in terms of dimensional accuracy, as demonstrated using the example of the bipolar plates. This process can realize more dimensionally accurate components with simple features than the other AM process, and the application of the design restrictions and process parameters was successful.

7.2 Mold manufacturing

For the subsequent manufacture of a functional product, a design automation system was developed that implements the infill structures in the desired variation from Chapter 5.4 into the CAD model. The design automation with the mold design and the measurements are available at [205]. Together with the stress prediction from Chapter 5.4, the optimum structure can be automatically integrated into any CAD design. This was done as an example for the production of an MIM injection mold for the production of regular serpentine bipolar plates. It should be noted that this is merely a demonstrator for which the identified design limits and process parameters used in the MEX/M process apply. Requirements for the injection mold, such as important features like ejector pins for removing the injection molded parts, draft angles or gate location of the mold, were neglected [206]. Figure 7.4 illustrates the injection mold.

Figure 7.4 MIM injection mold for the production of the serpentine bipolar plate: Green part (left) and sintered part (right).

The mold is designed with a percentage infill of 60 %, with a wall thickness of 1 mm and a 0° orientation for the angles in all dimensions. The relative shrinkage measured against the dimensions of the CAD model of the two injection molded parts is in the range of 15.14 % to 16.34 % in the x and y directions, and in the range of 13.97 % to 14.13 % in the z direction. It can be assumed that the use of infill structures increases the number of void spaces in the z direction, which counteracts the shrinkage. The average channel width of the injection mold BPP channels are smaller than the design measurement and are, on average, 0.896 mm (1 mm in the design) as measured with the Keyence VHX-S600E microscope. The percentage weight saving compared to the dense injection mold is 37.33 %, which was evaluated by analyzing the CAD model.

7.3 Cost estimations

To determine the unit costs of the BPP samples and the MIM injection molded parts, a cost model is used, incorporating both the acquisition costs and the material costs of the MEX/M and PBF-LB/M systems as data [207]. The cost application roughly estimates the manufacturing costs of the different processes, MEX/M, PBF-LB/M, BJT/M, and MIM, for a component based on the material used, and the component volume. The total costs are made up of the material costs, the machine costs, the post-processing costs, and the overhead costs. The material costs are determined by the density of the material, the number of components to be produced, the component volume, the material price, and a shrinkage factor for sinter-based processes [208]. The machine costs (K_M) take into account the process costs for the shaping, debinding, curing, sintering, and heat treatment steps and are defined as follows:

$$K_M = k_M \cdot (t_P + n_{Charge} \cdot t_R) \tag{7.1}$$

The process time t_P depends on the respective process step and is dependent on the geometric complexity and the selected material; different holding times and heating rates are therefore required for the sintering step. A fixed process time for the respective process step is used as a simplified assumption, neglecting the influencing factors mentioned [208]. t_R is the setup time for each machine, n_{Charge} is the number of batches, and k_M is the machine hourly rate, which consists of the acquisition costs A_M of the respective machine, the depreciation period t_d, and the percentage annual utilization δ_i:

$$k_M = \frac{A_M}{8.760 \cdot t_d \cdot \frac{\delta_i}{100}} \tag{7.2}$$

The number of batches n_{Charge} consists of the number of units n_B, and the batch size $X_{Charge,i}$:

$$n_{Charge} = \left\lceil \frac{n_B}{X_{Charge,i}} \right\rceil \tag{7.3}$$

The $X_{Charge,i}$ is calculated for the respective process step, and the length l_B, and the width b_B, of a rectangle are assumed as the component surface. Finer structures are calculated oversized in this way. The batch size is determined from the distance between the components x_B, and the dimension of the installation space, consisting of the width b_{Br} and length l_{Br}:

$$X_{Charge} = \left\lfloor \frac{b_{Br}}{b_B + x_B} \right\rfloor \cdot \left\lfloor \frac{l_{Br}}{l_B + x_B} \right\rfloor \tag{7.4}$$

Costs such as employee salaries, energy costs, and transportation costs are counted as overhead costs [208]. The cost of post-processing is approximated using factors ranging from 0 to 3. If no post-processing step is required, the factor 0 is used instead. With a factor of 1, post-processing such as the removal of support structures is required. Factor 2 is used if a functional surface, e.g. a bearing seat, is present, and a factor of 3 is used if several toleranced functional surfaces are present. For the cost comparison, the machine and material prices from the manufacturers were employed for the MEX/M process. The machines and material costs employed in the production of the BPPs were utilized for the PBF-LB/M process. For the BJT/M process, a DM P2500 for the shaping step, a TR60/LS for curing, a NA 120/45-1 for debinding, and a VHT 25/16-MO-1 for sintering were specified as machines. The same machines are used for debinding and sintering in the MIM process, with an Arburg Allrounder 470S assumed for shaping [208].

The dimensions of the bipolar plate from Chapter 7.1 are 33 mm in width and length, and 3 mm in height. The volume is, therefore, 3267 mm^3 with an assumed post-processing factor of 2. The cost comparison of MEX/M and PBF-LB/M shows that the MEX/M process is more expensive for this component, regardless of the number of pieces. This is mainly due to the fact that the material costs are 4.6 times higher than for the PBF-LB/M process. The cost of a bipolar plate in the MEX/M process is €168, and €413 for the BJT/M process. Up to a quantity of 12 components, the process is more cost-effective than the BJT/M process. From 13 components, it is therefore more expensive. As in the other comparison, the reason for this is the higher starting material costs as well as the limited installation space of the sintering process. The furnace assumed for the BJT/M process has larger dimensions, so that more components can be sintered at the same time. For the MIM process, costs of €50,000 were assumed for one injection mold [208]. Due to the high initial costs, the MIM process is more expensive than the MEX/M for this component up to a quantity of 2152. One half of the MIM injection mold has the dimensions of 76.82 mm in length, 57.46 mm in width, and a height of 33.44 mm with a volume of 63.93 cm^3. The other half simply has a different height of 26.6 mm and a volume of 69.22 cm^3 and is added to these dimensions. Manufacturing the part using the MEX/M process with an infill of 60 % costs €873 and is therefore 57 times cheaper than the assumed injection mold costs from [208].

8. Multi-material approach in the MEX/M process

From the previous chapters, it has been demonstrated that the MEX/M can produce functional products cost-effectively compared to other sinter-based AM processes. This chapter explores the further potential of multi-material manufacturing and is based on publication [109] (see Appendix B.7).

AISI 316L and AISI 17-4PH samples were selected for the study of the multi-material application due to their magnetic and non-magnetic properties. The filament originates from the same manufacturer as described in Chapter 3.3 and has identical particle sizes. A TGA test also revealed that similar binder systems have been used for both filament materials. The benefit of this is that shaping and debinding occur under the same boundary conditions as for AISI 316L, which have been examined in Chapters 4 and 5. Thirty-two cubes with an edge length of 10 mm were manufactured for the analysis. The process parameters for shaping are defined in Chapter 4.2.1 and were kept constant for both materials. Eight cubes each of 316L and 17-4PH were manufactured without a combination of materials. Another eight cubes consisted of the first 25 layers of 316L, followed by 25 layers of 17-4PH. The last eight cubes consisted of 25 layers of 17-4PH followed by 25 layers of 316L. For the sintering process, maximum temperatures of 1286 °C, 1310 °C, 1320 °C, and 1330 °C were investigated using 99.9 % argon as the atmosphere.

The examination of the shrinkage in the various multi-material constellations did not reveal any unusual deformations, indicating that the multi-material test specimens were produced successfully. At a temperature of 1286 °C, the sintered monomaterials 316L and 17-4PH exhibit less shrinkage in the x- and y-directions than in the z-direction. Starting at a sintering temperature of 1310 °C, a change in shrinkage was observed in both materials, with shrinkage in the x- and y-directions greater than in the z-direction. The multi-material combinations in both constellations do not follow this trend. Up to the sintering temperature of 1310 °C, the shrinkage in the z-direction remained larger than in the x- and y-directions. The position of the materials appears to play a negligible role in the influence of the sintering temperature, as the multi-material samples exhibited less shrinkage in all dimensions than the monomaterials 316L and 17-4PH. The shrinkage for the 316L/17-4PH constellation is in the range of 10.98 % to 11 % for the x- and y-directions at a sintering temperature of 1320 °C. In the z-direction, the shrinkage is approximately 11.7 %. For the 17-4PH/316L constellation, the shrinkage in the x- and y-directions is 11.79 %, and in the z-direction, it is 12.43 %. The determined shrinkage thus fulfills the requirements of the MPIF 35 MIM standard.

The density was measured optically from the SEM images by determining the number of pores in the micrographs. At the sintering temperature of 1330 °C, the optical density for 316L is 97.4 % and for 17-4PH 97.6 %. However, since this is an optical density determination, further optical evaluation procedures are required to determine the overall

K.Asami, *Analysis of Sinter-Based Additive Manufacturing for Functional Products*,
Light Engineering für die Praxis, https://doi.org/10.1007/978-3-032-28433-4_8

density. The samples showed occasional voids at the interface of the two materials. Based on this measurement, it can be stated that the production of multi-material samples was successful and that the samples have a high density and fulfill the MIM standard. The production of the multi-material samples showed no major deformations, meaning that functional components should be feasible without any post-processing step. The determined shrinkage of the multi-material samples was consistently smaller in all dimensions than that of the monomaterials 316L and 17-4PH. No adaptation to the equipment, such as powder layering techniques, co-injection in the MIM process or two recoaters in the PBF-LB/M process, is required to manufacture multi-material components. As a result, the MEX/M process has a general cost advantage and can therefore be easily and simply used for the production of multi-material components.

9. Evaluation, conclusions, and outlook

This section summarizes the results of the work in this thesis and addresses the research hypotheses. Furthermore, it presents an outlook on future research directions.

9.1 Key findings, limitations, and discussion

In this thesis, the process parameters for shaping, debinding, and sintering were investigated in the first step. For the embedding of the results in the current scientific state-of-the-art, it is confirmed that the used filament and the influencing factors in the shaping process have a direct effect on the sintered components [31,49–53]. Through the process parameter investigation and the variation of the layer height, it was observed that the surface roughness increases with greater layer height (Table A.1.1.). The statistical evaluations from the ANOVA and regression analysis were used to identify an optimal set of process parameters. These process parameters were then optimized with regard to increasing green part density. The extrusion multiplier, in particular, has a direct influence on the green part density, confirming the state of the art [63]. The investigation of the debinding process led to a selection of parameters that made debinding feasible. The TGA analysis successfully determined appropriate holding periods for the thermal debinding process, which aligns with existing findings for common binder types and is also applicable to this filament type [77]. The sintering furnace and atmosphere used revealed that higher sintering temperatures lead to a lower relative density and inferior mechanical properties. In particular, the maximum temperatures of 1320 °C to 1380 °C could not achieve the MIM standard of relative density and tensile strength as in other sources [31]. In this machine configuration, the limits of the achievable sintering quality become apparent, and the maximum temperature of 1286 °C occasionally meets the MIM requirements in terms of relative density and tensile strength. One way to overcome these process-related limitations in this study is through post-treatment of the sintered samples using the HIP process, which can increase the tensile strength of the 316L material by up to 35 % and reduce porosity to as low as 0.3 vol% [73]. However, a different filament was utilized, so the transferability of the results needs to be examined.

The investigation of the design features across the entire process chain enabled the identification of restrictions and the development of design guidelines, ensuring that dimensionally accurate components can be manufactured in the desired geometric configuration. Although machine manufacturers already have recommendations for the production of design features for the MEX/M process [122], there is a lack of detailed information on deviations from these recommendations due to design requirements. The guidelines developed and the design features examined in this thesis provide additional information on potential geometric deviations in the design and thus enable an estimation of any necessary post-processing. Through the development of these guidelines, the research gap could be closed, and depending on the component to be manufactured, a

K.Asami, *Analysis of Sinter-Based Additive Manufacturing for Functional Products*,
Light Engineering für die Praxis, https://doi.org/10.1007/978-3-032-28433-4_9

decision can be made about which sinter-based AM process — such as BJT/M or the MEX/M process — is to be used to achieve the desired accuracy. A comparison with the design guidelines in the BJT/M, for example, indicates that free walls with the same height to width ratio of less than 12 can be produced from a wall thickness of 0.05 mm [117]. With the MEX/M process, a minimum wall thickness of 1.6 mm is required. In the BJT/M, the cylinders can be manufactured from a diameter of 0.2 mm [117], whereas in the MEX/M, this is only feasible from 0.8 mm. Boreholes can be produced from 0.2 mm in the BJT/M process [117] and from 0.68 mm in the MEX/M process. These restrictions highlight the opportunities for using the AM processes. As a possible application, individual spare parts that require lower resolution in terms of dimensional accuracy can be manufactured more cost-effectively using the MEX/M process (see Chapter 7.3). It should be noted that these guidelines are only valid for the machines, material, and process parameters used. Nonetheless, the methodology can generally be applied to all other machine configurations, materials, and process parameters. The use of a finer nozzle is expected to enable more finely realizable design features but, conversely, increases the time required for the printing process.

The simulative investigation of the lattice structures with the honeycomb and grid infill has led to the insight that a component with an infill volume of 60 % can theoretically be realized. A weight saving of up to 40 % is therefore feasible under known load situations. Furthermore, the holistic investigation of the infill structures was capable of identifying the load situations in the orientations and wall thicknesses. A simulative comparison of the maximum stress in a grid structure with 60% infill volume, 1 mm wall thickness, and 0° angular orientation along all dimensions, with a topology-optimized bending specimen under the same simulation setup, dimensions, and load, shows that the maximum stress in the component with infill is only 3.87% higher than in the topology-optimized result. The material saved is also 16.43 % less in the topology-optimized case than in the grid infill design. For the honeycomb structures with the same wall thicknesses and orientation, the maximum stress is 13 % higher and the material savings are 49.93 % lower than the topology optimization result. It is evident that the grid structures achieved a better results in terms of material saving, while the honeycomb structures performed worser in terms of lower maximum stresses. This example demonstrates that, with knowledge of the occurring loads and the given boundary conditions, an optimization can be carried out instead of topology optimization. Furthermore, this approach allows for material savings without altering the outer contour of the component, which may be necessary due to integration requirements with other components. A simulative and experimental validation of the tensile test with the same material in the PBF-LB/M process indicated that the global stiffness is overestimated when assuming a linear elastic model for the simulation, and can be used as an approximation [209]. Since a linear elastic model was also used, but the material parameters differ from the simulations in this thesis, an experimental validation is necessary. Furthermore, not all orientations and wall thicknesses for the honeycomb and grid structures have been simulated, meaning a full evaluation is still pending.

Various ML models were developed and examined using the simulation data of the infill, indicating that the GBR model has the lowest error (RMSE). With knowledge of the load situations, the ML model can be used directly, eliminating the simulation effort and time required in topology optimization. Furthermore, the transfer of the optimization model into a CAD file is no longer necessary. By using the parametric design, the infill structures can be selected in terms of shape, wall thickness, and orientation, and integrated into the design once the prediction has been made. Additionally, in scenarios where the infill structures cannot be oriented in the direction of loading due to installation situations, it is now possible to estimate the maximum stress without a simulation step. Therefore, the results of the study could contribute to ML-based design optimization. The methodology demonstrates that a prediction is feasible with the available input. However, the limits of the predictions are clearly recognizable as the RMSE is still high. In order to reduce the errors in the prediction, more simulation data is required. The results were validated in the PBF-LB/M process with subsequent heat treatment. It is expected that validation in the MEX/M process will also be conducted if the material properties are comparable to those in the PBF-LB/M process. As already mentioned, these can be achieved with the subsequent HIP post-processing.

The ML-based prediction of the quality parameters density and surface roughness was examined. It is possible to predict the surface roughness of the green parts on the downskin area. By knowing the surface roughness in the as-built condition, an estimation of the potential post-processing effort on the components can be made. This approach allows for the reduction of process time and costs. There are design-related cases in which post-processing is not feasible; therefore, the predicted roughness serves as a boundary condition for the application. A comparison of the roughness prediction in the FDM process shows with an R^2 value of 0.96 and an MSE of 0.05 µm a high performance using the extreme gradient boosting (XGBoost) model [210]. It should be noted that only tensile specimens were produced with different process parameters. The measured surface roughness data are in the range of 5.99 µm to 10.6 µm, allowing for such an accurate prediction. This indicates the limited ML performance with the MLP model was achieved due to the large scatter of the measurement data and the small number of data points. A more extensive experimental data collection is expected to achieve a higher and more robust ML performance.

Robust approaches already exist in AM for density prediction with little data, which can achieve an error of 0.91 % with the MLP model [211]. The prediction of density using the Bag model was successful. By selecting the varied input parameters, it was found that the percentage infill correlates most strongly with the density output. The remaining parameters have little influence on the output data. The model can therefore be used for the density prediction of hollow structures, which has so far been investigated experimentally for the MEX/M process in only a few cases. The predictions allow for the determination of the density parameters for porous structures with a low error and without high experimental effort.

BPPs have already been manufactured and examined in the PBF-LB/M process [212–214]. Specifically, the channel width [213] and honeycomb structures [214] in BPPs using 316L material were investigated, resulting in a minimum channel width of 1.25 mm being manufacturable. The BPPs examined were designed and manufactured with a standard thickness of 1 mm. The optical examination of the standard deviation showed that a channel width of 1 mm is already feasible for both processes. However, more extensive investigations, such as examining the cross-sectional area of the channel, must be carried out in order to obtain statistical validation. Additionally, wine-glass-shaped channels were examined in the cross-sectional area, which was not used in this thesis. For the honeycomb structures, droplet-shaped powder outlets have to be designed in the PFB-LB/M process to remove the enclosed powder in the post-processing step [214]. In the MEX/M process, no powder accumulation was observed due to the stacking of BPPs [203], thus negating the need for additional design elements like powder outlets. The functionality of the MEX/M BPP has not yet been examined in the area of application and must be tested due to the existing porosity. The environmental conditions of hydrogen need to be taken into account, and the corrosion resistance of the material must be investigated. Since it was not possible to produce small design features in the BPP design with the MEX/M process, a smaller nozzle can potentially realize smaller design features. For this purpose, the methodology for developing the design guidelines must be repeated. The benefit here is the direct comparability, as the same material, process parameters, equipment, and design elements are used.

A demonstrator was successfully produced for the manufacture of the injection mold with infill structures. Approaches already exist for the production of MIM tools in the MEX/M process with the integration of infill patterns [215,216]. An advantage of injection molds produced with the MEX/M process is the feasible integration of complex cooling channels, which conventionally consist only of straight channels due to manufacturing constraints [215]. These were not taken into account during the demonstration phase of the production and must be investigated in the future. The design automation [205] in this thesis enables the stress analysis (Chapter 5.3) and the prediction of the maximum stress (Chapter 5.4), as well as a prediction of the density (Chapter 6), to be estimated in applications before the component is manufactured. As a result, injection molds can be equipped with the desired variation of the infill structure, and an investigation of the stress situation can be conducted without simulation effort, resulting in a manufacturable injection mold. The prediction of the density with the infill parameters, the process parameters, and a sintering temperature of 1286 °C using the Bag model described in Chapter 6.3 gives a relative density of 80.95 %. It is clear that this does not correspond to the MIM standard. A comparison with the state of the art (Chapter 2.1.1.4) shows that the density at the applied sintering temperature is higher than the density of the infill sample with 75 % infill volume at a sintering temperature of 1380 °C, produced with the Ultrafuse filament, which has a density of approx. 75 % (see Figure 2.5). However, the prediction needs to be validated experimentally. The clamping force in the MIM process depends on the machine used and starts at 50 kN for micro injection molding [217]. The comparison with Chapter 5.3 reveals that the maximum stresses occur in the pressure simulations at a load of 50 kN in Figure 5.26. The maximum stress at the given load for the grid structures and an infill of 60 % used in this mold design is approximately 62 MPa and theoretically withstands the given loads. Other materials, such as 17-4PH, are usually used for injection molding [215]. The methodology and simulations were succesfully demonstrated on the material 316L and must, therefore, be repeated with other materials in order to determine the stress conditions and weight savings. Finally, it must be determined experimentally whether a green part can be manufactured with this mold.

The investigation of the process parameters and different configurations for the multi-material components successfully demonstrated that multi-material components can be manufactured in the MEX/M process without additional costs for the equipment and achieve a density in accordance with the MIM standard. Further multi-material investigations of 316L and 17-4 PH in the MEX/M process were carried out in [218] and confirm the findings from Chapter 8. Hexagonal multi-material samples could be produced without defects. An essential prerequisite for this is precise printing of the different materials in a component otherwise sintering failure occur due to the different shrinkage rates of the materials [218]. This makes it possible to produce multi-material BPPs, for example, to avoid the need for a coating process or to combine the advantages of corrosion resistance and higher tensile strength without requiring a complex joining process. In the case of the injection mold, the infill structures can be produced with the material 17-4PH and the contour surface with the material 316L to achieve higher strength.

9.2 Conclusions

Compared to other AM technologies, the MEX/M process, as a sinter-based AM technology, offers a significant advantage in that no loose powder is used in the manufacture of functional products, and multi-material components are feasible without any adaptation in the machine setup. Furthermore, it is one AM process that is capable of realizing closed porous structures. The MEX/M process leverages from the existing knowledge of the MIM process, particularly in the steps of debinding and sintering. However, this technology still lacks a suitable application that will gain industrial acceptance. The aim of this research work was to identify the right application to facilitate wider industrial adoption in the future.

H1: Primary research hypothesis

Functional products can be manufactured through the systematic development of process parameters and design guidelines in the MEX/M process.

The material and process parameter investigations in Chapter 4, as well as the investigation of the design elements in Chapters 5.1 and 5.2, have been successfully used for the manufacture of a functional BPP product in Chapter 7.1. A comparison of the basic serpentine structure with the manufactured counterpart in the PBF-LB/M showed a reduction in the mean standard deviation of 52.3 %. The process parameter development and investigation for the multi-material production of 316L and 17-4PH enabled the successful manufacture of multi-material components without significant deformation. It was also found that the bonding points exhibited good adhesion. The optical evaluation of the density revealed a high relative density of more than 97 %, which indicates that the MIM standard has been achieved. The research hypothesis can therefore be confirmed, and functional products can be realized through the development of the design guidelines and the process parameters.

H2: Secondary research hypothesis

By investigating lattice structures and ML approaches, functional MEX/M products can be optimized with minimal effort.

The development of a parametric design of infill structures and the simulative evaluation in Chapter 5.3 have demonstrated that component weight can be reduced based on the applied loads without altering the contour geometry of the components. The development and implementation of ML models in Chapter 5.4 have demonstrated that a prediction with an RMSE of 67.31 MPa is feasible using the GBR model, thereby allowing the optimization effort for weight reduction at a given load to be completely avoided. The results were successfully applied in Chapter 7.2 for the manufacture of a BPP injection mold, which can simulatively withstand the stresses in the application of micro injection molding, and a closed porous geometry can be realized with a weight reduction of 37.33 % for one half of the mold. Therefore, the research question can be affirmed, demonstrating that components can be effectively optimized using lattice-based infill structures and ML approaches.

H3: Tertiary research hypothesis

The prediction of target parameters such as surface roughness and density is possible using ML in the MEX/M method with minimal experimental data.

The development and implementation of the ML models in Chapter 6, with a small amount of experimental data, have demonstrated that the MLP model is suitable for the prediction of surface roughness for the downskin surface, with an accuracy of 75.71 % and an RMSE of 32.92 μm in the test. The surface roughness of the upskin surface exhibited an RMSE of 138.99 μm in the test and an accuracy of 39.26 %. Therefore, no robust prediction can be made based on these experimental data. The Bagging model provided robust density predictions, achieving a RMSE of 1.88% in relative density and an accuracy of 97.44% on the test data set. Therefore, the research question can therefore be partially answered positively. It is possible to predict the quality parameters density and surface roughness of the downskin area with the given data set. For robust predictions of the roughness in the upskin surface, a larger data set is required.

9.3 Prospects for future research

The research results obtained have contributed to the identification of potential applications in the sinter-based AM process, the MEX/M process. The use of ML models enabled the validation that both process and component optimization are feasible for this manufacturing approach. These findings can serve as a foundation for future research and contribute to meeting quality requirements in both the manufacturing process and the final application. Especially in the production of MIM injection molded parts, the MEX/M process chain is suitable as a complementary solution and can reduce manufacturing and delivery times, and efficiently produce functional products and prototypes.

The investigation of multi-material production in the MEX/M process in Chapter 8 further demonstrated that this process is a key technology for the simple and fast manufacture of multi-material components. This insight opens up further potential applications, such as the development of multi-material components with different infill patterns and must be explored in the future.

10. References

[1] R. Guchhait, B. Sarkar, A decision-making problem for product outsourcing with flexible production under a global supply chain management, International Journal of Production Economics 272 (2024) 109230. https://doi.org/10.1016/j.ijpe.2024.109230.

[2] P. Hou, Y. Zhao, Y. Li, Strategic analysis of supplier integration and encroachment in an outsourcing supply chain, Transportation Research Part E: Logistics and Transportation Review 177 (2023) 103238. https://doi.org/10.1016/j.tre.2023.103238.

[3] R.K. Bachar, S. Bhuniya, S.K. Ghosh, A. AlArjani, E. Attia, Md.S. Uddin, B. Sarkar, Product outsourcing policy for a sustainable flexible manufacturing system with reworking and green investment, MBE 20 (2022) 1376–1401. https://doi.org/10.3934/mbe.2023062.

[4] A. Fischer, S. Gebauer, E. Khavkin, 3D-Druck im Unternehmen: Entscheidungsmodelle, Best Practices und Anwendungsbeispiele: am Beispiel Fused Layer Modeling (FLM), Hanser, München, 2018.

[5] L. Waalkes, Potenzialerschließung und -bewertung der sinterbasierten Kolbenextrusion, Springer Berlin Heidelberg, Berlin, Heidelberg, 2023. https://doi.org/10.1007/978-3-662-66883-2.

[6] A. Ullah, K. Asami, L. Holtz, T. Röver, K. Azher, K. Bartsch, C. Emmelmann, A Machine Learning Approach for Mechanical Component Design Based on Topology Optimization Considering the Restrictions of Additive Manufacturing, JMMP 8 (2024) 220. https://doi.org/10.3390/jmmp8050220.

[7] Damilola Oluwaseun Ogundipe, Sodiq Odetunde Babatunde, Emmanuel Adeyemi Abaku, AI AND PRODUCT MANAGEMENT: A THEORETICAL OVERVIEW FROM IDEA TO MARKET, Int. j. Manag. Entrep. Res 6 (2024) 950–969. https://doi.org/10.51594/ijmer.v6i3.965.

[8] ISO/ASTM 52900:2023 - Additive manufacturing — General principles - Part positioning, coordinates and orientation, (2023).

[9] W. Muhammad, A.P. Brahme, O. Ibragimova, J. Kang, K. Inal, A machine learning framework to predict local strain distribution and the evolution of plastic anisotropy & fracture in additively manufactured alloys, International Journal of Plasticity 136 (2021) 102867. https://doi.org/10.1016/j.ijplas.2020.102867.

[10] A. Olleak, Z. Xi, Calibration and Validation Framework for Selective Laser Melting Process Based on Multi-Fidelity Models and Limited Experiment Data, Journal of Mechanical Design 142 (2020) 081701. https://doi.org/10.1115/1.4045744.

[11] J.R. Deneault, J. Chang, J. Myung, D. Hooper, A. Armstrong, M. Pitt, B. Maruyama, Toward autonomous additive manufacturing: Bayesian optimization on a 3D printer, MRS Bulletin 46 (2021) 566–575. https://doi.org/10.1557/s43577-021-00051-1.

[12] F. Ciccone, A. Bacciaglia, A. Ceruti, Optimization with artificial intelligence in additive manufacturing: a systematic review, J Braz. Soc. Mech. Sci. Eng. 45 (2023) 303. https://doi.org/10.1007/s40430-023-04200-2.

[13] P.V. Osswald, S.K. Mustafa, C. Kaa, P. Obst, M. Friedrich, M. Pfeil, D. Rietzel, G. Witt, Optimization of the production processes of powder-based additive manufacturing technologies by means of a machine learning model for the temporal

K.Asami, *Analysis of Sinter-Based Additive Manufacturing for Functional Products*, Light Engineering für die Praxis, https://doi.org/10.1007/978-3-032-28433-4_10

prognosis of the build and cooling phase, Prod. Eng. Res. Devel. 14 (2020) 677–691. https://doi.org/10.1007/s11740-020-00987-4.

[14] C. Silbernagel, A. Aremu, I. Ashcroft, Using machine learning to aid in the parameter optimisation process for metal-based additive manufacturing, RPJ 26 (2019) 625–637. https://doi.org/10.1108/RPJ-08-2019-0213.

[15] D.S. Shin, C.H. Lee, U. Kühn, S.C. Lee, S.J. Park, H. Schwab, S. Scudino, K. Kosiba, Optimizing laser powder bed fusion of Ti-5Al-5V-5Mo-3Cr by artificial intelligence, Journal of Alloys and Compounds 862 (2021) 158018. https://doi.org/10.1016/j.jallcom.2020.158018.

[16] S. Liu, A.P. Stebner, B.B. Kappes, X. Zhang, Machine learning for knowledge transfer across multiple metals additive manufacturing printers, Additive Manufacturing 39 (2021) 101877. https://doi.org/10.1016/j.addma.2021.101877.

[17] M. Kuehne, K. Bartsch, B. Bossen, C. Emmelmann, Predicting melt track geometry and part density in laser powder bed fusion of metals using machine learning, Progress in Additive Manufacturing (2023). https://doi.org/10.15480/882.4950.

[18] J. Zhu, H. Zhou, C. Wang, L. Zhou, S. Yuan, W. Zhang, A review of topology optimization for additive manufacturing: Status and challenges, Chinese Journal of Aeronautics 34 (2021) 91–110. https://doi.org/10.1016/j.cja.2020.09.020.

[19] S.M.E. Sepasgozar, A. Shi, L. Yang, S. Shirowzhan, D.J. Edwards, Additive Manufacturing Applications for Industry 4.0: A Systematic Critical Review, Buildings 10 (2020) 231. https://doi.org/10.3390/buildings10120231.

[20] K. Bartsch, Digitalization of design for support structures in laser powder bed fusion of metals, Springer Nature Switzerland, Cham, 2023. https://doi.org/10.1007/978-3-031-22956-5.

[21] M. Munsch, M. Schmidt-Lehr, E. Wycisk, AMPOWER REPORT 2024 "Additive Manufacturing Market Report," AMPOWER GmbH & Co. KG, 2024. https://additive-manufacturing-report.com/.

[22] ISO/ASTM 52911-1:2019, Additive manufacturing — Design, Part 1: Laser-based powder bed fusion of metals, (2019).

[23] ISO/ASTM 52900:2021 (en) Additive manufacturing — General principles — Fundamentals and vocabulary, (2021).

[24] I. Gibson, D. Rosen, B. Stucker, M. Khorasani, Additive Manufacturing Technologies, Springer International Publishing, Cham, 2021. https://doi.org/10.1007/978-3-030-56127-7.

[25] Y. Zhang, S. Bai, M. Riede, E. Garratt, A. Roch, A comprehensive study on fused filament fabrication of Ti-6Al-4V structures, Additive Manufacturing 34 (2020) 101256. https://doi.org/10.1016/j.addma.2020.101256.

[26] M. Munsch, M. Schmidt-Lehr, E. Wycisk, Metal Additive Manufacturing with sinter-based technologies, (2018).

[27] M. Mun, M. Schmidt-Lehr, E. Wycisk, Metal Additive Manufacturing Report 2019, AMPOWER Report (2019).

[28] R. Matsuzaki, T. Kanatani, A. Todoroki, Multi-material additive manufacturing of polymers and metals using fused filament fabrication and electroforming, Additive Manufacturing 29 (2019) 100812. https://doi.org/10.1016/j.addma.2019.100812.

[29] M. Annoni, H. Giberti, M. Strano, Feasibility Study of an Extrusion-based Direct Metal Additive Manufacturing Technique, Procedia Manufacturing 5 (2016) 916–927. https://doi.org/10.1016/j.promfg.2016.08.079.

[30] Y. Thompson, J. Gonzalez-Gutierrez, C. Kukla, P. Felfer, Fused filament fabrication, debinding and sintering as a low cost additive manufacturing method of

316L stainless steel, Additive Manufacturing 30 (2019) 100861. https://doi.org/10.1016/j.addma.2019.100861.

[31] C. Suwanpreecha, A. Manonukul, A Review on Material Extrusion Additive Manufacturing of Metal and How It Compares with Metal Injection Moulding, Metals 12 (2022) 429. https://doi.org/10.3390/met12030429.

[32] N. Warrier, K.H. Kate, Fused filament fabrication 3D printing with low-melt alloys, Prog Addit Manuf 3 (2018) 51–63. https://doi.org/10.1007/s40964-018-0050-6.

[33] A. Gebhardt, Additive Fertigungsverfahren: additive Manufacturing und 3D-Drucken für Prototyping - Tooling - Produktion, 5., neu bearbeitete und erweiterte Auflage, Hanser, München, 2016.

[34] S. Hazra, Gourab Ganguly, Subhajit Chattopadhyay, Slicing on the Tessellated Model for the 3D Printing Process based on the Variable Thickness Layers, JME 19 (2024) 082–088. https://doi.org/10.37255/jme.v19i3pp082-088.

[35] V. Kuznetsov, A. Solonin, O. Urzhumtsev, R. Schilling, A. Tavitov, Strength of PLA Components Fabricated with Fused Deposition Technology Using a Desktop 3D Printer as a Function of Geometrical Parameters of the Process, Polymers 10 (2018) 313. https://doi.org/10.3390/polym10030313.

[36] G. Singh, J.-M. Missiaen, D. Bouvard, J.-M. Chaix, Copper additive manufacturing using MIM feedstock: adjustment of printing, debinding, and sintering parameters for processing dense and defectless parts, Int J Adv Manuf Technol 115 (2021) 449–462. https://doi.org/10.1007/s00170-021-07188-y.

[37] A. Bellini, L. Shor, S.I. Guceri, New developments in fused deposition modeling of ceramics, Rapid Prototyping Journal 11 (2005) 214–220. https://doi.org/10.1108/13552540510612901.

[38] A. Manonukul, W. Likityingwara, P. Rungkiatnawin, N. Muenya, S. Amoranan, W. Kittinantapol, S. Surapunt, Study of Recycled and Virgin Compounded Metal Injection Moulded Feedstock for Stainless Steel 630, JMMP 1 (2007) 411–420. https://doi.org/10.1299/jmmp.1.411.

[39] H. Valkenaers, F. Vogeler, E. Ferraris, A. Voet, J.-P. Kruth, A Novel Approach to Additive Manufacturing: Screw Extrusion 3D-Printing, in: Proceedings of the 10th International Conference on Multi-Material Micro Manufacture, Research Publishing Services, 2013: pp. 235–238. https://doi.org/10.3850/978-981-07-7247-5-359.

[40] J. Jacob, D. Pejak Simunec, A.E.Z. Kandjani, A. Trinchi, A. Sola, A Review of Fused Filament Fabrication of Metal Parts (Metal FFF): Current Developments and Future Challenges, Technologies 12 (2024) 267. https://doi.org/10.3390/technologies12120267.

[41] R.M. German, Sintering theory and practice, Wiley, New York Chichester, 1996.

[42] J. Jiang, X. Xu, J. Stringer, Support Structures for Additive Manufacturing: A Review, JMMP 2 (2018) 64. https://doi.org/10.3390/jmmp2040064.

[43] R.M. German, A. Bose, Injection molding of metals and ceramics, Metal Powder Industries Federation, Princeton, N.J., U.S.A, 1997.

[44] F. Klocke, Fertigungsverfahren 5: Gießen und Pulvermetallurgie, 5. Aufl. 2018, Springer Berlin Heidelberg, Berlin, Heidelberg, 2018. https://doi.org/10.1007/978-3-662-54728-1.

[45] P. Vervoort, M. Martens, Secondaries for metal injection molding (MIM), in: Handbook of Metal Injection Molding, Elsevier, 2019: pp. 173–194. https://doi.org/10.1016/B978-0-08-102152-1.00010-6.

[46] D. Pejak Simunec, A. Sola, Emerging Research in Conductive Materials for Fused Filament Fabrication: A Critical Review, Adv Eng Mater 24 (2022) 2101476. https://doi.org/10.1002/adem.202101476.

[47] J.C. Horvath, R. Cameron, Mastering 3D printing: a guide to modeling, printing, and prototyping, Second edition, Apress, New York, 2020.

[48] O. Miclette, R. Côté, V. Demers, V. Brailovski, Material extrusion additive manufacturing of low-viscosity metallic feedstocks: Performances of the plunger-based approach, Additive Manufacturing 60 (2022) 103252. https://doi.org/10.1016/j.addma.2022.103252.

[49] D. Godec, S. Cano, C. Holzer, J. Gonzalez-Gutierrez, Optimization of the 3D Printing Parameters for Tensile Properties of Specimens Produced by Fused Filament Fabrication of 17-4PH Stainless Steel, Materials 13 (2020) 774. https://doi.org/10.3390/ma13030774.

[50] M. Condruz, A. Paraschiv, C. Puscasu, HEAT TREATMENT INFLUENCE ON HARDNESS AND MICROSTRUCTURE OF ADAM MANUFACTURED 17-4 PH, V (2018) 39–45.

[51] J. Gonzalez-Gutierrez, S. Cano, S. Schuschnigg, C. Kukla, J. Sapkota, C. Holzer, Additive Manufacturing of Metallic and Ceramic Components by the Material Extrusion of Highly-Filled Polymers: A Review and Future Perspectives, Materials 11 (2018) 840. https://doi.org/10.3390/ma11050840.

[52] M.Á. Caminero, A. Romero Gutiérrez, J.M. Chacón, E. García-Plaza, P.J. Núñez, Effects of fused filament fabrication parameters on the manufacturing of 316L stainless-steel components: geometric and mechanical properties, RPJ 28 (2022) 2004–2026. https://doi.org/10.1108/RPJ-01-2022-0023.

[53] T. Kurose, Y. Abe, M.V.A. Santos, Y. Kanaya, A. Ishigami, S. Tanaka, H. Ito, Influence of the Layer Directions on the Properties of 316L Stainless Steel Parts Fabricated through Fused Deposition of Metals, Materials 13 (2020) 2493. https://doi.org/10.3390/ma13112493.

[54] J. Allum, Proof of full-strength bonding in material extrusion additive manufacturing, PhD Thesis, Loughborough University, 2021.

[55] M. Zhou, X. Zhou, L. Si, P. Chen, M. Li, Y. Zhang, H. Zhou, Modeling of bonding strength for Fused Filament Fabrication considering bonding interface evolution and molecular diffusion, Journal of Manufacturing Processes 68 (2021) 1485–1494. https://doi.org/10.1016/j.jmapro.2021.06.064.

[56] X. Ang, J.Y. Tey, W.H. Yeo, K.P.Y. Shak, A review on metallic and ceramic material extrusion method: Materials, rheology, and printing parameters, Journal of Manufacturing Processes 90 (2023) 28–42. https://doi.org/10.1016/j.jmapro.2023.01.077.

[57] W. Hassan, M.A. Farid, A. Tosi, K. Rane, M. Strano, The effect of printing parameters on sintered properties of extrusion-based additively manufactured stainless steel 316L parts, Int J Adv Manuf Technol 114 (2021) 3057–3067. https://doi.org/10.1007/s00170-021-07047-w.

[58] S. Atatreh, M.S. Alyammahi, H. Vasilyan, T. Alkindi, R.A. Susantyoko, Evaluation of the infill design on the tensile properties of metal parts produced by fused filament fabrication, Results in Engineering 17 (2023) 100954. https://doi.org/10.1016/j.rineng.2023.100954.

[59] H. Dave, J. Davim, Fused Deposition Modeling Based 3D Printing 2021, (n.d.).

[60] M. Quarto, M. Carminati, G. D'Urso, Density and shrinkage evaluation of AISI 316L parts printed via FDM process, Materials and Manufacturing Processes 36 (2021) 1535–1543. https://doi.org/10.1080/10426914.2021.1905830.

[61] G. Singh, J.-M. Missiaen, D. Bouvard, J.-M. Chaix, Copper extrusion 3D printing using metal injection moulding feedstock: Analysis of process parameters for green density and surface roughness optimization, Additive Manufacturing 38 (2021) 101778. https://doi.org/10.1016/j.addma.2020.101778.

[62] D. Syrlybayev, B. Zharylkassyn, A. Seisekulova, M. Akhmetov, A. Perveen, D. Talamona, Optimisation of Strength Properties of FDM Printed Parts—A Critical Review, Polymers 13 (2021) 1587. https://doi.org/10.3390/polym13101587.

[63] J. Ghorbani, P. Koirala, Y.-L. Shen, M. Tehrani, Eliminating voids and reducing mechanical anisotropy in fused filament fabrication parts by adjusting the filament extrusion rate, Journal of Manufacturing Processes 80 (2022) 651–658. https://doi.org/10.1016/j.jmapro.2022.06.026.

[64] R. Srinivasan, K. Nirmal Kumar, A. Jenish Ibrahim, K.V. Anandu, R. Gurudhevan, Impact of fused deposition process parameter (infill pattern) on the strength of PETG part, Materials Today: Proceedings 27 (2020) 1801–1805. https://doi.org/10.1016/j.matpr.2020.03.777.

[65] E. Moritzer, C.L. Elsner, C. Schumacher, Investigation of METAL-POLYMER composites manufactured by fused deposition modeling with regard to process parameters, Polymer Composites 42 (2021) 6065–6079. https://doi.org/10.1002/pc.26285.

[66] T. Rosnitschek, A. Seefeldt, B. Alber-Laukant, T. Neumeyer, V. Altstädt, S. Tremmel, Correlations of Geometry and Infill Degree of Extrusion Additively Manufactured 316L Stainless Steel Components, Materials 14 (2021) 5173. https://doi.org/10.3390/ma14185173.

[67] I. Ait-Mansour, N. Kretzschmar, S. Chekurov, M. Salmi, J. Rech, Design-dependent shrinkage compensation modeling and mechanical property targeting of metal FFF, Prog Addit Manuf 5 (2020) 51–57. https://doi.org/10.1007/s40964-020-00124-8.

[68] B. Liu, Y. Wang, Z. Lin, T. Zhang, Creating metal parts by Fused Deposition Modeling and Sintering, Materials Letters 263 (2020) 127252. https://doi.org/10.1016/j.matlet.2019.127252.

[69] P. Singh, V.K. Balla, A. Gokce, S.V. Atre, K.H. Kate, Additive manufacturing of Ti-6Al-4V alloy by metal fused filament fabrication (MF3): producing parts comparable to that of metal injection molding, Prog Addit Manuf 6 (2021) 593–606. https://doi.org/10.1007/s40964-021-00167-5.

[70] J. Gonzalez-Gutierrez, S. Cano, J.V. Ecker, M. Kitzmantel, F. Arbeiter, C. Kukla, C. Holzer, Bending Properties of Lightweight Copper Specimens with Different Infill Patterns Produced by Material Extrusion Additive Manufacturing, Solvent Debinding and Sintering, Applied Sciences 11 (2021) 7262. https://doi.org/10.3390/app11167262.

[71] C. Suwanpreecha, P. Seensattayawong, V. Vadhanakovint, A. Manonukul, Influence of Specimen Layout on 17-4PH (AISI 630) Alloys Fabricated by Low-Cost Additive Manufacturing, Metall Mater Trans A 52 (2021) 1999–2009. https://doi.org/10.1007/s11661-021-06211-x.

[72] Y. Abe, T. Kurose, M. Santos, Y. Kanaya, A. Ishigami, S. Tanaka, H. Ito, Effect of Layer Directions on Internal Structures and Tensile Properties of 17-4PH Stainless Steel Parts Fabricated by Fused Deposition of Metals, Materials 14 (2021) 243. https://doi.org/10.3390/ma14020243.

[73] Y. Wang, L. Zhang, X. Li, Z. Yan, On hot isostatic pressing sintering of fused filament fabricated 316L stainless steel – Evaluation of microstructure, porosity, and tensile properties, Materials Letters 296 (2021) 129854. https://doi.org/10.1016/j.matlet.2021.129854.
[74] P. Singh, V.K. Balla, S.V. Atre, R.M. German, K.H. Kate, Factors affecting properties of Ti-6Al-4V alloy additive manufactured by metal fused filament fabrication, Powder Technology 386 (2021) 9–19. https://doi.org/10.1016/j.powtec.2021.03.026.
[75] M. S. Alsoufi, M. W. Alhazmi, D. K. Suker, T. A. Alghamdi, R. A. Sabbagh, M. A. Felemban, F. K. Bazuhair, Experimental Characterization of the Influence of Nozzle Temperature in FDM 3D Printed Pure PLA and Advanced PLA+, AJME 7 (2019) 45–60. https://doi.org/10.12691/ajme-7-2-1.
[76] P. Das, R. Chandran, R. Samant, S. Anand, Optimum Part Build Orientation in Additive Manufacturing for Minimizing Part Errors and Support Structures, Procedia Manufacturing 1 (2015) 343–354. https://doi.org/10.1016/j.promfg.2015.09.041.
[77] D.F. Heaney, Handbook of metal injection molding, Woodhead Publishing, 2018.
[78] C. Kukla, J. Gonzalez-Gutierrez, I. Duretek, S. Schuschnigg, C. Holzer, Effect of particle size on the properties of highly-filled polymers for fused filament fabrication, in: AIP Conference Proceedings, AIP Publishing, 2017. https://doi.org/10.1063/1.5016795.
[79] M.-W. Wu, Z.-K. Huang, C.-F. Tseng, K.-S. Hwang, Microstructures, mechanical properties, and fracture behaviors of metal-injection molded 17-4PH stainless steel, Met. Mater. Int. 21 (2015) 531–537. https://doi.org/10.1007/s12540-015-4369-y.
[80] A. Manonukul, S. Songkuea, P. Moonchaleanporn, M. Tange, Effect of weld line positions on the tensile deformation of two-component metal injection moulding, Int J Miner Metall Mater 24 (2017) 1384–1393. https://doi.org/10.1007/s12613-017-1531-1.
[81] M.-S. Huang, H.-C. Hsu, Effect of backbone polymer on properties of 316L stainless steel MIM compact, Journal of Materials Processing Technology 209 (2009) 5527–5535. https://doi.org/10.1016/j.jmatprotec.2009.05.011.
[82] L. Castro, S. Merino, B. Levenfeld, A. Várez, J.M. Torralba, Mechanical properties and pitting corrosion behaviour of 316L stainless steel parts obtained by a modified metal injection moulding process, Journal of Materials Processing Technology 143–144 (2003) 397–402. https://doi.org/10.1016/S0924-0136(03)00402-3.
[83] D.F. Heaney, T.W. Mueller, P.A. Davies, Mechanical properties of metal injection moulded 316L stainless steel using both prealloy and master alloy techniques, Powder Metallurgy 47 (2004) 367–373. https://doi.org/10.1179/003258904225020855.
[84] P. Singh, Q. Shaikh, V.K. Balla, S.V. Atre, K.H. Kate, Estimating Powder-Polymer Material Properties Used in Design for Metal Fused Filament Fabrication (DfMF3), JOM 72 (2020) 485–495. https://doi.org/10.1007/s11837-019-03920-y.
[85] X. Kan, D. Yang, Z. Zhao, J. Sun, 316L FFF binder development and debinding optimization, Mater. Res. Express 8 (2021) 116515. https://doi.org/10.1088/2053-1591/ac3b15.
[86] M.A. Wagner, J. Engel, A. Hadian, F. Clemens, M. Rodriguez-Arbaizar, E. Carreño-Morelli, J.M. Wheeler, R. Spolenak, Filament extrusion-based additive manufacturing of 316L stainless steel: Effects of sintering conditions on the

microstructure and mechanical properties, Additive Manufacturing 59 (2022) 103147. https://doi.org/10.1016/j.addma.2022.103147.
[87] K. Kudo, K. Shinagawa, H. Miura, Effect of α + β region sintering on the mechanical properties of injection molded Ti-6Al-4V compacts, Mechanical Engineering Journal 5 (2018) 17-00686-17–00686. https://doi.org/10.1299/mej.17-00686.
[88] O.M. Ferri, T. Ebel, R. Bormann, Influence of surface quality and porosity on fatigue behaviour of Ti–6Al–4V components processed by MIM, Materials Science and Engineering: A 527 (2010) 1800–1805. https://doi.org/10.1016/j.msea.2009.11.007.
[89] I. Subuki, M.H. Ismail, A. Amir, M.A. Omar, Effect of Stearic Acid on Rheological Properties of 316L Feedstock for Metal injection Moulding, Jurnal Teknologi 59 (2012). https://doi.org/10.11113/jt.v59.2587.
[90] S. Cano, J. Gonzalez-Gutierrez, J. Sapkota, M. Spoerk, F. Arbeiter, S. Schuschnigg, C. Holzer, C. Kukla, Additive manufacturing of zirconia parts by fused filament fabrication and solvent debinding: Selection of binder formulation, Additive Manufacturing 26 (2019) 117–128. https://doi.org/10.1016/j.addma.2019.01.001.
[91] G. Herranz, C. Berges, J. Hidalgo, Material Extrusion Additive Technologies: Benefits, Challenges, and Niche Applications, in: Additive Manufacturing - Present and Sustainable Future, Materials and Applications [Working Title], IntechOpen, 2025. https://doi.org/10.5772/intechopen.1008718.
[92] M. Sadaf, M. Bragaglia, F. Nanni, A simple route for additive manufacturing of 316L stainless steel via Fused Filament Fabrication, Journal of Manufacturing Processes 67 (2021) 141–150. https://doi.org/10.1016/j.jmapro.2021.04.055.
[93] R.K. Enneti, V.P. Onbattuvelli, O. Gulsoy, K.H. Kate, S.V. Atre, Powder-binder formulation and compound manufacture in metal injection molding (MIM), in: Handbook of Metal Injection Molding, Elsevier, 2019: pp. 57–88. https://doi.org/10.1016/B978-0-08-102152-1.00005-2.
[94] P. Singh, Materials-processing relationships for metal fused filament fabrication of Ti-6Al-4V alloy., University of Louisville, 2020. https://doi.org/10.18297/etd/3388.
[95] A. Das, E.L. Gilmer, S. Biria, M.J. Bortner, Importance of Polymer Rheology on Material Extrusion Additive Manufacturing: Correlating Process Physics to Print Properties, ACS Appl. Polym. Mater. 3 (2021) 1218–1249. https://doi.org/10.1021/acsapm.0c01228.
[96] C. Duty, C. Ajinjeru, V. Kishore, B. Compton, N. Hmeidat, X. Chen, P. Liu, A.A. Hassen, J. Lindahl, V. Kunc, What makes a material printable? A viscoelastic model for extrusion-based 3D printing of polymers, Journal of Manufacturing Processes 35 (2018) 526–537. https://doi.org/10.1016/j.jmapro.2018.08.008.
[97] Sintering and Grain Growth, in: Ceramic Materials, Springer New York, New York, NY, 2007: pp. 427–443. https://doi.org/10.1007/978-0-387-46271-4_24.
[98] E.P.M. Association, others, Metal Injection Moulding: A Manufacturing Process for Precision Engineering Components, European Powder Metallurgy Association: Shrewsbury, UK (2013).
[99] S. Riecker, J. Clouse, T. Studnitzky, O. Andersen, B. Kieback, Fused deposition modeling–opportunities for cheap metal AM, in: European Congress and Exhibition on Powder Metallurgy. European PM Conference Proceedings, The European Powder Metallurgy Association, 2016: pp. 1–6.

[100] J. Gonzalez-Gutierrez, D. Godec, C. Kukla, T. Schlauf, C. Burkhardt, C. Holzer, Shaping, debinding and sintering of steel components via fused filament fabrication, in: 16th International Scientific Conference on Production Engineering, 2017: pp. 8–10.

[101] K. Gante Lokesha Renukaradhya, Metal filament 3D printing of SS316L: Focusing on the printing process, (2019).

[102] J. Damon, S. Dietrich, S. Gorantla, U. Popp, B. Okolo, V. Schulze, Process porosity and mechanical performance of fused filament fabricated 316L stainless steel, RPJ 25 (2019) 1319–1327. https://doi.org/10.1108/RPJ-01-2019-0002.

[103] D. Jiang, F. Ning, Anisotropic deformation of 316L stainless steel overhang structures built by material extrusion based additive manufacturing, Additive Manufacturing 50 (2022) 102545. https://doi.org/10.1016/j.addma.2021.102545.

[104] M.Á. Caminero, A. Romero, J.M. Chacón, P.J. Núñez, E. García-Plaza, G.P. Rodríguez, Additive manufacturing of 316L stainless-steel structures using fused filament fabrication technology: mechanical and geometric properties, RPJ 27 (2021) 583–591. https://doi.org/10.1108/RPJ-06-2020-0120.

[105] D. Jiang, F. Ning, Additive Manufacturing of 316L Stainless Steel by a Printing-Debinding-Sintering Method: Effects of Microstructure on Fatigue Property, Journal of Manufacturing Science and Engineering 143 (2021) 091007. https://doi.org/10.1115/1.4050190.

[106] M. Mousapour, M. Salmi, L. Klemettinen, J. Partanen, Feasibility study of producing multi-metal parts by Fused Filament Fabrication (FFF) technique, Journal of Manufacturing Processes 67 (2021) 438–446. https://doi.org/10.1016/j.jmapro.2021.05.021.

[107] F. Cerejo, D. Gatões, M.T. Vieira, Optimization of metallic powder filaments for additive manufacturing extrusion (MEX), Int J Adv Manuf Technol 115 (2021) 2449–2464. https://doi.org/10.1007/s00170-021-07043-0.

[108] M.A. Wagner, A. Hadian, T. Sebastian, F. Clemens, T. Schweizer, M. Rodriguez-Arbaizar, E. Carreño-Morelli, R. Spolenak, Fused filament fabrication of stainless steel structures - from binder development to sintered properties, Additive Manufacturing 49 (2022) 102472. https://doi.org/10.1016/j.addma.2021.102472.

[109] K. Asami, J.M. Crego Lozares, A. Ullah, B. Bossen, L. Clague, C. Emmelmann, Material extrusion of metals: Enabling multi-material alloys in additive manufacturing, Materials Today Communications 38 (2024) 107889. https://doi.org/10.1016/j.mtcomm.2023.107889.

[110] Metal Powder Industries Federation, Materials Standards for Metal Injection Molded Parts, (2016). ISBN 978-1-943694-05-1.

[111] E.A. Olevsky, R.M. German, Effect of gravity on dimensional change during sintering—I. Shrinkage anisotropy, Acta Materialia 48 (2000) 1153–1166. https://doi.org/10.1016/S1359-6454(99)00368-7.

[112] J. Kranz, Methodik und Richtlinien für die Konstruktion von laseradditiv gefertigten Leichtbaustrukturen, Springer Berlin Heidelberg, Berlin, Heidelberg, 2017. https://doi.org/10.1007/978-3-662-55339-8.

[113] M. Kamal, G. Rizza, Design for metal additive manufacturing for aerospace applications, in: Additive Manufacturing for the Aerospace Industry, Elsevier, 2019: pp. 67–86. https://doi.org/10.1016/B978-0-12-814062-8.00005-4.

[114] C. Emmelmann, D. Herzog, J. Kranz, Design for laser additive manufacturing, in: Laser Additive Manufacturing, Elsevier, 2017: pp. 259–279. https://doi.org/10.1016/B978-0-08-100433-3.00010-5.

[115] D. Herzog, K. Asami, C. Scholl, C. Ohle, C. Emmelmann, A. Sharma, N. Markovic, A. Harris, Design guidelines for laser powder bed fusion in Inconel 718, Journal of Laser Applications 34 (2022) 012015. https://doi.org/10.2351/7.0000508.

[116] J. Kranz, D. Herzog, C. Emmelmann, Design guidelines for laser additive manufacturing of lightweight structures in TiAl6V4, Journal of Laser Applications 27 (2015) S14001. https://doi.org/10.2351/1.4885235.

[117] H. Blunk, A. Seibel, Design guidelines for metal binder jetting, Prog Addit Manuf 9 (2024) 725–732. https://doi.org/10.1007/s40964-023-00475-y.

[118] DIN EN ISO/ASTM 52910:2020-02, Additive Fertigung_- Konstruktion_- Anforderungen, Richtlinien und Empfehlungen (ISO/ASTM 52910:2018); Deutsche Fassung EN_ISO/ASTM 52910:2019, (n.d.). https://doi.org/10.31030/3111254.

[119] G.A. Adam, Systematische Erarbeitung von Konstruktionsregeln für die additiven Fertigungsverfahren Lasersintern, Laserschmelzen und Fused Deposition Modeling, Shaker, 2015.

[120] P. Wang, J. Song, M.L.S. Nai, J. Wei, Experimental analysis of additively manufactured component and design guidelines for lightweight structures: A case study using electron beam melting, Additive Manufacturing 33 (2020) 101088.

[121] M. Mani, P. Witherell, H. Jee, Design Rules for Additive Manufacturing: A Categorization, in: Volume 1: 37th Computers and Information in Engineering Conference, American Society of Mechanical Engineers, Cleveland, Ohio, USA, 2017: p. V001T02A035. https://doi.org/10.1115/DETC2017-68446.

[122] Markforged, Inc., Design Guide for 3D Printing with Metals, (2025). https://3dologie.com/wp-content/uploads/2020/06/Markforged-Metal-Design-Guide.pdf?srsltid=AfmBOopWaJZtY_zSlmhMj_CFV8agFZDrx-LUCLVMumywJFlT5Hdhoror (accessed March 23, 2025).

[123] T.M. Mitchell, Machine Learning, McGraw-Hill, New York, 1997.

[124] J. Frochte, Maschinelles Lernen: Grundlagen und Algorithmen in Python, 3., überarbeitete und erweiterte Auflage, Hanser, München, 2021. https://doi.org/10.3139/9783446463554.

[125] K. Chatzilygeroudis, I. Hatzilygeroudis, I. Perikos, Machine Learning Basics, in: P. Eslambolchilar, A. Komninos, M. Dunlop (Eds.), Intelligent Computing for Interactive System Design, 1st ed., ACM, New York, NY, USA, 2021: pp. 143–193. https://doi.org/10.1145/3447404.3447414.

[126] A.F.A.H. Alnuaimi, T.H.K. Albaldawi, An overview of machine learning classification techniques, BIO Web Conf. 97 (2024) 00133. https://doi.org/10.1051/bioconf/20249700133.

[127] K. Bakshi, K. Bakshi, Considerations for artificial intelligence and machine learning: Approaches and use cases, in: 2018 IEEE Aerospace Conference, IEEE, Big Sky, MT, 2018: pp. 1–9. https://doi.org/10.1109/AERO.2018.8396488.

[128] S.V. Wawre, S.N. Deshmukh, Classification using Machine Learning Techniques, in: 2016. https://api.semanticscholar.org/CorpusID:29754985.

[129] A.K.A. Kumar, Predicting Diabetes using Machine Learning Technique, International Journal of Advance Research, Ideas and Innovations in Technology 6 (2020) 37–43. https://api.semanticscholar.org/CorpusID:221761820.

[130] Karthikeyan P., K. Velswamy, P. Harshavardhanan, Rajagopal R., JeyaKrishnan V., Velliangiri S., Machine Learning Techniques Application: Social Media, Agriculture, and Scheduling in Distributed Systems, in: S. Velayutham (Ed.),

Advances in Computational Intelligence and Robotics, IGI Global, 2020: pp. 380–401. https://doi.org/10.4018/978-1-5225-9902-9.ch020.

[131] C.-F. Tsai, Y.-H. Hu, Empirical comparison of supervised learning techniques for missing value imputation, Knowl Inf Syst 64 (2022) 1047–1075. https://doi.org/10.1007/s10115-022-01661-0.

[132] I. Goodfellow, Y. Bengio, A. Courville, Deep Learning, MIT Press, 2016.

[133] A. Géron, Hands-on machine learning with Scikit-Learn, Keras, and TensorFlow: concepts, tools, and techniques to build intelligent systems, Second edition, O'Reilly, Beijing Boston Farnham Sebastopol Tokyo, 2019.

[134] R. Kruse, C. Borgelt, F. Klawonn, C. Moewes, M. Steinbrecher, P. Held, Multi-Layer Perceptrons, in: Computational Intelligence, Springer London, London, 2013: pp. 47–81. https://doi.org/10.1007/978-1-4471-5013-8_5.

[135] H. Zhu, Multi-layered perceptron and its applications in biotechnology, TNS 20 (2023) 159–165. https://doi.org/10.54254/2753-8818/20/20230753.

[136] Y. Hu, Comparison and Analysis of the Effectiveness of Linear Regression, Decision Tree, and Random Forest Models for Health Insurance Premium Forecasting, AEMPS 79 (2024) 347–353. https://doi.org/10.54254/2754-1169/79/20241754.

[137] N.S.S.V.S. Rao, S.J.J. Thangaraj, Flight Ticket Prediction using Random Forest Regressor Compared with Decision Tree Regressor, in: 2023 Eighth International Conference on Science Technology Engineering and Mathematics (ICONSTEM), IEEE, Chennai, India, 2023: pp. 1–5. https://doi.org/10.1109/ICONSTEM56934.2023.10142260.

[138] D. Gao, Health Insurance Premium Prediction Based on Multiple Machine Learning Algorithms, ACE 100 (2025) 145–152. https://doi.org/10.54254/2755-2721/2025.20593.

[139] C. Li, Machine Learning-Based Models for Accurate Car Prices Prediction, HBEM 40 (2024) 416–421. https://doi.org/10.54097/9zcpv779.

[140] A. Murphy, C. Moore, Random forest (machine learning), in: Radiopaedia.Org, Radiopaedia.org, 2019. https://doi.org/10.53347/rID-67772.

[141] F.J. Livingston, Implementation of Breiman's Random Forest Machine Learning Algorithm, in: 2005. https://api.semanticscholar.org/CorpusID:59894003.

[142] P. Probst, Hyperparameters, tuning and meta-learning for random forest and other machine learning algorithms, in: 2019. https://api.semanticscholar.org/CorpusID:201710457.

[143] C. Zhang, P. Zhong, M. Liu, Q. Song, Z. Liang, X. Wang, Hybrid Metric K-Nearest Neighbor Algorithm and Applications, Mathematical Problems in Engineering 2022 (2022) 1–15. https://doi.org/10.1155/2022/8212546.

[144] K.-L. Du, M.N.S. Swamy, Fundamentals of Machine Learning, in: Neural Networks and Statistical Learning, Springer London, London, 2014: pp. 15–65. https://doi.org/10.1007/978-1-4471-5571-3_2.

[145] F. Bodendorf, J. Franke, A machine learning approach to estimate product costs in the early product design phase: a use case from the automotive industry, Procedia CIRP 100 (2021) 643–648. https://doi.org/10.1016/j.procir.2021.05.137.

[146] K. Asami, S. Roth, M. Krukenberg, T. Röver, D. Herzog, C. Emmelmann, Predictive modeling of lattice structure design for 316L stainless steel using machine learning in the L-PBF process, Journal of Laser Applications 35 (2023) 042046. https://doi.org/10.2351/7.0001174.

[147] B. Everitt, A. Skrondal, The Cambridge dictionary of statistics, 4th edition, Cambridge University Press, Cambridge New York, 2010.

[148] L. Jin, X. Zhai, K. Wang, K. Zhang, D. Wu, A. Nazir, J. Jiang, W.-H. Liao, Big data, machine learning, and digital twin assisted additive manufacturing: A review, Materials & Design 244 (2024) 113086. https://doi.org/10.1016/j.matdes.2024.113086.

[149] L. Meng, B. McWilliams, W. Jarosinski, H.-Y. Park, Y.-G. Jung, J. Lee, J. Zhang, Machine Learning in Additive Manufacturing: A Review, JOM 72 (2020) 2363–2377. https://doi.org/10.1007/s11837-020-04155-y.

[150] A. Farrag, Y. Yang, N. Cao, D. Won, Y. Jin, Physics-Informed Machine Learning for metal additive manufacturing, Prog Addit Manuf 10 (2025) 171–185. https://doi.org/10.1007/s40964-024-00612-1.

[151] X. Yao, S.K. Moon, G. Bi, A hybrid machine learning approach for additive manufacturing design feature recommendation, RPJ 23 (2017) 983–997. https://doi.org/10.1108/RPJ-03-2016-0041.

[152] H. Ko, P. Witherell, Y. Lu, S. Kim, D.W. Rosen, Machine learning and knowledge graph based design rule construction for additive manufacturing, Additive Manufacturing 37 (2021) 101620. https://doi.org/10.1016/j.addma.2020.101620.

[153] D. Wu, Y. Wei, J. Terpenny, Predictive modelling of surface roughness in fused deposition modelling using data fusion, International Journal of Production Research 57 (2019) 3992–4006. https://doi.org/10.1080/00207543.2018.1505058.

[154] J. Zhang, P. Wang, R.X. Gao, Deep learning-based tensile strength prediction in fused deposition modeling, Computers in Industry 107 (2019) 11–21. https://doi.org/10.1016/j.compind.2019.01.011.

[155] F. Wang, S. Fathizadan, F. Ju, K. Rowe, N. Hofmann, Print surface thermal modeling and layer time control for large-scale additive manufacturing, IEEE Transactions on Automation Science and Engineering 18 (2020) 244–254.

[156] Z. Zhu, N. Anwer, Q. Huang, L. Mathieu, Machine learning in tolerancing for additive manufacturing, CIRP Annals 67 (2018) 157–160. https://doi.org/10.1016/j.cirp.2018.04.119.

[157] N.A. Fountas, N.M. Vaxevanidis, Optimization of fused deposition modeling process using a virus-evolutionary genetic algorithm, Computers in Industry 125 (2021) 103371. https://doi.org/10.1016/j.compind.2020.103371.

[158] E. Luis, H.M. Pan, S.L. Sing, A.K. Bastola, G.D. Goh, G.L. Goh, H.K.J. Tan, R. Bajpai, J. Song, W.Y. Yeong, Silicone 3D Printing: Process Optimization, Product Biocompatibility, and Reliability of Silicone Meniscus Implants, 3D Printing and Additive Manufacturing 6 (2019) 319–332. https://doi.org/10.1089/3dp.2018.0226.

[159] M. Malviya, K. Desai, Build Orientation Optimization for Strength Enhancement of FDM Parts Using Machine Learning based Algorithm, CADandA 17 (2019) 783–796. https://doi.org/10.14733/cadaps.2020.783-796.

[160] N. Després, E. Cyr, P. Setoodeh, M. Mohammadi, Deep Learning and Design for Additive Manufacturing: A Framework for Microlattice Architecture, JOM 72 (2020) 2408–2418. https://doi.org/10.1007/s11837-020-04131-6.

[161] N. Hertlein, P.R. Buskohl, A. Gillman, K. Vemaganti, S. Anand, Generative adversarial network for early-stage design flexibility in topology optimization for additive manufacturing, Journal of Manufacturing Systems 59 (2021) 675–685. https://doi.org/10.1016/j.jmsy.2021.04.007.

[162] J. Huang, T.-H. Kwok, C. Zhou, W. Xu, Surfel convolutional neural network for support detection in additive manufacturing, Int J Adv Manuf Technol 105 (2019) 3593–3604. https://doi.org/10.1007/s00170-019-03792-1.

[163] G. Liu, Y. Xiong, D.W. Rosen, Multidisciplinary design optimization in design for additive manufacturing, Journal of Computational Design and Engineering 9 (2021) 128–143. https://doi.org/10.1093/jcde/qwab073.

[164] J.D. Alejandrino, R.S.I. Concepcion, S.C. Lauguico, R.R. Tobias, L. Venancio, D. Macasaet, A.A. Bandala, E.P. Dadios, A Machine Learning Approach of Lattice Infill Pattern for Increasing Material Efficiency in Additive Manufacturing Processes, IJMERR (2020) 1253–1263. https://doi.org/10.18178/ijmerr.9.9.1253-1263.

[165] Z. Zhang, J. Femi-Oyetoro, I. Fidan, M. Ismail, M. Allen, Prediction of Dimensional Changes of Low-Cost Metal Material Extrusion Fabricated Parts Using Machine Learning Techniques, Metals 11 (2021) 690. https://doi.org/10.3390/met11050690.

[166] N. Khan, A. Riccio, A systematic review of design for additive manufacturing of aerospace lattice structures: Current trends and future directions, Progress in Aerospace Sciences 149 (2024) 101021. https://doi.org/10.1016/j.paerosci.2024.101021.

[167] K. Asami, S. Roth, M. Krukenberg, C. Emmelmann, Design Optimization of Additively Manufactured Components Using Simulation-Based Analysis of Infill Structures, in: Volume 2: Advanced Design and Information Technologies, American Society of Mechanical Engineers, New Orleans, Louisiana, USA, 2023: p. V002T02A026. https://doi.org/10.1115/IMECE2023-113065.

[168] C. Pan, Y. Han, J. Lu, Design and Optimization of Lattice Structures: A Review, Applied Sciences 10 (2020) 6374. https://doi.org/10.3390/app10186374.

[169] D. Bhate, Four Questions in Cellular Material Design, Materials 12 (2019) 1060. https://doi.org/10.3390/ma12071060.

[170] T.A. Schaedler, W.B. Carter, Architected Cellular Materials, Annu. Rev. Mater. Res. 46 (2016) 187–210. https://doi.org/10.1146/annurev-matsci-070115-031624.

[171] L.J. Gibson, M.F. Ashby, B.A. Harley, Cellular materials in nature and medicine, Cambridge University Press, 2010.

[172] L.J. Gibson, Cellular solids, Mrs Bulletin 28 (2003) 270–274.

[173] Y. Tang, Y.F. Zhao, A survey of the design methods for additive manufacturing to improve functional performance, Rapid Prototyping Journal 22 (2016) 569–590. https://doi.org/10.1108/RPJ-01-2015-0011.

[174] F. Teng, Y. Sun, S. Guo, B. Gao, G. Yu, Topological and Mechanical Properties of Different Lattice Structures Based on Additive Manufacturing, Micromachines 13 (2022) 1017. https://doi.org/10.3390/mi13071017.

[175] O. Al-Ketan, R.K. Abu Al-Rub, Multifunctional Mechanical Metamaterials Based on Triply Periodic Minimal Surface Lattices, Adv Eng Mater 21 (2019) 1900524. https://doi.org/10.1002/adem.201900524.

[176] S. AlMahri, R. Santiago, D.-W. Lee, H. Ramos, H. Alabdouli, M. Alteneiji, Z. Guan, W. Cantwell, M. Alves, Evaluation of the dynamic response of triply periodic minimal surfaces subjected to high strain-rate compression, Additive Manufacturing 46 (2021) 102220. https://doi.org/10.1016/j.addma.2021.102220.

[177] H.A. Schwarz, Gesammelte mathematische abhandlungen, American Mathematical Soc., 1972.

[178] M. Leary, Design of titanium implants for additive manufacturing, in: Titanium in Medical and Dental Applications, Elsevier, 2018: pp. 203–224. https://doi.org/10.1016/B978-0-12-812456-7.00009-3.

[179] Y. Chen, M.-H. Fu, A novel three-dimensional auxetic lattice meta-material with enhanced stiffness, Smart Mater. Struct. 26 (2017) 105029. https://doi.org/10.1088/1361-665X/aa819e.

[180] Y. Xu, D. Zhang, S. Hu, R. Chen, Y. Gu, X. Kong, J. Tao, Y. Jiang, Mechanical properties tailoring of topology optimized and selective laser melting fabricated Ti6Al4V lattice structure, Journal of the Mechanical Behavior of Biomedical Materials 99 (2019) 225–239. https://doi.org/10.1016/j.jmbbm.2019.06.021.

[181] J. León-Becerra, O.A. González-Estrada, J. Quiroga, Effect of Relative Density in In-Plane Mechanical Properties of Common 3D-Printed Polylactic Acid Lattice Structures, ACS Omega 6 (2021) 29830–29838. https://doi.org/10.1021/acsomega.1c04295.

[182] H. Zhang, Y. Wang, Z. Kang, Topology optimization for concurrent design of layer-wise graded lattice materials and structures, International Journal of Engineering Science 138 (2019) 26–49. https://doi.org/10.1016/j.ijengsci.2019.01.006.

[183] Y.M. Hailu, A. Nazir, S.-C. Lin, J.-Y. Jeng, The Effect of Functional Gradient Material Distribution and Patterning on Torsional Properties of Lattice Structures Manufactured Using MultiJet Fusion Technology, Materials 14 (2021) 6521. https://doi.org/10.3390/ma14216521.

[184] J. Plocher, A. Panesar, Effect of density and unit cell size grading on the stiffness and energy absorption of short fibre-reinforced functionally graded lattice structures, Additive Manufacturing 33 (2020) 101171. https://doi.org/10.1016/j.addma.2020.101171.

[185] H.S. Abdulhadi, A. Mian, Effect of strut length and orientation on elastic mechanical response of modified body-centered cubic lattice structures, Proceedings of the Institution of Mechanical Engineers, Part L: Journal of Materials: Design and Applications 233 (2019) 2219–2233. https://doi.org/10.1177/1464420719841084.

[186] L.P.S. Hartanti, Haniel, P.W. Anggoro, R. Ismail, J. Jamari, A.P. Bayuseno, Optimization Fused Deposition Modeling Parameter of Interference Screw using Full Factorial Design, in: 2023 IEEE International Biomedical Instrumentation and Technology Conference (IBITeC), IEEE, Yogyakarta, Indonesia, 2023: pp. 1–6. https://doi.org/10.1109/IBITeC59006.2023.10390976.

[187] M.T. Zaky, Effect of solvent debinding variables on the shape maintenance of green molded bodies, Journal of Materials Science 39 (2004) 3397–3402. https://doi.org/10.1023/B:JMSC.0000026942.64551.97.

[188] A.R. Boccaccini, N.P. Bansal, eds., Ceramics and composites processing methods, John Wiley & Sons, Hoboken, N.J, 2012. https://doi.org/10.1002/9781118176665.

[189] R. Telle, ed., Keramik, Springer Berlin Heidelberg, 2007. https://doi.org/10.1007/978-3-540-49469-0.

[190] W. Weißbach, M. Dahms, C. Jaroschek, Werkstoffkunde: Strukturen, Eigenschaften, Prüfung, 19., vollst. überarb. u. erw. Aufl. 2015, Springer Vieweg, Wiesbaden, 2015. https://doi.org/10.1007/978-3-658-03919-6.

[191] N. Kurgan, Effects of sintering atmosphere on microstructure and mechanical property of sintered powder metallurgy 316L stainless steel, Materials & Design (1980-2015) 52 (2013) 995–998. https://doi.org/10.1016/j.matdes.2013.06.035.

[192] P.S. Schütz, Schadensanalytik Vorgangsweise, Fraktographie und Beispiele, Diplomarbeit, Montanuniversität Leoben, 2015. https://pureadmin.unileoben.ac.at/ws/portalfiles/portal/4717158/AC12703554.pdf.

[193] European Powder Metallurgy Association, ed., Developing the powder metallurgy future: powder production, consolidation technologies, materials, applications, tools for improving PM: World PM2022 Congress & Exhibition: proceedings: 9-13 October 2022, Lyon Convention Centre, Lyon, France, European Powder Metallurgy Association (EPMA), Shrewsbury, United Kingdom, 2022.

[194] M.K. Asami, D. Herzog, B. Bossen, L. Geyer, C. Klemp, C. Emmelmann, Design Guidelines For Green Parts Manufactured With Stainless Steel In The Filament Based Material Extrusion Process For Metals (MEX/M), (2022). https://doi.org/10.15480/882.9046.

[195] K. Asami, M.P.R. Medapati, T. Rakow, T. Röver, C. Emmelmann, Design Guidelines for Material Extrusion of Metals (MEX/M), JETA 3 (2025) 15. https://doi.org/10.3390/jeta3020015.

[196] M.K. Asami, M.P.R. Medapati, T. Rakow, T. Röver, C. Emmelmann, Supplementary Data: Design Guidelines for Material Extrusion of Metals (MEX/M), (2025). https://doi.org/10.15480/882.14720.

[197] E28 Committee, Test Methods of Compression Testing of Metallic Materials at Room Temperature, (2018). https://doi.org/10.1520/E0009-09.

[198] ISO 7438:2020(en) Metallic materials — Bend test, (2020).

[199] M.K. Asami, Parametric Design of Infillstructures as CAD Data for the publication entitled: Design Optimization of Additively Manufactured Components Using Simulation-Based Analysis of Infill Structures, (2024). https://doi.org/10.15480/882.9622.

[200] M.K. Asami, S. Roth, Additional Data: ML-Algorithm Source Code; for Publication: PREDICTIVE MODELING OF LATTICE STRUCTURE DESIGN FOR 316L STAINLESS STEEL USING MACHINE LEARNING IN THE L-PBF PROCESS, (2024). https://doi.org/10.15480/882.8838.

[201] K. Asami, M. Kuehne, T. Röver, C. Emmelmann, Application of Machine Learning in Predicting Quality Parameters in Metal Material Extrusion (MEX/M), Metals 15 (2025) 505. https://doi.org/10.3390/met15050505.

[202] M.K. Asami, M. Kuehne, Machine Learning Models and Data for the Application of Machine Learning in Predicting Quality Parameters in Metal Material Extrusion (MEX/M), (2025). https://doi.org/10.15480/882.14880.

[203] K. Asami, S. Roth, J. Hünting, T. Röver, C. Emmelmann, Metallic Bipolar Plate Production Through Additive Manufacturing: Contrasting MEX/M and PBF-LB/M Approaches, JETA 3 (2025) 12. https://doi.org/10.3390/jeta3020012.

[204] M.K. Asami, S. Roth, J. Hünting, T. Röver, C. Emmelmann, Metallic Bipolar Plate Production Through Additive Manufacturing: Contrasting MEX/M and PBF-LB/M Approaches, (2025). https://doi.org/10.15480/882.14567.

[205] M.K. Asami, S. Prathveraj, P. Prathamesh, T. Röver, C. Emmelmann, Automatic Infill Generation with Honeycomb and Grid Structures in MIM Mold Design for Metallic Bipolar Plates, (2025). https://doi.org/10.15480/882.14884.

[206] D.F. Heaney, Designing for metal injection molding (MIM), in: Handbook of Metal Injection Molding, Elsevier, 2012: pp. 29–49. https://doi.org/10.1533/9780857096234.1.29.

[207] M.K. Asami, D. Herzog, T. Deutschmann, T. Röver, I. Kelbassa, C. Emmelmann, Cost Application for the Publication: Methodology for Cost Estimation Using

Characteristic Factors in Additive Manufacturing, (2024). https://doi.org/10.15480/882.9688.

[208] K. Asami, D. Herzog, T. Deutschmann, T. Röver, I. Kelbassa, C. Emmelmann, Methodology for Cost Estimation Using Characteristic Factors in Additive Manufacturing, Journal of the Japan Society of Powder and Powder Metallurgy 72 (2025) S75–S82. https://doi.org/10.2497/jjspm.14B-T6-02.

[209] T. Goldmann, W.-C. Huang, S. Rzepa, J. Džugan, R. Sedláček, M. Daniel, Additive Manufacturing of Honeycomb Lattice Structure—From Theoretical Models to Polymer and Metal Products, Materials 15 (2022) 1838. https://doi.org/10.3390/ma15051838.

[210] A. Mishra, V.S. Jatti, E.M. Sefene, S. Paliwal, Explainable Artificial Intelligence (XAI) and Supervised Machine Learning-based Algorithms for Prediction of Surface Roughness of Additively Manufactured Polylactic Acid (PLA) Specimens, Applied Mechanics 4 (2023) 668–698. https://doi.org/10.3390/applmech4020034.

[211] B. Bossen, M. Kuehne, O. Kristanovski, C. Emmelmann, Data-driven density prediction of AlSi10Mg parts produced by laser powder bed fusion using machine learning and finite element simulation, Journal of Laser Applications (2023). https://doi.org/10.15480/882.8765.

[212] G. Yang, J. Mo, Z. Kang, Y. Dohrmann, F.A. List, J.B. Green, S.S. Babu, F.-Y. Zhang, Fully printed and integrated electrolyzer cells with additive manufacturing for high-efficiency water splitting, Applied Energy 215 (2018) 202–210. https://doi.org/10.1016/j.apenergy.2018.02.001.

[213] K. Lin, J. Qiao, K. Shi, W. Dong, D. Gu, Laser powder bed fusion of micro-channels for the application of proton exchange membrane fuel cell bipolar plates, CIRP Journal of Manufacturing Science and Technology 43 (2023) 193–204. https://doi.org/10.1016/j.cirpj.2023.01.007.

[214] K. Lin, Y. Xu, D. Gu, J. Shan, K. Shi, W. Zhang, Laser Powder Bed Fusion of Multifunctional Bio-inspired Vertical Honeycomb Sandwich Structures: For the Application of Lightweight Bipolar Plates of Proton Exchange Membrane Fuel Cells, Chin. J. Mech. Eng. 37 (2024) 102. https://doi.org/10.1186/s10033-024-01088-4.

[215] Desktop Metal, Inc., Injection molds with conformal cooling channels, (2022). https://www.desktopmetal.com/resources/metal-3d-printing-tooling (accessed March 27, 2025).

[216] Desktop Metal, Inc., Innovative Plastics: Better injection molds with metal 3D printing, (2022). https://www.desktopmetal.com/resources/innovative-plastics-injection-molding (accessed March 27, 2025).

[217] K. Nishiyabu, Micro Metal Powder Injection Molding, in: J. Wang (Ed.), Some Critical Issues for Injection Molding, InTech, 2012. https://doi.org/10.5772/34498.

[218] J. Hidalgo, C. Berges, G. Herranz, Mesostructured emerging 316L-17-4PH composite steel parts by material extrusion for the property maximization through an efficient microstructural design, Int J Adv Manuf Technol 135 (2024) 3343–3362. https://doi.org/10.1007/s00170-024-14659-5.

Appendix A

A.1 Parameter variation of the shaping process

Table A.1.1 Parameter varation for the influence identification of the target values.

No.	Temperature [°C]	Speed [mm/s]	Laxer height [mm]	Extrusion Multiplier	Density [g/cm³]	Surface roughness Sa [µm]	Surface roughness Sz [µm]	Dimension [mm]		
								x	y	z
1	120	35	0.4	1,0	4.75	27.95	152.1	10.15	10.12	10.26
2	120	35	0.4	1,2	4.59	58.91	323.68	10.50	10.51	10.56
3	120	35	0.4	1.35	5.04	52.71	261.68	11.19	11.17	10.72
4	120	35	0.2	1	4.63	22.99	154.39	10.12	10.15	10.27
5	120	35	0.2	1.2	4.56	22.8	164.14	10.19	10.09	10.18
6	120	35	0.2	1.35	5.07	43.53	297.45	11.06	11.05	11.06
7	120	70	0.4	1	4.71	35.28	241.44	10.14	10.18	10.27
8	120	70	0.4	1.2	4.54	35.29	272.78	10.15	10.21	10.24
9	120	70	0.4	1.35	4.73	50.42	265.93	10.27	10.34	10.28
10	120	70	0.2	1	4.68	24.84	163.29	9.95	10.03	10.18
11	120	70	0.2	1.2	4.99	23.8	163.38	10.07	10.15	10.19
12	120	70	0.2	1.35	5.11	22.48	139.16	10.36	10.44	10.23
13	150	35	0.4	1	4.69	38.08	237.67	10.10	10.17	10.26
14	150	35	0.4	1.2	5.06	37.51	248.76	10.42	10.44	10.22
15	150	35	0.4	1.35	5.12	47.52	266.31	11.05	11.11	10.27
16	150	35	0.2	1	4.57	28.67	174.81	10.28	10.17	10.19
17	150	35	0.2	1.2	5.06	20.52	149.31	10.34	10.40	10.30
18	150	35	0.2	1.35	5.15	24.42	184.07	10.67	10.81	10.26
19	150	70	0.4	1	4.62	36.65	226.24	10.12	10.05	10.20
20	150	70	0.4	1.2	5.03	52.05	299.56	10.39	10.33	10.17
21	150	70	0.4	1.35	5.09	65.94	362.21	11.07	11.16	10.29
22	150	70	0.2	1	4.62	28.69	194.79	10.17	10.29	10.23
23	150	70	0.2	1.2	5.04	25.86	167.27	10.29	10.40	10.21
24	150	70	0.2	1.35	5.11	26.5	163.41	10.99	10.90	10.26
25	135	35	0.4	1	4.55	34.9	191.04	10.17	10.08	10.08
26	135	35	0.4	1.2	4.96	33.95	196.45	10.31	10.23	10.12
27	135	35	0.4	1.35	5.12	56.24	293.59	10.68	10.71	10.22
28	135	35	0.2	1	4.41	39.16	254.1	10.10	10.17	10.19
29	135	35	0.2	1.2	4.93	28.25	159.4	10.17	10.27	10.15
30	135	35	0.2	1.35	4.82	45.81	209.8	10.77	10.80	10.21
31	135	70	0.4	1	4.57	39.08	250.28	10.05	9.95	10.11
32	135	70	0.4	1.2	5.03	36.98	236.17	10.30	10.22	10.11
33	135	70	0.4	1.35	5.09	60.31	313.04	10.76	10.80	10.23
34	135	70	0.2	1	4.55	24.33	221.94	10.12	10.20	10.15
35	135	70	0.2	1.2	5.09	25.43	153.24	10.33	10.41	10.16
36	135	70	0.2	1.35	5.08	22.97	137.33	10.85	10.92	10.29
37	120	30	0.4	1	4.52	33.42	192.75	10.11	10.03	10.12
38	120	40	0.4	1.2	4.70	44.51	228.46	10.28	10.18	10.13
39	120	50	0.4	1.35	4.79	35.2	232.27	10.39	10.30	10.14
40	120	60	0.2	1	4.13	19.22	129.03	9.98	10.08	10.12
41	120	80	0.2	1.35	4.82	33.99	188.4	10.71	10.73	10.26
42	125	30	0.4	1	4.47	31.6	185	10.12	10.05	10.18
43	125	40	0.4	1.2	4.84	41.63	221.15	10.41	10.32	10.21
44	125	50	0.4	1.35	4.85	50.97	282.25	10.94	10.87	10.30
45	125	60	0.2	1	4.26	16.4	97.51	10.02	10.11	10.16
46	125	70	0.2	1.2	4.74	19.62	131.48	10.20	10.31	10.19
47	125	80	0.2	1.35	4.80	32.96	194.84	10.65	10.69	10.26
48	130	30	0.4	1	4.40	39.55	198.56	10.13	10.00	10.12
49	130	40	0.4	1.2	4.84	45.8	245.37	10.34	10.29	10.15
50	130	50	0.4	1.35	4.87	56.34	323.07	10.87	10.82	10.28
51	130	60	0.2	1	4.26	15.95	108.55	9.99	10.10	10.14
52	130	70	0.2	1.2	4.69	17.45	110.21	10.17	10.30	10.15

K.Asami, *Analysis of Sinter-Based Additive Manufacturing for Functional Products*, Light Engineering für die Praxis, https://doi.org/10.1007/978-3-032-28433-4

53	130	80	0.2	1.35	4.84	45.04	237.88	10.68	10.73	10.23
54	135	30	0.4	1	4.44	32.49	187.52	10.11	9.99	10.10
55	135	40	0.4	1.2	4.80	46.72	251.4	10.39	10.32	10.15
56	135	50	0.4	1.35	4.85	50.5	337.69	10.80	10.72	10.26
57	135	60	0.2	1	4.32	14.74	90.49	9.97	10.07	10.13
58	135	80	0.2	1.35	4.82	48.11	249.31	10.64	10.70	10.21
59	140	30	0.4	1	4.61	33.49	199.15	10.12	10.03	10.14
60	140	40	0.4	1.2	4.87	38.34	242.48	10.51	10.45	10.25
61	140	50	0.4	1.35	4.91	66.93	386.93	11.03	11.14	10.45
62	140	60	0.2	1	4.36	17.68	117.73	9.97	10.20	10.17
63	140	70	0.2	1.2	4.79	18.71	116.04	10.21	10.30	10.18
64	140	80	0.2	1.35	4.81	63.05	319.48	10.91	11.02	10.30
65	145	30	0.4	1	4.43	43.2	245.86	10.20	10.04	10.09
66	145	40	0.4	1.2	4.80	30.81	180.72	10.49	10.39	10.20
67	145	50	0.4	1.35	4.86	44.48	254.51	10.99	10.90	10.32
68	145	60	0.2	1	4.38	22.38	172.78	10.10	10.04	10.16
69	145	70	0.2	1.2	4.81	20.16	114.47	10.15	10.22	10.09
70	145	80	0.2	1.35	4.87	24.44	145.65	10.93	10.98	10.20
71	150	30	0.4	1	4.44	50.6	199.45	10.14	10.15	10.05
72	150	40	0.4	1.2	4.70	46.4	215.91	10.47	10.40	10.10
73	150	50	0.4	1.35	4.85	46.76	249.88	10.91	10.84	10.21
74	150	60	0.2	1	4.42	22.74	114.17	10.20	10.09	10.18
75	150	80	0.2	1.35	4.91	34.91	153.76	10.90	10.62	10.40

A.2 Geometric dimensions of the tensile specimens

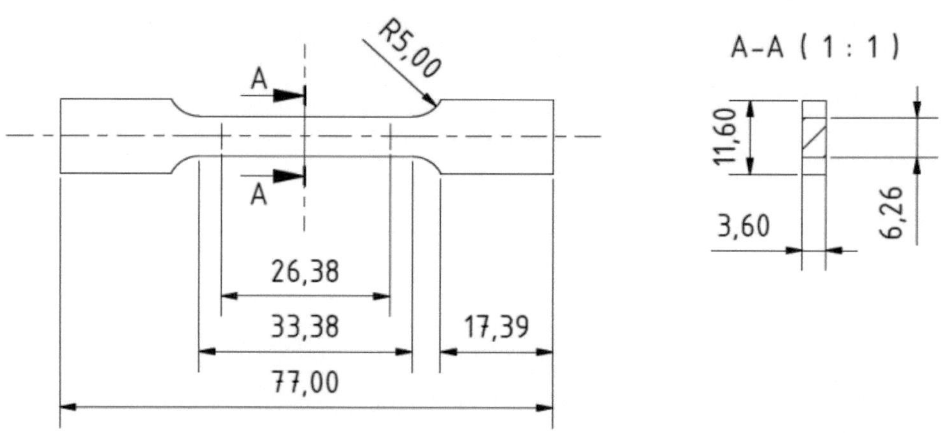

Figure A.1 Geometrical dimension of the tensile specimen.

Appendix B

B.1 Design Guidelines For Green Parts Manufactured With Stainless Steel In The Filament Based Material Extrusion Process For Metals (MEX/M)

The summary of the findings can be found in Chapter 5.1.

K.Asami, *Analysis of Sinter-Based Additive Manufacturing for Functional Products*, Light Engineering für die Praxis, https://doi.org/10.1007/978-3-032-28433-4

World PM2022 - Session 85 : AM Sinter based Technologies - Material Extrusion Applications

Manuscript refereed by Dr Christian Kukla (Montanuniversitaet Leoben, Austria)

Design Guidelines For Green Parts Manufactured With Stainless Steel In The Filament Based Material Extrusion Process For Metals (MEX/M)

Karim Asami[1] karim.asami@tuhh.de; Dirk Herzog[1] dirk.herzog@tuhh.de; Bastian Bossen[1] bastian.bossen@tuhh.de; Leon Geyer[1] leon.geyer@tuhh.de; Clarissa Klemp[1] clarissa.klemp@tuhh.de, Claus Emmelmann[1] c.emmelmann@tuhh.de

[1]Hamburg University of Technology, Institute of Laser and System Technologies, Harburger Schloßstr. 28, 21079 Hamburg, Germany

Abstract

Filament-based material extrusion (MEX/M) presents a rapid and inexpensive alternative to e.g. metal injection molding, particularly for prototype production. The filament consists of metal powder in a plastic matrix and is melted and applied layer by layer until a so-called green body is created. These green parts are subsequently debinded and sintered at high temperatures to form a dense metal component. It is crucial to identify the material-specific and process-specific limits in order to be able to manufacture true to size. This paper therefore develops design guidelines for the MEX/M process for green part manufacturing for the widely used austenitic stainless steel AISI 316L (1.4404). Basic geometrical features such as walls, boreholes and overhangs are studied and influencing factors on the dimensional accuracy are assessed. Based on the results, recommendations for part design are presented.

Introduction

The filament-based material extrusion (MEX/M, also known Metal Fused-Filament Fabrication as per ISO 52900) has become popular over the last years in different fields such as functional prototyping and small scale fabrication [1]. One reason is the existing knowledge of the metal injection molding (MIM) and the fused deposition modelling (FDM) processes. Since the MEX/M process consists of the combination of both MIM and FDM processes it is easy to transfer and to add knowledge that has been previously accumulated. The main advantage compared to MIM is that no mold is needed and the time for its creation can be neglected. Therefore, the limitations in the design complexity of MEX/M-parts are not restricted due to the degree of freedom and the limited shape [2]. Another reason for the popularity is the larger availability of different filament-based materials and machines provided by different suppliers such as BASF SE [3], Virtual foundry [4], Anycubic [5], PT&A [6], Desktop Metal [7], Triditive [8] and Markforged [9]. The MEX/M process consists of three process steps: printing, debinding and sintering. While the debinding and sintering have been extensively studied through the injection molding process (see [10,11]) there hardly exist design studies in the printing step of MEX/M process [12]. Therefore, this paper investigates design features to develop a design catalogue for an accurate component design.

Materials and Methods

For this investigation different design features are introduced. The design limitations and recommendations identified in this paper are dependent on the material, the machine, the particular process parameters and environmental conditions. The basic geometrical elements used for the tests are typical features which occur in a number of industrial parts and have successfully been applied to established design guidelines for other AM processes [13,14,15]. The geometric properties are changed parameter wise to identify the manufacturing limits. Table 1 shows the investigated specimens.

Table 1: Developed specimens for the investigation.

Unsupported Walls	Horizontal Overhangs
	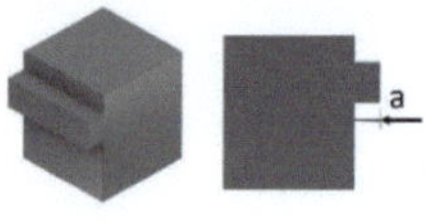
Width (mm): 10/24/32 Height (mm): 0.4/0.6/0.8/1/4.8/6.4/9.6/10/12/14.4/16/19.2 Thickness (mm): 0.1/0.2/.../0.5/0.7/0.8/1/1.2/1.6	Width, Thickness (mm): 5 Height (mm): 6 Overhang length a (mm): 0.3/0.5/1/2 Overhang height (mm): 3

Inclined Walls

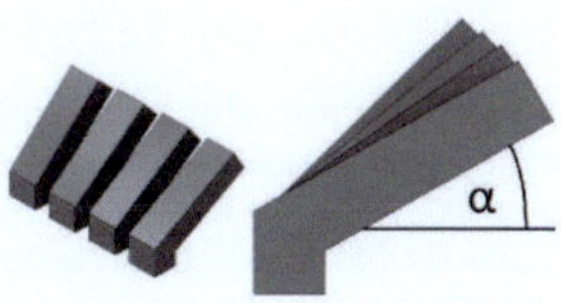

Width (mm): 10
Thickness (mm): 15
Height (mm): 20
Overhang angle α (°): 15,25,30,35,40,45,50

Bridges

Width (mm): 5
Total Height (mm): 8
Total length (mm): 10+ b
Bridge distance b (mm): 0.3/0.5/1/2/3/5

Cylindrical rods

Diameter (mm): 0.8/1.2/1.6/2/3/4/5
Height (mm): 3,4,…,8/10/15/20/30

Vertical gaps

Width, Thickness (mm): 12
Height (mm): 10
Gap width (mm): 0.1, 0.2, …, 0.5

Horizontal boreholes

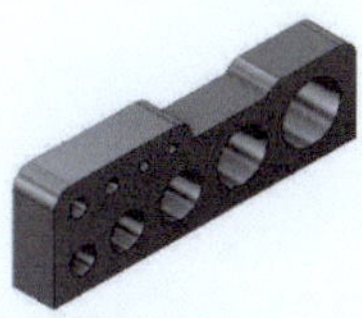

Width (mm): 5
Diameter (mm): 0.8, 1, 1.5, 2, 3, 4, 5, 6, 7

Vertical boreholes

Height (mm): 5
Diameter (mm): 0.8, 1, 1.5, 2, 2.5, 3

The specimens are manufactured with AISI 316L (1.4404) filament provided by the company PT&A (Dresden, Germany) with the starting powder composition illustrated in table 2 are provided by the manufacturer. The filament diameter is 2.85 mm with a tolerance of 0.05 mm.

Table 2: starting powder composition for 316-L (1.4404) in weight %

Cr	Ni	Mo	Mn	Si	P	S	C	F
16.8	10.1	2.2	1.38	0.74	0.02	0.018	0.004	bal.

The components of the binder are not known. According to the manufacturer, debinding with acetone as solvent is recommended.

The test specimens are produced with the MEX/M process on a Renkforce RF 2000 (Conrad Electronic SE, Hirschau, Germany) 3D printer with 0.4 mm nozzle diameter using Simplify3D (Cincinnati, USA) slicer software. The extruder of this printer has been optimized for constant metal filament feeding. The process parameters are chosen to avoid typical printing errors known in FDM such as stringing [16], oozing [16], bridging [17], elephant foot [18], etc. These are the main influence factors for accurate printing, displayed in table 3. The distance and roughness measurements have been performed with a Keyence VK-X150 (Osaka, Japan) microscope, a caliper gauge and a feeler gauge. The specimens (green parts) have been measured at the edges and in the center of each feature. The roughness measurements for the inclined walls were performed on the downskin area, since this area is critical for a free overhang, where the material can sink. Line measurements have been made for the surface according to the method presented in [15].

Table 3: Printing parameters

Extrusion temperature (°C)	135
Heat bed temperature (°C)	40
Printing speed (mm/s)	2100
Extrusion multiplier	1.35
Skirt layer	2
Retraction distance (mm)	4.5
Layer height (mm)	0.2

Results and Discussion

The results from the experiments are given in figures 1-7 in the form of a comparison of the actual geometrical values with the respective target values. Each diagram shows the modeled parameter on the x-axis and the printed quantity as a percentage on the y-axis. For the overhang measurements the overhang angles are plotted over the mean roughness values Rz of the downskin area. Typical defects are additionally visualized using light microscope and photographic images.

Unsupported walls

Figure 1 (left) shows the percentual measurement results of modeled and fabricated heights for different lengths and widths. Furthermore, the differently developed widths at fixed length and height are shown (right). Walls with a width and length of 10 mm are created with a height between 0.4 mm and 4.8 mm. The walls with a width of 1.2 mm are created with a height between 7.2 mm and 14.4 mm. Walls with a width of 1.6 mm were in the range of 6.4 mm to 19.2 mm. All samples could be built and reached >95% of the designed target height with a height higher than 0.4 mm. For the samples with a width of 1.2 mm, the created height tends to decrease at 12 mm and 14.4 mm (gray bars). Figure 1 below shows different target thicknesses for fixed wall widths and lengths of 10 mm each. All specimens have been built up thicker than designed. From a thickness of 1 mm the over-extrusion is less than 20%. The thickness of 0.7 mm has been manufactured nearly as designed.

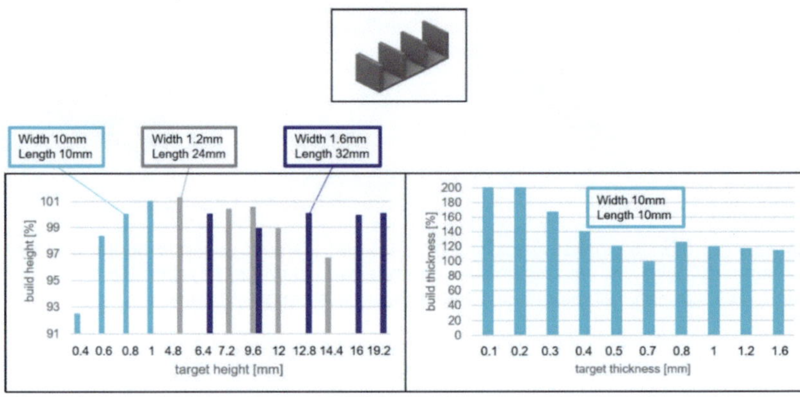

Figure 1: Height (left) and thickness (right) measurements for the unsupported walls.

Overhang, inclined walls and bridge features

The horizontal overhangs could only be produced up to an overhang length of 0.5 mm with a slight material droop in of up to 2 layers in height. Specimens with overhangs exceeding this value show disproportionate deformation of the overhangs. Figure 2 depicts an example of the defects of the straight overhangs, where the lowest filament layer sinks. Also, trying to use a different layer path in printing did not prevent the overhang from sinking.

Figure 2: Representation of the 0.5 mm (left), 1.0 mm (middle) and 2.0 mm (right) overhang and defects

Figure 3 illustrates a diagram of the average roughness values on the downskin surfaces, dependent of the overhang angle. The 0° overhang angle could not be produced. Due to the adhesion during printing with the build platform, overhangs of 15° are manufacturable. The average roughness Rz for the angle 15° is 376.6 µm, for 30° 423.6 µm. Therefore, it can be assumed that the average roughness value for 25° of 227.7 µm can be regarded as an outlier due to measurement inaccuracy. Starting at an overhang angle of more than 30° the average roughness of the downskin surface decreases noticeably. The surface for the overhang angle of 35° has a mean roughness Rz of 179,94 µm and for an overhang angle of 50° a mean roughness of 110.45 µm.

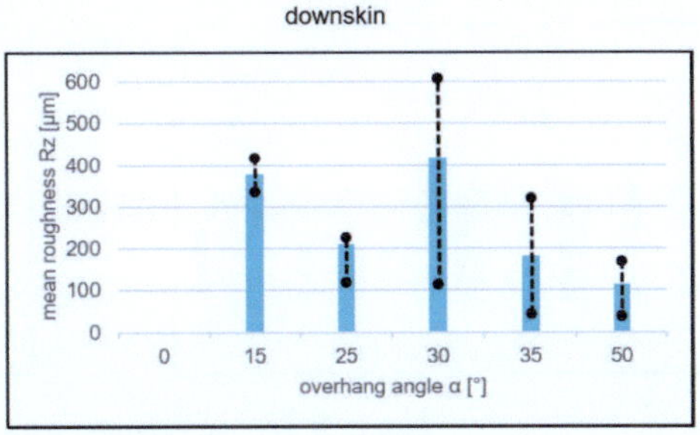

Figure 3: Mean Roughness measurements of the downskin surface for different overhang angle α.

The investigated bridge structures could all be manufactured and are depicted in figure 4. However, deformations are visible for bridge distances greater than 2 mm. With an unsupported bridge distance of 2 mm, it can be observed that individual filament strands sink distinctly. A microscope image of the 1.0 mm bridge distance illustrates the tendency of the lowest layer to bend already at this distance. (Fig. 4 bottom)

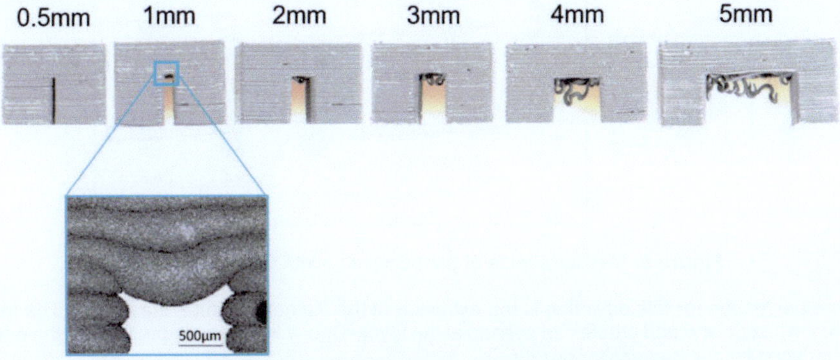

Figure 4: Investigated different bridge size geometries.

Cylindrical rods

The cylindrical rods could be manufactured for a thickness greater than 3 mm and are illustrated in Figure 5 (right). Narrower rods were not producible and tend to deform. In figure 5 (right), it can be seen that the percentage diameter of the created samples is constantly at 97% of the designed thickness. In Figure 5 on the left, different heights of the cylindrical rods have been manufactured. The small rods have a diameter manufactured with a height of 8 mm and the tall rods up to a height of 30 mm. All test geometries were successfully built. It can be observed that all specimens are higher than designed and approach the designed height incrementally. This deviation appears to decrease with increasing height, which could be a result of an over extrusion from the chosen slicer, the layer height and printing parameters.

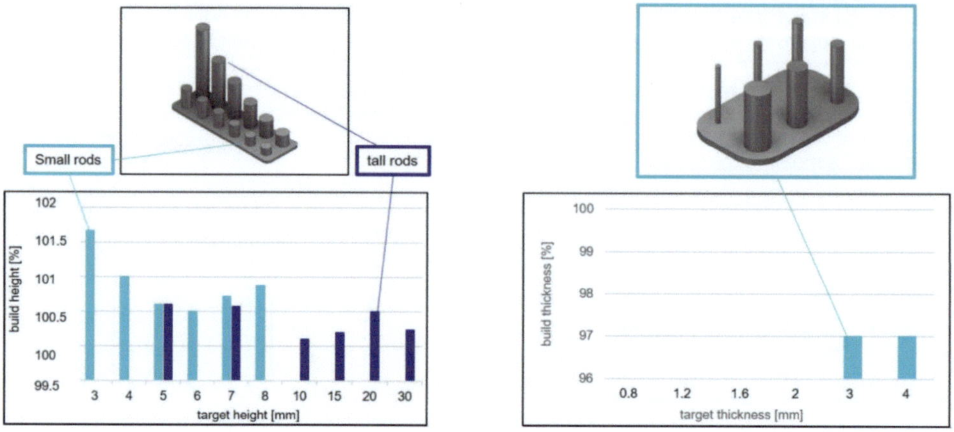

Figure 5: Measurements of the cylindrical rods.

Vertical gaps

The vertical gaps could be manufactured with a gap width of 0.3 mm to 0.5 mm. However, it can be observed that the widths of 0.5 mm are 43% smaller than the modeled widths. (fig. 6)

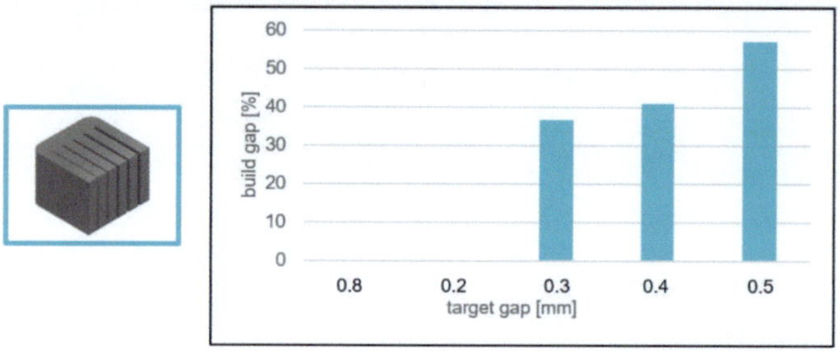

Figure 6: Measurements of the achieved width for vertical gaps.

One possible reason for this deviation is the extrusion of the filament through the nozzle. The material swells at the nozzle end and creates an overlap at the layer edge. It is hereby expected that the deviation can be reduced by using a smaller nozzle.

Boreholes

The diagram in Figure 7 displays the deviation of the diameter from the design in vertical and horizontal orientation. All boreholes could be manufactured in the vertical orientation except the borehole with a diameter of 0.8 mm. At a diameter size of 5 mm in vertical orientation no deviation from the design occurs. The diameters < 5 mm have all been manufactured smaller than designed. The reason for this is potentially the same as for the vertical gaps, where the material swells at the nozzle end and creates an overlap and the effective diameter is reduced.

The horizontal boreholes have been manufactured in the range of 1 mm – 4 mm. In this range the produced boreholes are significantly smaller than the designed. At a diameter of 4 mm, the layer of material in the overhang area sinks and a build-up is not possible. To overcome this failure support structures are needed.

World PM2022 - Session 85 : AM Sinter based Technologies - Material Extrusion Applications

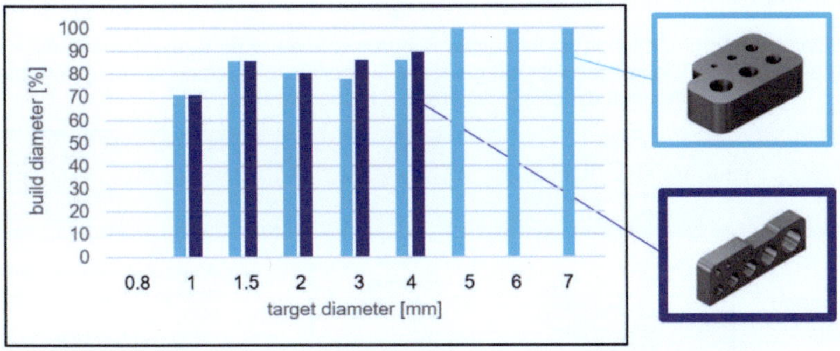

Figure 7: Borehole measurements of the diameter in different orientation.

Discussion

The results of the measurements allowed the development of design recommendations for AISI 316L (1.4404). For the design recommendations, the minimum deviations from the design have been defined as limits. This corresponds to a comparison to the design guidelines for piston-based metal printing [19] with an identical nozzle parameter and material. The results presented show some similarities e.g. vertical boreholes with a diameter of 1 mm can be manufactured and height limitations for unsupported walls in dependence of thickness. However, the results in [19] also show some disadvantages compared to filament-based extrusion. For example, cylindrical rods larger than 5 mm are to be constructed, whereas in the results shown above a build-up is already possible from 3 mm diameter. Furthermore, inclined walls can be produced from an angle of 35° and in piston-based only from an angle of 50°.
A further comparison with the design recommendations Markedforged [20] and Desktop Metal [21] indicates a transferability to the guidelines of this paper. Using the same 0.4 mm nozzle diameter, both guidelines give a recommendation for the cylindrical rods of 3 mm. The horizontal overhangs should be a maximum of 0.5 mm. The minimum wall thickness of the unsupported walls is 1.0 mm and the inclined walls should have a downskin angle of at least 40°. It can be assumed that the machine conditions have a smaller impact on the printable design than the material and process parameters.

Conclusion

Various design features have been manufactured in the MEX/M process using AISI 316L (1.4404). The experimental study of the features allows to identify the limitations for manufacture. Horizontal overhangs and bridges could be produced without a significant deviation with a length of 0.5 mm and 1.0 mm. Unsupported walls with a thickness of 1.6 mm are not limited by height. The inclined walls above 35° have a mean roughness of 310 µm. Cylindrical rods could be manufactured with 3 mm of the diameter, vertical boreholes 4 mm in diameter, while horizontal boreholes in the range of 1 mm to 4 mm were manufacturable. Vertical gaps larger than 0.5 mm show a higher design accuracy. The identified limits to the specimens allow a development of design guidelines for each feature.
With the help of the design recommendations, the knowledge about the process steps can be completed and a successful component design can be realized. Further research is planned for the extension of the design guideline for the debinding and sintering process steps for 316L (1.4404) and other materials. Also, demonstrators are being developed that will improve the mechanical properties related to design accuracy. The guideline is intended to ensure repeatability and save time in the production of a functional component.

World PM2022 - Session 85 : AM Sinter based Technologies - Material Extrusion Applications

Appendix Design Guideline Catalogue

Table 4: Design recommendation for the tested features with a nozzle size of 0.4 mm for AISI 316L (1.4404) after [13,15]. It is noticed that the guideline is specific to machine set-up, parameters and material especially with regard to the identified limits.

Geometry Feature		Recommendation	
Name	Illustration	Value	Comment
unsupported walls		Wall thickness >1.6 mm Height thickness ratio < 12	• Produced thickness max. 15% higher for thickness of 1.6 mm (for 0.4 mm nozzle size) • Height to thickness ratio <12 for thicknesses <1.6 mm to prevent bending
Vertical gaps		Width > 0.5 mm	• Negative width deviation of max. 43% of total width
Inclined walls	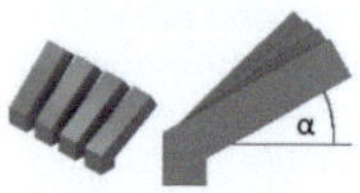	Overhang Angle ≥ 35°	• Manufacturable from 15° • **Downskin:** mean roughness max. ~610 μm for <35° and ~310 μm for 35° angle
Horizontal overhang	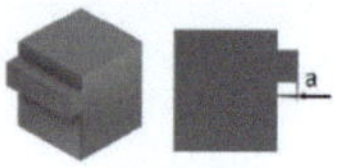	a ≤ 0.5 mm	• Overhang length > 0.5 mm must be supported to prevent material fall-in
Horizontal bridges		b ≤ 1.0 mm	• Bridge distance of 1.0 mm manufacturable • Greater bridge length should be supported to prevent the material from sinking
Cylindrical rods		Diameter ≥ 3 mm	• Positive height deviation max. 1.5% of total height
Horizontal boreholes		Diameter 1 mm – 4 mm	• No support structure necessary • For Diameter > 4 mm only with supportstructures manufacturable
Vertical boreholes		Diameter ≥ 4 mm	• For diameter in a range of 1 mm - 4 mm borehole size <25% smaller than total diameter • Diameter ≥ 5 mm produced borehole size meets the design

References

[1] Sargini, M.I.M.; Masood, S.H.; Palanisamy, S.; Jayamani, E.; Kapoor, A. Additive manufacturing of an automotive brake pedal by metal fused deposition modelling. Mater. Today Proc. **2021,** 45, 4601–4605.
[2] Khorasani, M.; Ghasemi, A.; Rolfe, B.; Gibson, I. Additive manufacturing a powerful tool for the aerospace industry. Rapid Prototyp. J. **2021,** 28, 87–100.
[3] BASF Ultrafuse, Available online https://www.ultrafusefff.com/product-category/metal/ultrafuse-316l/ (accessed on 26 April 2022)
[4] Virtual foundry, Available online https://filament2print.com/gb/45_the-virtual-foundry (accessed on 26 April 2022)
[5] Anycubic, Available online https://www.anycubic.com/products/316l-metal-filament-1-75mm (accessed on 26 April 2022)
[6] PT&A, Available online http://www.pt-a.de/ (accessed on 26 April 2022)
[7] Desktop Metal, Available online https://www.desktopmetal.com/products/studio (accessed on 26 April 2022)
[8] Triditive, Available online https://www.triditive.com (accessed on 26 April 2022)
[9] Markedforged, Available online https://markforged.com (accessed on 26 April 2022)
[10] Heaney, D.F. Handbook of Metal Injection Molding; Woodhead Publishing: Sawston, UK, 2018
[11] European Powder Metallurgy Association. Metal Injection Moulding: A Manufacturing Process for Precision Engineering Components; European Powder Metallurgy Association: Shrewsbury, UK, 2013
[12] C. Suwanpreecha, A. Manonukul, A review on material extrusion additive manufacturing of metal and how it compares with metal injection moulding, J 2022, 48-49
[13] J. Kranz, D. Herzog, and C. Emmelmann, Design guidelines for laser additive manufacturing of lightweight structures in TiAl6V4, J. Laser Appl. 27, S14001 (**2015**).
[14] L. Rebaioli and I. Fassi, A review on benchmark artifacts for evaluating the geometrical performance of additive manufacturing processes, Int. J. Adv. Manuf. Technol. 93, 2571–2598 (**2017**).
[15] D. Herzog, K. Asami, C. Scholl, C. Ohle, C. Emmelman, A. Sharma, N. Markovic, and A. Harris, Design guidelines for laser powder bed fusion in Inconel 718 (2021)
[16] Simplify3D, Available online https://www.simplify3d.com/support/print-quality-troubleshooting/stringing-or-oozing/ (accessed on 20 April 2022
[17] Simplify3D, Available online https://www.simplify3d.com/support/print-quality-troubleshooting/poor-bridging/ (accessed on 20 April 2022)
[18] All3DP, Available online https://all3dp.com/2/elephant-s-foot-3d-printing-problem-easy-fixes/ (accessed on 20 April 2022)
[19] C. Klemp, Systematic development of design rules for the production of green parts by means of piston-based feedstock extrusion, master thesis, 27 January 2021, Fraunhofer IAPT, TUHH iLAS
[20] Metal X design guideline, Available online https://assets-global.website-files.com/5b782089dfe0db977c49f302/60480ac3b3b559811a63ab4a_Markforged%20Metal%20Design%20Guide.pdf (accessed on 21 April 2022)
[21] BMD Design guide, Available online https://www.desktopmetal.com/uploads/Desktop-Metal-BMD-Design-Guide.pdf (accessed on 21 April 2022)

B.2 Design Guidelines for Material Extrusion of Metals (MEX/M)

The summary of the findings can be found in Chapter 5.2.

Journal of
Experimental and
Theoretical Analyses

Article

Design Guidelines for Material Extrusion of Metals (MEX/M)

Karim Asami, Mehar Prakash Reddy Medapati, Titus Rakow, Tim Röver and Claus Emmelmann

MDPI

https://doi.org/10.3390/jeta3020015

Journal of *Experimental and Theoretical Analyses*

MDPI

Article

Design Guidelines for Material Extrusion of Metals (MEX/M)

Karim Asami [1,2,*], Mehar Prakash Reddy Medapati [1,2], Titus Rakow [1], Tim Röver [1,2] and Claus Emmelmann [1,2]

1 Institute of Laser and Systems Technology, Hamburg University of Technology, Am Schwarzenberg-Campus 1, 21073 Hamburg, Germany; mehar.medapati@tuhh.de (M.P.R.M.); titus.rakow@tuhh.de (T.R.); tim.roever@tuhh.de (T.R.); c.emmelmann@tuhh.de (C.E.)
2 Institute of Laser and Systems Technology, Harburger Schloßstraße 28, 21079 Hamburg, Germany
* Correspondence: karim.asami@tuhh.de

Abstract: This study introduced a systematic framework to develop practical design guidelines specifically for filament-based material extrusion of metals (MEX/M), an additive manufacturing (AM) process defined by ISO/ASTM 52900. MEX/M provides a cost-efficient alternative to conventional manufacturing methods, which is particularly valuable for rapid prototyping. Although AM offers significant design flexibility, the MEX/M process imposes distinct geometric and process constraints requiring targeted optimization. The research formulates and validates design guidelines tailored for the MEX/M using an austenitic steel 316L (1.4404) alloy filament. The feedstock consists of a uniform blend of 316L stainless steel powder and polymeric binder embedded within a thermoplastic matrix, extruded and deposited layer by layer. Benchmark parts were fabricated to examine geometric feasibility, such as minimum printable wall thickness, feature inclination angles, borehole precision, overhang stability, and achievable resolution of horizontal and vertical gaps. After fabrication, the as-built (green-state) components undergo a two-step thermal post-processing treatment involving binder removal (debinding), followed by sintering at elevated temperatures to reach densification. Geometric accuracy was quantitatively assessed through a 3D scan by comparing the manufactured parts to their original CAD models, allowing the identification of deformation patterns and shrinkage rates. Finally, the practical utility of the developed guidelines was demonstrated by successfully manufacturing an impeller designed according to the established geometric constraints. These design guidelines apply specifically to the machine and filament type utilized in this study.

Keywords: material extrusion of metals (MEX/M); design guidelines; stainless steel 316L (1.4404)

check for updates

Academic Editor: Marco Rossi

Received: 10 April 2025
Revised: 17 May 2025
Accepted: 23 May 2025
Published: 28 May 2025

Citation: Asami, K.; Medapati, M.P.R.; Rakow, T.; Röver, T.; Emmelmann, C. Design Guidelines for Material Extrusion of Metals (MEX/M). *J. Exp. Theor. Anal.* **2025**, *3*, 15. https://doi.org/10.3390/jeta3020015

1. Introduction

Additive manufacturing (AM) technologies have revolutionized the manufacturing landscape by facilitating the direct fabrication of complex geometries from digital models through a layer-by-layer process. Unlike conventional techniques such as milling and casting, AM enables the creation of intricate structures which were previously unattainable with traditional subtractive or formative methods. This capability makes AM particularly advantageous for lightweight, biomimetic, and topology-optimized components that are relevant in the aerospace, biomedical, and automotive industries [1]. Additionally, AM allows the consolidation of multiple parts into single monolithic structures, significantly reducing material usage, assembly complexity, and overall production costs [2]. Among AM technologies, filament-based material extrusion of metals (MEX/M), also known as metal fused-filament fabrication (metal FFF) according to ISO/ASTM 52900 https://www.

iso.org/obp/ui/#iso:std:iso-astm:52900:ed-2:v1:en, 3 May 2025, has attracted significant attention due to its distinctive advantages [2,3]. In MEX/M, a filament composed of metal powder and polymer binder is extruded through a heated nozzle and deposited layer-wise to form a "green" component. Subsequently, the part undergoes thermal processing to achieve final densification. Compared to selective laser melting (SLM), direct metal laser sintering (DMLS), and electron beam melting (EBM) AM methods, MEX/M generates less material waste and mitigates the safety risks associated with the handling of fine metal powders, providing a safer and more cost-effective alternative for metal part production [2].

Furthermore, MEX/M systems are typically compact, versatile, and cost-efficient, expanding their applicability to diverse industrial fields, including aerospace and biomedical sectors [3,4]. However, the process has some drawbacks, such as a low relative density, which does not meet the standards of the metal injection moulding process [2,3]. To meet this standard, the acquisition of an expensive process chain that requires, among other things, a powerful sintering furnace negates the cost-effectiveness compared to other AM processes. Further process-related drawbacks are the low production speed, which is slower due to the multi-stage process; higher surface roughness, which depends on the nozzle diameter [3]; and the limitation of the component size due to the sintering process. These disadvantages make it necessary to investigate process parameters and design aspects in order to identify suitable applications for the MEX/M process.

In this study, the focus is on the production of metal components using austenitic stainless steel 316L (1.4404) via MEX/M [2,3]. 316L stainless steel is chosen for its superior corrosion resistance, high tensile strength, excellent weldability, biocompatibility, and mechanical stability, making it suitable for biomedical implants, chemical processing equipment, and marine structures [5–7]. The capability of 316L stainless steel to fabricate complex geometries with high structural integrity is particularly beneficial for additive manufacturing [8]. Despite its advantages, the MEX/M process introduces several challenges, including minimum printable wall thickness, overhang stability, feature resolution, and significant dimensional shrinkage during the sintering phase [9,10]. Additionally, anisotropic properties resulting from layer-wise deposition affect the mechanical behaviour of final components, highlighting the need for optimized design guidelines [11–13]. Previous studies, such as that of Rebaioli [14], emphasize the necessity of developing comprehensive geometric benchmarks tailored specifically for different AM technologies and materials. Desktop metal has established design guidelines specific to its bound metal deposition (BMD) AM technology [15]. Consequently, the primary objective of this research is to establish a systematic set of design guidelines specific to MEX/M constraints and capabilities. These guidelines are developed with the following aims:

- Assist designers in optimizing part geometries for enhanced manufacturability in the MEX/M process.
- Evaluate the impact of process parameters on shrinkage behaviour, dimensional accuracy, and geometric fidelity.
- Validate the potential of MEX/M for producing high-density metal components.
- Increase technology acceptance and reduce experimental effort and material costs [16]

2. Materials and Methods

In the following section, selected geometric features representative of typical industrial components are described. These features have been effectively integrated into established additive manufacturing (AM) design guidelines [14,16–18]. The guidelines presented here are formulated based on the specific machine parameters and processing conditions used in a controlled environment.

For this investigation, the 316L (1.4404) filament was supplied by PT + A (Dresden, Germany) [19]. The initial powder composition, along with the recommended debinding and sintering parameters, were specified in the datasheet provided by the PT + A, as summarized in Table 1 and illustrated in Figure 1. The filament diameter was 2.85 ± 0.05 mm, and the metal powder exhibited a size distribution between 2 µm and 15 µm. The feedstock was composed of 316L stainless steel powder particles uniformly mixed with a polymeric binder system. This mixture was embedded within a thermoplastic matrix. The chemical composition of the primary and secondary backbone binder is not disclosed by the filament supplier. Benchmark specimens were fabricated to evaluate geometric constraints, such as minimum printable wall thickness, feature inclination angles, borehole precision, overhang stability, and resolution of horizontal and vertical gaps. Table 2 summarizes the process parameters applied in this study to produce green parts. The specific geometric features and their dimensions evaluated in this study are listed in Table 3. The manufactured samples with the resulting geometric deviations can be observed in Table A1 in the Appendix A. The slicing and toolpath generation were performed using Simplify3D software Version 4.1.2 (Simplify3D, Cincinnati, OH, USA), and optimized printing parameters, derived from prior studies, were applied to achieve consistent extrusion and optimal interlayer adhesion. Benchmark specimens were printed using a Renkforce RF 2000 3D printer (Conrad Electronic SE, Hirschau, Germany), employing a direct-drive extruder with constant filament feeding.

Table 1. Material composition for 316L (1.4404), reproduced from [19].

Material	316L
Cr	16.8
Ni	10.1
Mo	2.2
Mn	1.38
Si	0.74
P	0.02
S	0.018
C	0.004
Fe	bal.
Cu	-
N	-
O	-

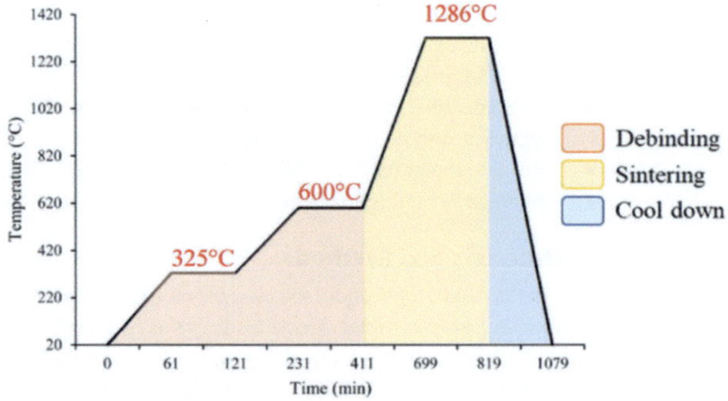

Figure 1. Thermal cycle of the debinding and sintering process adapted from [17].

Table 2. MEX process parameters [18].

Extrusion temperature (°C)	135
Heat bed temperature (°C)	40
Printing speed (mm/s)	2100
Extrusion multiplier	1.35
Skirt layer	2
Retraction distance (mm)	4.5
Layer height (mm)	0.2

Table 3. Selected design features.

Unsupported walls (WoS)  *Thickness (mm): 0.2/0.4/0.6/0.8/1/1.5/2* *Length (mm): 38.5* *Width (mm): 20* *Height (mm): 23*	**Supported horizontal overhangs (ÜZ)**  *Overhang gap (mm): 1/2/3/4/5* *Height (mm): 23* *Width (mm):3*
Supported walls (WmS)  *Thickness (mm): 2/1.5/1/0.8/0.6/0.4/0.2* *Length (mm): 38.5* *Width (mm): 20* *Height (mm): 23*	**Horizontal and vertical boreholes (BH-V and BH-H)** 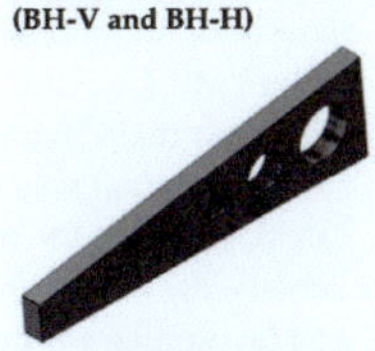 *Bore diameter (mm): 0.4/0.6/0.8/1/2/3/4/6/8*
Vertical cylinder (ZV) 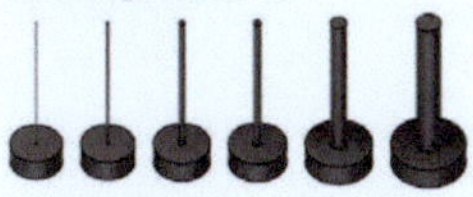 *Bar Ø (mm): 8/6/4/3/2/1/0.8/0.4/0.2* *Height (mm): 20*	**Hollow cylinders (ZH-4)**  *Thickness (mm): 0.4/0.8/1.6/2.4/3.2/4* *Height (mm): 23*
Hollow cylinders (ZH-2)  *Thickness (mm): 0.4/0.8/1.6/2.4/3.2/4* *Height (mm):23*	**Teardrops features (TD)**  *Angle (degree): 60/80/100/120* *Diameter (mm): 6*
Horizontal overhangs (HO)  *Overhang distance (mm): 0.5/1/1.5/2/3/4*	**Inclined walls (DW)** 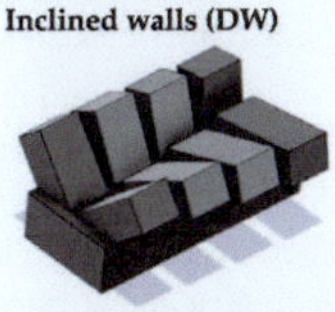 *Incline angle (degree): 10/20/30/40/50/60/70/80*

Chemical debinding of the green-state components was carried out using acetone as the solvent-based debinding agent. The debinding procedure was performed at a controlled temperature at 40 °C for 12 h, maintaining a sample-to-solvent volume ratio of 8:1 according to the filament manufacturer recommended guidelines. Post-debinding analysis indicated an average mass loss of around 5%, consistent with the supplier's provided data. Following chemical debinding, specimens underwent thermal debinding and sintering processes in a high-temperature ExSO90 furnace (Aim3D GmbH, Rostock, Germany). These processes were integrated into a single continuous thermal cycle, with peak sintering temperatures reaching 1286 °C. This temperature was selected based on prior studies to achieve final part densities greater than 95% [17]. To prevent oxidation during thermal processing, a high-purity argon atmosphere (99.9% purity) with a steady gas flow of 1 L/min was maintained throughout the cycle. The thermal debinding consisted of two distinct isothermal stages: an initial stage at 325 °C for 1 h, followed by a second stage at 600 °C for 3 h. The heating rate was controlled at 300 K/h until the first holding temperature was achieved and was subsequently reduced to 150 K/h, optimizing binder removal while minimizing thermal stresses.

2.1. Density and Porosity Measurement

Measurements of the mass loss from chemical debinding and density were conducted based on Archimedes principle, utilizing a precision analytical balance (MJ-300 series, Chyo YMC Co., Ltd., Kyoto, Japan). Porosity was calculated using measurements of dry specimen mass in air (m_B) and the immersed mass in fluid ($m_{B,Fl}$). The density of the component (ρ_B) was then derived from these values using Equation (1). Porosity (p) was calculated from the measured density according to Equation (2). Ethanol, with a known density, $\rho_{Fl} = 0.81\frac{g}{cm^3}$, was chosen as the immersion fluid due to its low surface tension and favourable wetting characteristics, reducing potential measurement inaccuracies from trapped air bubbles.

$$\rho_B = \rho_{Fl} \cdot \frac{m_B}{m_B - m_{B,Fl}} \tag{1}$$

$$p = \frac{\rho_W - \rho_B}{\rho_W} \tag{2}$$

2.2. Dimensional Analysis and Shrinkage Qualification

Dimensional measurements were conducted along the X, Y, and Z axes using a digital Vernier calliper and a Keyence VK-X150 microscope (Osaka, Japan). Each specimen was measured at multiple points, including edges and central regions, to accurately capture variations caused by non-uniform shrinkage. Surface roughness was specifically evaluated on downskin areas of inclined walls, as these regions often experience material sagging and present significant challenges in maintaining dimensional stability during sintering. The Keyence digital microscope used in this study comprises a VHX-7020 digital camera, VH-Z20T lens, VHS-S600E holder, and VHX-970F control unit with an integrated display. Magnification adjustments between 20X and 200X facilitated detailed visual inspections, geometric measurements, and surface roughness evaluations.

2.3. Geometric Accuracy Assessment Using 3D Scanning

Conventional measurement tools were inadequate for accurately capturing the intricate details of the demonstrator impeller produced using the MEX/M process. To address this challenge, a Keyence VL-500 3D coordinate measurement system, offering a measurement accuracy of ±10 μm, was utilized for high-resolution scanning of the component's geometry and surface texture. Scanning operations, data acquisition, and subsequent analysis were managed using Keyence VL-500 Application Series software Version 2.3.6.74 [20].

The scanning parameters specified in Table 4 were uniformly applied to all design features, in both green and sintered states. Due to variations in surface reflectivity at different stages of processing, scanner brightness settings were manually calibrated for each component to enhance data quality. Dimensional analysis of basic geometric features, such as circular elements, was carried out directly within the software using the "Cylinder" function located under the "3D Measurement—Basic Measuring Elements" tab. For measuring distances and angles between planar surfaces, reference planes were first established, after which the "Distance" or "Angle" functions under the "3D Measurement Between Two Planes" tab were employed. In addition to fundamental geometric measurements, the 3D scan data provided supplemental information for assessing shrinkage-induced deformations. This integrated method ensured accurate evaluation of both linear and volumetric shrinkage throughout the manufacturing process.

Table 4. Keyence VL 500 settings.

Size	Abbreviation [Entity]
Change magnification	High
Measurement method	Stitch
Full auto	Manual
Resolution	Fine
Measurement area	Standard
Rotation degree	360°
Rotation segment	60°

Dimensional deviations were visualized by comparing the 3D scan data with proportionally scaled CAD (S-CAD) models using the VL-500 series software. Initially, the point clouds were manually aligned to the STL models using defined reference planes, followed by a precise alignment using the software's built-in "fit" function. Subsequently, dimensional deviations were determined through the "3D compare" function, allowing quantification of tolerance differences between the scanned data and the original CAD geometries. To ensure accurate and meaningful comparisons, the CAD models were scaled according to experimentally derived shrinkage factors. This methodology facilitated reliable assessments of dimensional accuracy, clear identification of deformation patterns, and thorough validation of the shrinkage characteristics inherent to the MEX/M process.

3. Results and Discussion

3.1. Shrinkage Behaviour (WmS, BH, ZHs, and TD)

Shrinkage in the MEX/M process is primarily attributed to the sintering stage [17]. This study systematically investigates the shrinkage behaviour to establish reference values that can be utilized to compensate for dimensional changes during these stages. By incorporating these reference values, components can be initially designed to meet target dimensions and subsequently scaled in slicing software to account for the predicted shrinkage. To evaluate shrinkage behaviour, the external dimensions of the components were measured in three distinct process states: green (as-built), debinded, and sintered. By comparing these measurements, dimensional changes that occurred throughout each stage of the manufacturing process were quantified.

The results of this analysis are presented in Figure 2 and Table 5. Substantial dimensional shrinkage primarily occurs during the sintering stage. During the chemical debinding phase, parts exhibited an average shrinkage of 1.8% ± 0.2% along the X and Y axes, while shrinkage along the Z axis was lower, measured at 0.8% ± 0.3%. After transitioning from the green state to the sintered state, shrinkage intensified, with reduction factors of 0.845 $\pm$ 0.004 in the X and Y directions and 0.851 $\pm$ 0.002 in the Z direction.

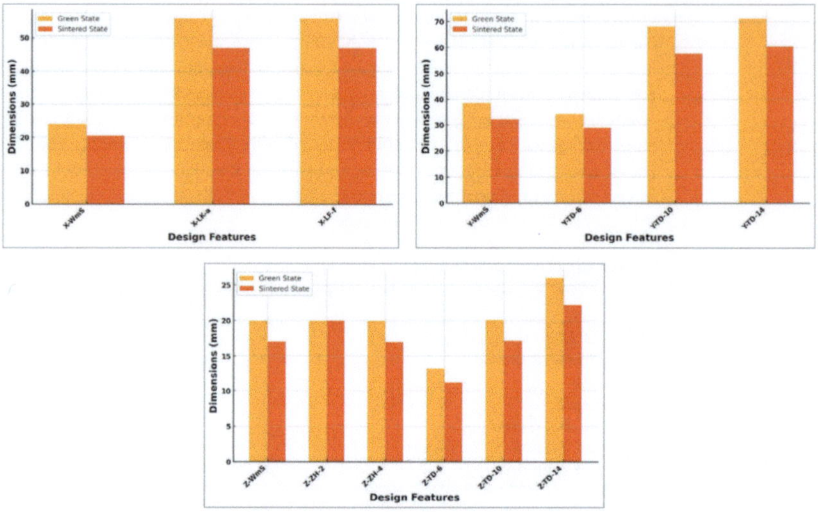

Figure 2. Comparison of shrinkage of design features in X, Y and Z directions (the first letter denotes the axis; WmS: supported walls; LK,a: boreholes manufactured in a vertical direction (BH-V); LK,f: boreholes manufactured flat on the building platform (BH-H);TD: teardrop borehole; ZH: hollow cylinder. The last number denotes the diameter of the feature cavity).

Table 5. Shrinkage behaviour in X, Y and Y axis of design features.

	Design Feature	Green State (mm)	Debinded State (mm)	Sintered State (mm)	Shrinkage (%)
X	WmS	23.93	23.62	20.40	14.8
	LK-a (BH-V)	55.79	54.63	46.94	15.9
	LF-f (BH-H)	55.79	54.82	47.01	15.8
Y	WmS	38.40	37.57	32.22	16.1
	TD-6	34.17	33.56	28.90	15.4
	TD-10	68.05	66.93	57.68	15.2
	TD-14	71.08	69.86	60.33	15.1
Z	WmS	19.93	19.92	17.07	14.3
	ZH-2	19.93	19.80	19.97	14.9
	ZH-4	19.93	19.78	16.97	14.9
	TD-6	13.16	12.96	11.15	15.3
	TD-10	20.10	19.92	17.13	14.8
	TD-14	25.97	25.76	22.13	14.8

This increase in shrinkage was accompanied by greater dimensional variability, as reflected in the higher standard deviations observed post-sintering. Unlike trends reported in previous studies [21,22], shrinkage along the Z axis was lower than in the X and Y directions. This deviation is likely due to the relatively small dimensions and fine features of the fabricated geometries, which contributed to a more isotropic shrinkage behaviour. The experimentally determined shrinkage factors were used to scale the CAD models for dimensional comparisons against the sintered components (S-CAD models). To account for the shrinkage observed during slicing, the recommended scaling factors are 1.183 for the X and Y axes and 1.176 for the Z axis. These adjustments are essential to ensure dimensional accuracy in the final sintered parts.

3.2. Dimensional Deviations of Design Features

3.2.1. Unsupported and Supported Wall Features (WoS and WmS):

Wall thicknesses exceeding 0.6 mm were successfully fabricated without the use of sidewall support, as shown in Figure 3a in the green (as-built) state. The specimen with a nominal wall thickness of 0.6 mm exhibited undersizing of up to 0.23 mm at its upper right corner. Furthermore, the presence of a visible staircase effect in Figure 3 highlights inconsistencies in filament extrusion, indicating non-uniform material deposition during the MEX/M process.

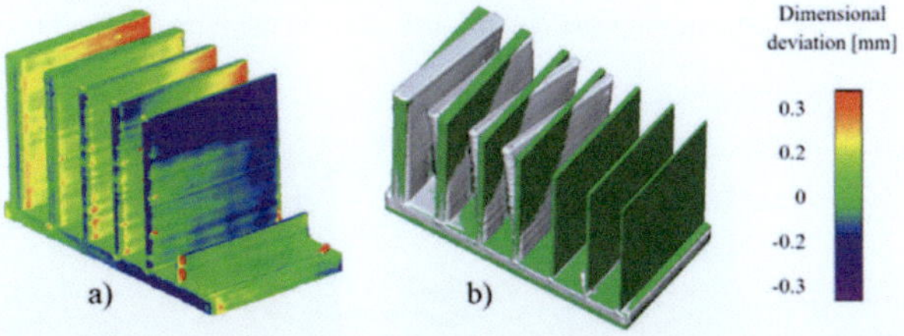

Figure 3. Three-dimensional scan results of unsupported wall (WoS) specimens, (**a**) green state, (**b**) debinding state.

After chemical debinding, pronounced dimensional deviations were observed in samples lacking sidewalls, with deformation becoming more severe during the debinding stage, as illustrated in Figure 4. The STL model, scaled to account for shrinkage, is shown in green, while the corresponding 3D scan data are presented in grey in Figure 3b. In general, wall thicknesses of 1.6 mm or more could be reliably fabricated [23]. The defects observed during debinding are attributed to anisotropic diffusion mechanisms, which lead to uneven binder removal and subsequent wall deformation [10].

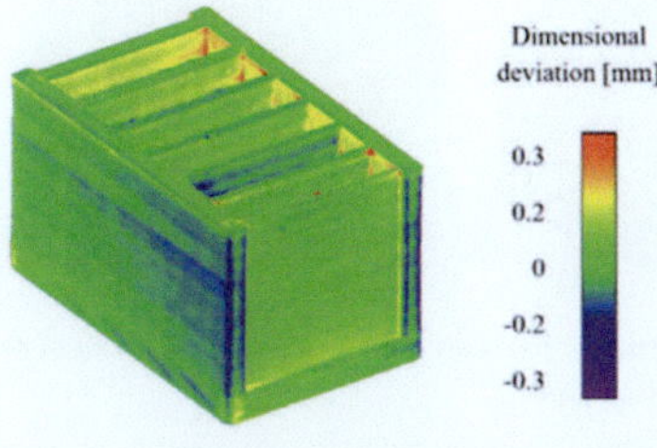

Figure 4. Three-dimensional scan results of supported wall (WmS) specimens in green state.

Incorporating sidewalls allows for the successful fabrication of walls as thin as 0.2 mm. A comparison between the CAD model and 3D scan data (Figure 4) indicates only minor dimensional deviations, aside from the naturally rounded edges attributed to the nozzle diameter. However, despite applying the shrinkage compensation factor to the STL model, deviations following the debinding process were particularly persistent in the side and rear walls, which display outward expansion. G-code analysis revealed that these 2 mm thick walls lacked internal infill, resulting in structural voids. This absence of internal support is likely responsible for the observed deformations, even in regions where internal connections were present.

After the debinding process, increased dimensional deviations were observed in several regions, even though the STL file had been adjusted to compensate for expected shrinkage. Outward deformation was especially evident in the sidewalls and rear wall, with the effect being most pronounced in walls with a nominal thickness of 2 mm. A detailed examination of the slicer-generated G-code revealed that these walls were printed without any internal infill, leading to the formation of structural voids. The absence of internal support likely contributed to the observed deformations, despite the walls being partially connected at their midsections. Furthermore, some deviations can be attributed to limitations in the accuracy of the shrinkage prediction model used during pre-processing. A comparison of deformation patterns before and after sintering, shown in Figures 5 and 6, indicates that while the overall geometry remained largely consistent, minor reductions in distortion were noted, particularly in the right-side and rear wall regions.

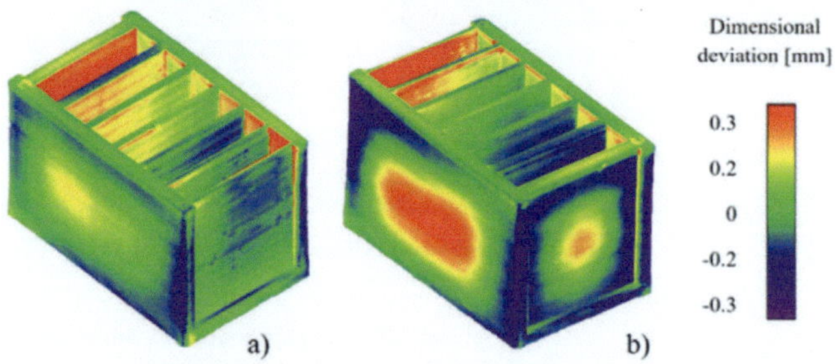

Figure 5. Three-dimensional scan of support wall specimens (WmS) in debinded state. (**a**) Front view. (**b**) Rear view.

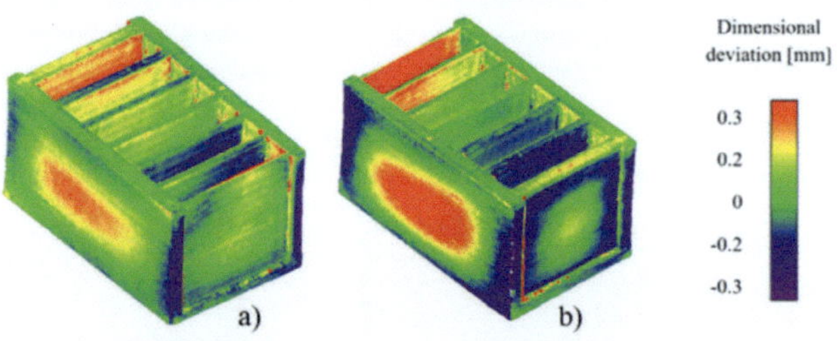

Figure 6. Three-dimensional scan of sintered support wall specimens (WmS) in debinded state. (**a**) Front view. (**b**) Rear view.

The measured wall thicknesses of the samples are presented in Figure 7. Although walls with a nominal thickness below 0.2 mm were successfully fabricated, they exhibited actual thicknesses of approximately 0.4 mm, due to the nozzle size used. In the MEX/M process, the actual dimensions of walls closely matched the nominal values of 0.4 mm and 0.6 mm, which was attributable to the single-track fabrication method employed for both wall types. This fabrication approach results in a relatively high standard deviation of 0.10 mm around a mean dimensional deviation of 0.09 mm for sintered MEX/M components.

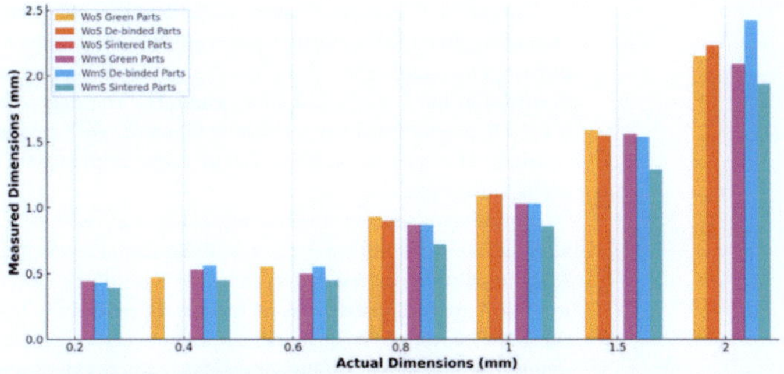

Figure 7. Dimensional deviation of the supported (WmS) and unsupported walls (WoS) samples in Green, debinded, and sintered states.

3.2.2. Cylinders and Hollow Cylinders with Different Thicknesses (ZV And ZH-2, ZH-4):

Cylindrical structures with nominal diameters of 0.2 mm and 0.4 mm could not be fabricated, as these fine features were excluded during the slicing stage. As shown in Figure 8a, cylinders with smaller diameters in the green state displayed notably higher dimensional deviations than those with larger cross-sections, suggesting a direct relationship between reduced cross-sectional area and increased geometric instability.

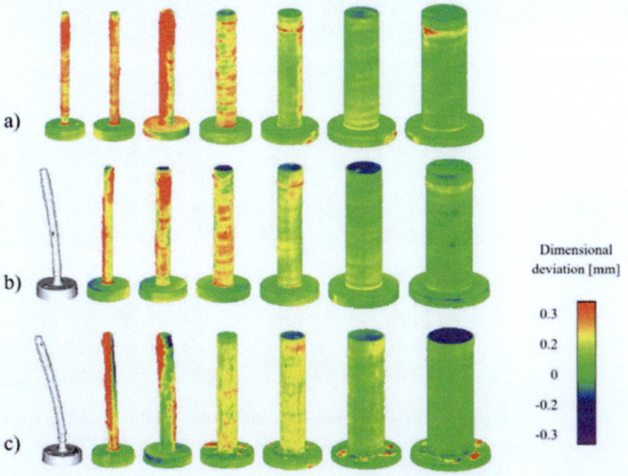

Figure 8. Three-dimensional scan results of cylindrical specimens [ZV]. (**a**) Green state; (**b**) debinded; (**c**) sintered.

The cylinder with a nominal diameter of 0.8 mm exhibited dimensional deviations of up to 0.42 mm, whereas the largest deviations reaching approximately 0.5 mm were observed in cylinders with diameters of 1 mm and 2 mm. Across all samples, the interface between the cylinder and its base was consistently rounded, leading to localized oversizing. The layer-specific dimensional variations were likely due to extrusion inconsistencies, indicating non-uniform material deposition during the printing process. After the chemical debinding stage, the smallest cylinder showed noticeable deformation (Figure 8b).

In contrast, geometric distortions in the other cylinders remained largely unchanged, suggesting that these deformations originated during the MEX/M fabrication phase and persisted thereafter. During the subsequent sintering process (Figure 8c), further bending occurred in the 1 mm and 2 mm cylinders. The degree of curvature was found to be inversely proportional to the cylinder diameter, with smaller cylinders exhibiting greater bending. The cylinder with the largest diameter displayed undersizing of up to 0.19 mm on its top surface.

In other cases, the undersizing evident in the brown parts was partially reduced after sintering, indicating that some dimensional recovery occurred during densification. Additionally, deviations observed at the base of the sintered cylinders were attributed to optical artefacts, likely caused by surface reflectivity interfering with the accuracy of the 3D scan system. A clear correlation between component diameter and dimensional oversizing is observed in the green-state samples shown in Figure 8, justifying a closer examination of this relationship. This trend remains consistent across both the green-state specimens in Figure 8a and the sintered counterparts in Figure 8c.

Hollow cylinders (ZHs) were produced with two nominal wall thicknesses, 2 mm and 4 mm, and included both uniform and varying wall thickness configurations, as shown in Figure 9a,b. In the green state, the outer hollow cylinders exhibited a slight inward deflection toward the central axis, with this deformation being more pronounced in the larger-diameter specimens. During the debinding stage, this deflection became more severe, as illustrated in Figure 9c,d, indicating a loss of structural stability.

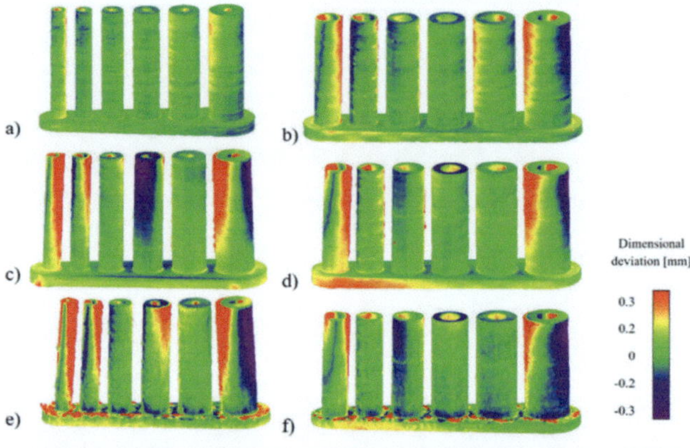

Figure 9. Three-dimensional scan results. Hollow cylinder specimens. **(a,c,e)** ZH-2 mm samples in green, debinded and sintered states, and **(b,d,f)** ZH-4 mm samples in green, debinded, and sintered states.

In the smaller samples, the central cylinder also showed significant tilting, which contributed to noticeable dimensional deviations, particularly in the upper regions. These deformation patterns were consistently observed in both ZH-2 and ZH-4 samples, especially in areas with a wall thickness of 0.2 mm. After sintering, as depicted in Figure 9e,f, the degree of tilt in the smaller cylinders increased markedly, while the dimensional deviations in the larger samples remained relatively unchanged. Form deviations at the base of the samples were also observed; however, these were primarily attributed to optical artefacts from surface reflectivity, rather than actual geometric distortions. The outer diameters of both solid and hollow cylinder specimens are summarized in Table 6.

Table 6. Inner diameter of the hollow cylinders.

	CAD (mm)	Green State (mm)	Debinded State (mm)	Sintered State (mm)
ZH-2 samples	2	1.55	1.4	1.26
	2	1.76	1.61	1.44
	2	1.71	1.59	1.45
	2	1.81	1.68	1.49
	2	1.73	1.63	1.46
	2	1.72	1.65	1.48
ZH-4 samples	4	3.66	3.48	2.63
	4	3.8	3.61	3.14
	4	3.81	3.61	3.12
	4	3.86	3.67	3.19
	4	3.76	3.53	3.05
	4	3.79	3.58	3.09

Due to structural damage incurred by the hollow cylinders during debinding, it was not possible to derive reliable shrinkage data between the green and debinded states. After accounting for shrinkage, the average dimensional deviation of the outer diameters was 0.09 mm, with a relatively high standard deviation of 0.14 mm. This variation is attributed to the accumulation of dimensional errors originating during green part fabrication, compounded by the inherent uncertainties in shrinkage behaviour during sintering. The green-state specimens alone showed an average dimensional deviation of 0.12 mm with a standard deviation of 0.17 mm, highlighting the overall process variability associated with the MEX/M technique. External dimensions of cylinder and hollow cylinder samples were summarized in Figure 10.

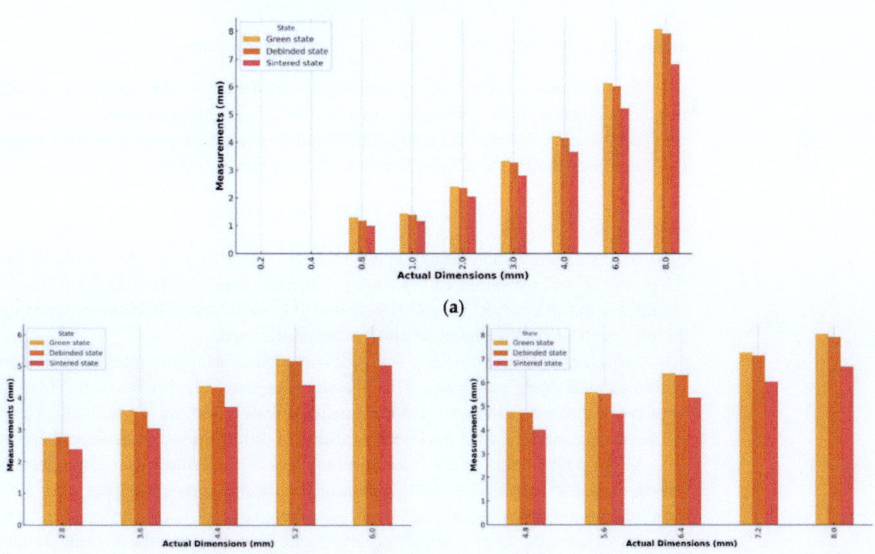

Figure 10. External dimensions for the samples in green, debinded, and sintered states (**a**) ZV, (**b**) ZH-2 and (**c**) ZH-4 samples.

To quantitatively assess this correlation, the dimensional deviations were plotted against the nominal diameters of the solid cylindrical samples, as illustrated in Figure 11. The observed linear relationship in the MEX/M-fabricated cylinders was further confirmed through a regression analysis, resulting in the following equation:

$$d_{abw} = -0.0461 \times d + 0.3827 \quad (3)$$

where d_{abw} represents the dimensional deviation and d denotes the cylinder diameter. The coefficient of determination R^2 of 0.874 confirms a strong linear relationship between these variables.

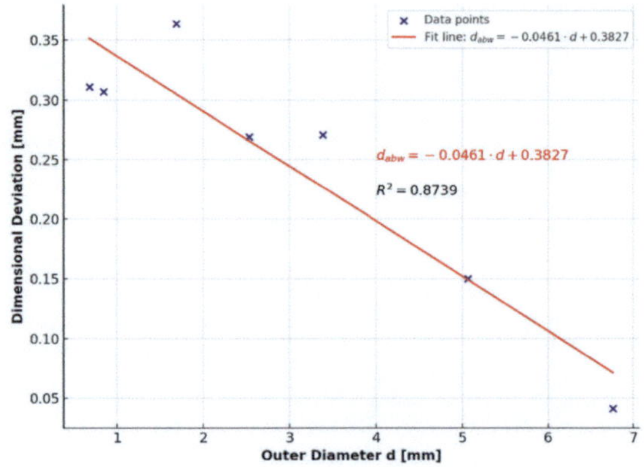

Figure 11. Dimensional deviation of the outer diameter of ZV samples.

The dimensional behaviour of the hollow cylinders, as shown in Figure 12, follows a trend consistent with the patterns observed in the ZV specimens. A linear regression model was used to describe the relationship between the outer diameter and the dimensional deviation for the ZH-2 and ZH-4 samples, and is given as follows:

$$d_{abw} = -0.0211 \times d + 0.0724 \quad (4)$$

The coefficient of determination ($R^2 = 0.644$) indicates a moderate correlation between the predicted and measured values, with a noticeable degree of data scatter. This variation, coupled with the lower R^2 value, is likely due to inconsistencies in cylinder wall thickness, which contribute to the observed dimensional deviations.

Cylinders with the smallest wall thickness exhibited pronounced dimensional deviations in the green part stage. This behaviour is primarily attributed to the mismatch between the nozzle diameter and the intended wall thickness. Specifically, the nozzle diameter exceeded the wall thickness, resulting in compromised dimensional accuracy.

As shown in Table 6, the internal diameters of the hollow cylinders demonstrate this deviation clearly. Due to the extent of the discrepancies, samples with a 0.2 mm wall thickness were excluded from the dimensional analysis to preserve the accuracy and integrity of the data. For the remaining ZH-2 and ZH-4 samples, the average dimensional deviation in internal diameter was measured at −0.24 mm, with a standard deviation of ±0.04 mm, indicating a consistent undersizing trend relative to the nominal design values.

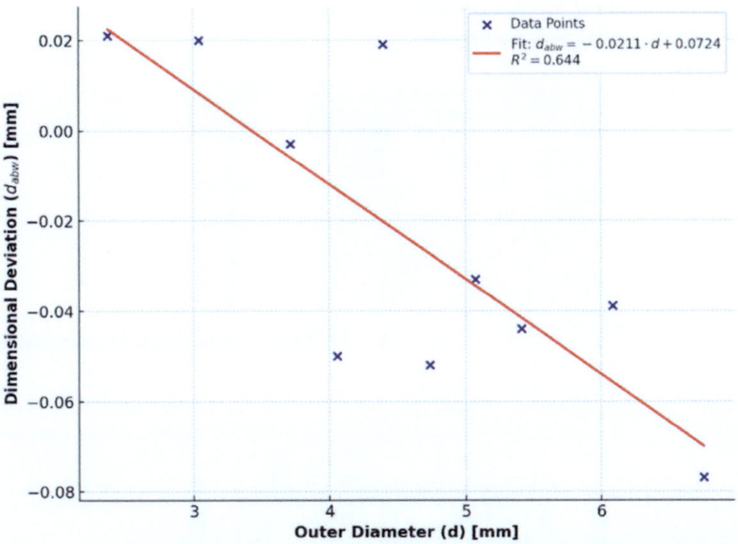

Figure 12. Dimensional deviation of the outer diameter of ZH-2 mm and ZH-4 mm samples.

3.2.3. Horizontal Overhangs (HOs):

Noticeable dimensional deviations were observed in single-sided supported overhangs starting from lengths as short as 1 mm. As shown in Figure 13, overhangs exceeding this threshold displayed pronounced layer sagging, resulting in significant geometric inaccuracies. To minimize such deviations, it is advisable to incline single-sided overhangs from the 1 mm mark onward during the green part stage or to implement support structures to improve stability. Additionally, rounding at the edges of the green parts was evident, and was primarily caused by the nozzle diameter, which in turn contributed to undersized features in the final components.

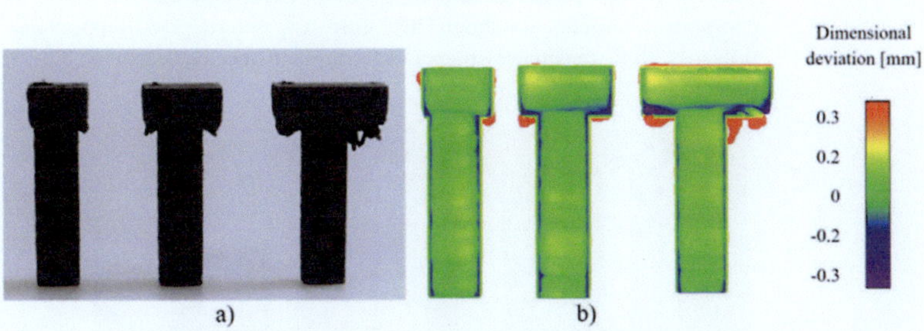

Figure 13. (**a**) Green (as-built) and (**b**) 3D scan results of HO samples.

The debinding and sintering processes were completed without major complications. Overhangs that were free of visible defects in the green state retained their shape well throughout post-processing, as illustrated by the comparison between Figures 13b and 14a. Microscopic analysis of the 0.5 mm and 1 mm overhangs (Figure 14a) showed that the

1 mm sample retained distinct sagging layers indicative of deformation originating during the initial printing phase.

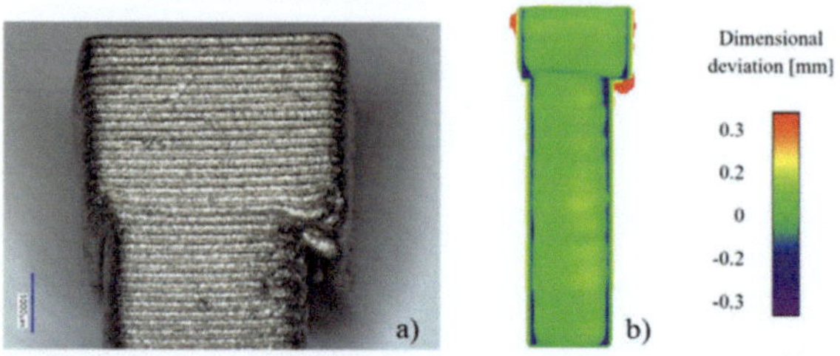

Figure 14. Sintered horizontal overhang (HO) sample. (**a**) Microscope image. (**b**) Three-dimensional scan results.

In the green-state specimens featuring double-sided supported overhangs, deformations became evident once the overhang length exceeded 2 mm, as illustrated in Figure 15.

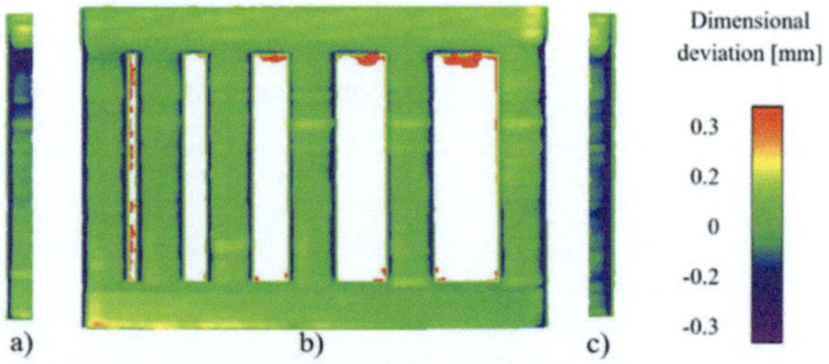

Figure 15. Three-dimensional scan results of green part of the horizontal overhang (HO) sample. (**a**) Right view. (**b**) Front view. (**c**) Side view.

After undergoing the chemical debinding process, the defects characteristic of the MEX/M process remained and were further exacerbated by additional warping at the component corners. As shown in Figure 16a, two diagonally opposite corners lifted upwards, while the other two experienced downward displacement.

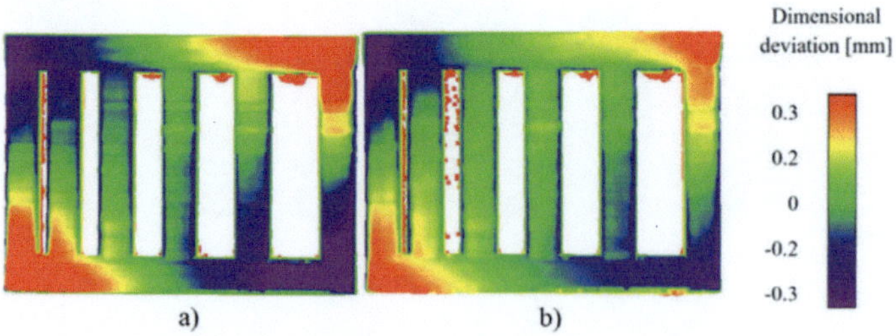

Figure 16. Three-dimensional scan results of horizontal overhangs (HOs) sample. (**a**) Debinded state. (**b**) Sintered state.

These distortions led to dimensional deviations, with elevated corners showing oversizing of up to 0.69 mm, while the depressed corners exhibited undersizing of up to 0.56 mm.

Although the sintering process helped reduce some of these deformations, they remained noticeable, with oversizing at the raised corners measuring 0.40 mm and undersizing at the lowered corners reaching 0.37 mm.

3.2.4. Boreholes and Teardrop Shapes (BH and TD)

Circular boreholes (BHs) and teardrop-shaped (TD) features were successfully produced in both vertical and horizontal orientations. However, specimens printed in the vertical orientation exhibited greater dimensional deviations compared to their horizontally printed counterparts.

The green-state vertical borehole (BH-V) specimens with diameters of 8 mm, 6 mm, and 4 mm showed visible layer sagging, leading to significant oversizing, as seen in Figure 17a. In contrast, the horizontally printed samples exhibited a slight curvature, characterized by raised edges and a depressed central region. After the debinding step, defects originating from the green-state fabrication remained evident, and further deformations developed in both orientations. As shown in Figure 17c,d, a convex distortion pattern was observed, marked by lower ends and an elevated centre in the borehole geometry.

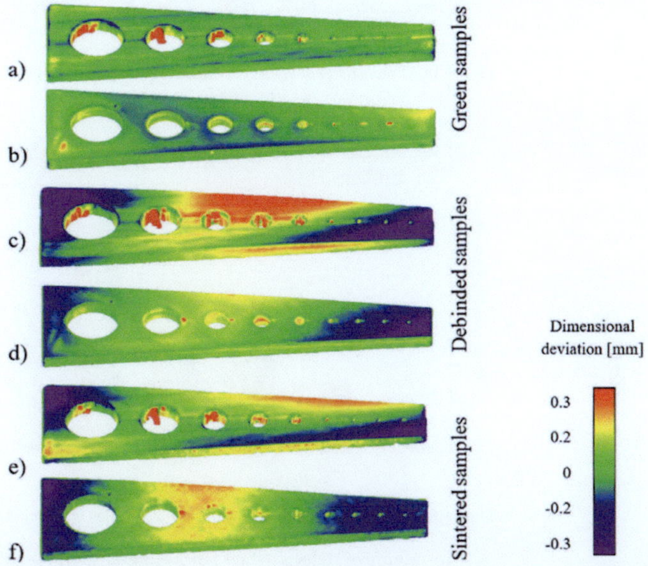

Figure 17. Three-dimensional scan results of the boreholes samples with vertical and horizontal build direction. **(a)** Green BH-V. **(b)** Green BH-H. **(c)** Debinded BH-V. **(d)** Debinded (BH-H). **(e)** Sintered BH-V. **(f)** Sintered BH-H.

In the vertically printed specimens (Figure 17c), the maximum oversize at the centre reached 0.35 mm, while undersizing was observed at 0.70 mm on the left side and 0.57 mm on the right. In comparison, the horizontally built borehole (BH) samples showed less deformation, with a central oversize of 0.17 mm. Following the sintering process, the convex deformation observed in the vertical specimens was partially reduced; the central oversize decreased to 0.24 mm, and the undersizing on both sides was reduced to approximately 0.40 mm. Conversely, in the horizontal BH specimen, the central oversize increased slightly to 0.22 mm after sintering. Figures 18 and 19 show the measured dimensions of BH samples fabricated in both horizontal and vertical orientations. Across all cases, the measured diameters were smaller than the nominal values. The average dimensional deviation was -0.37 mm, with a standard deviation of ± 0.11 mm, which is notably greater than the deviations observed in the internal diameters of the hollow cylinders.

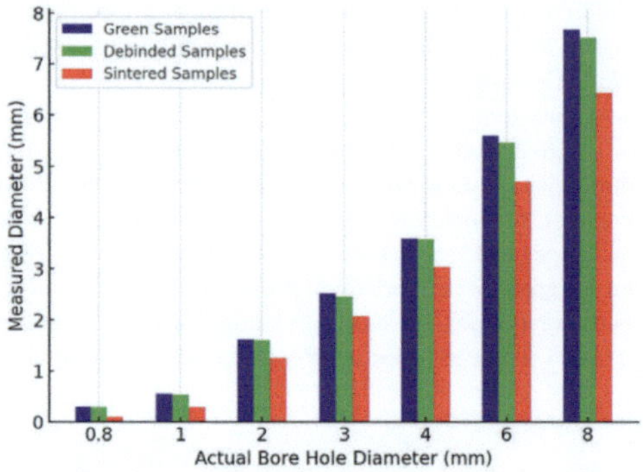

Figure 18. Dimensional deviation of the BH-H samples in green, debinded, and sintered states.

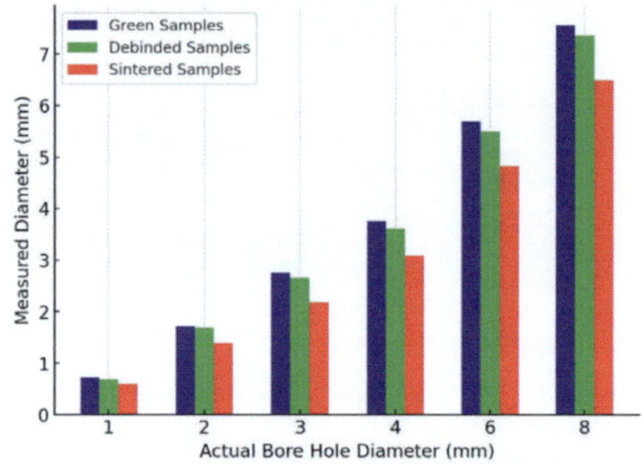

Figure 19. Dimensional deviation of the BH-V samples in green, debinded, and sintered states.

To minimize defects when fabricating larger boreholes aligned parallel with the x–y plane, an effective strategy is to incorporate an angular inclination in the upper regions of the holes [9]. This approach was applied in the TD specimens, which featured teardrop-shaped holes with diameters of 6 mm and 14 mm. A consistent shrinkage pattern was observed across all green-state parts, as shown in Figure 20. Notably, the outer edges of the holes exhibited oversizing along the surfaces facing the nearest component boundary, while no undersizing was detected on the opposite side. Additionally, Figure 20b reveals a horizontal stripe near the upper edge of the component, which was not observed during visual inspection and is therefore considered a measurement artefact.

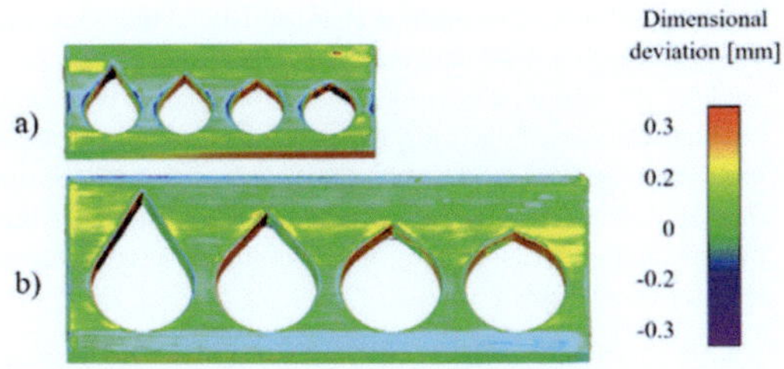

Figure 20. Three-dimensional scan result of TD samples in green state: (**a**) 6 mm and (**b**) 14 mm.

After the debinding process, further deformations were observed. In the specimen with a 6 mm diameter (Figure 21a), localized oversizing of up to 0.26 mm was measured at the corners. The 14 mm diameter specimen (Figure 21b) showed both oversizing—up to 0.19 mm—and undersizing, reaching up to 0.35 mm. These deformations became more pronounced following the sintering stage. As illustrated in Figure 22a,b, both undersizing and oversizing were evident across all of the examined specimens.

Figure 21. Three-dimensional scan result of TD samples in debinded state: (**a**) 6 mm (**b**) 14 mm.

Figure 22. Three-dimensional scan result of TD samples in sintered state: (**a**) 6 mm (**b**) 14 mm.

3.2.5. Inclined Walls (DWs):

Inclined wall features were designed with a range of angles from 0° to 90°, in 10° increments. The dimensional deviations observed in both the green and sintered states are shown in Figure 23. In the green parts, oversizing was evident in several regions, while undersizing was primarily found along the rounded edges of the components. Surfaces with shallower angles exhibited a pronounced stair-stepping effect, which was attributed to the layer-by-layer deposition characteristic of the MEX/M process. As shown in Table 7, downskin angles of 30° and above were successfully manufactured. However, a downskin angle of 10° could not be printed due to significant layer sagging. Dimensional deviations increased noticeably during the debinding stage, particularly for samples with lower inclination angles. Green-state specimens with shallow downskin angles exhibited the highest deviations, which further intensified after sintering. On average, the measured downskin angles deviated by approximately +4° from the nominal design values, with a standard deviation of ±1°.

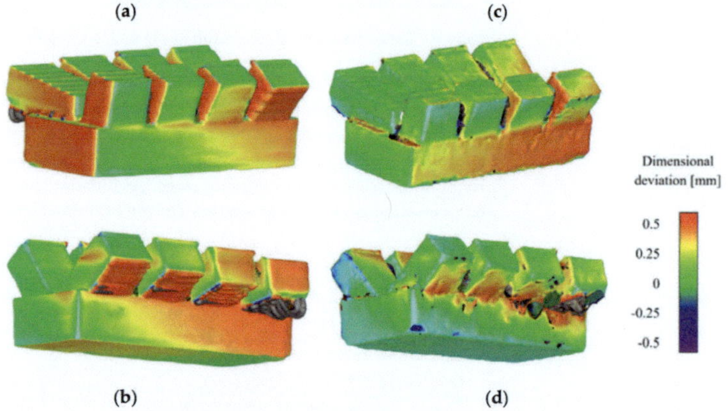

Figure 23. Three-dimensional scan results of the DW samples. (**a**,**b**) Green state. (**c**,**d**) Sintered state.

Table 7. Measured angle of the DW samples comparison.

CAD Data Angle (°)	Green Samples (°)	Sintered Samples (°)
10	-	-
20	24.1	25.7
30	32.6	34.2
40	42.0	45.2
50	47.5	53.1
60	60.3	63.8
70	69.6	73.6
80	80.1	83.1

3.3. Guidelines for Design and Production

The design guidelines derived from the analysis of the fabricated 316L samples using the MEX/M process are summarized in Table 8. The specified dimensions refer to the sintered components, under the assumption that the initial CAD models are scaled accordingly in the slicing software to account for process-induced shrinkage. In addition to nominal values, the expected dimensional deviations are provided in the Supplementary Data [22]. It is clear that these deviations are geometry-dependent, varying with the shape and structural complexity of each component. Moreover, due to deformation effects introduced during the debinding and sintering stages, the production of flat geometries is not recommended, as ensuring dimensional stability in such cases remains particularly difficult.

Table 8. Design guidelines of features printed via the MEX/M process using AISI 316L.

Geometry Feature		Recommendation	
Name	**Illustration**	**Value**	**Comment**
Unsupported walls (WoSs)		Wall thickness $\geq$ 1.5 mm	Maximum deviation of 0.4 mm
Supported walls (WmSs)		Wall thickness $\geq$ 0.2 mm	Dimensional deviation of 0.09 $\pm$ 0.1 mm
Vertical cylinder (ZV)	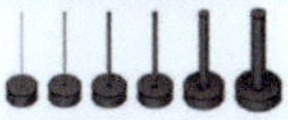	Minimum diameter $\geq$ 0.8 mm	Dimensional deviation of 0.24 $\pm$ 0.1 mm
Hollow cylinders (ZH-2) Hollow cylinders (ZH-4)		Cylinder thickness $\geq$ 0.34 mm	1. Dimensional deviation of the outer diameter $\pm$0.006 mm 2. Dimensional deviation 0.11 $\pm$ 0.02 mm
Horizontal overhangs (HOs)		Overhang < 0.5 mm	Without supports, the maximum overhang length is less than 0.5 mm
Supported horizontal overhangs (ÜZs)		b < 0.5 mm	A maximum of 0.5 mm is supported without any print issues
Vertical boreholes (BH-V)		0.84 mm $\leq$ diameter $\leq$ 2.53 mm	1. Dimensional dev -0.28 ± 0.04 mm 2. Diameter > 3 mm supports are needed; otherwise, use a teardrop shape
Horizontal boreholes (BH-H)		Diameter $\geq$ 0.68 mm	Dimensional dev -0.44 ± 0.09 mm
Teardrops features (TDs)		60° < Teardrops angle < 120° For diameter 6 mm and 14 mm	Maximum deviation of the overhang area 0.3 mm occurs due to staircase effect
Inclined walls (DWs)		Minimum downskin angle of 30°; otherwise, support is needed	Downskin angle $\geq$ 30° maximum deviation of 0.45 mm

3.4. Demonstrator:

To demonstrate the industrial applicability of the MEX/M additive manufacturing process, an impeller structure adapted and redesigned from the model in [24] was fabricated based on the design guidelines established in Chapter 3. The component incorporates several critical features, including angled walls exceeding 30°, boreholes with a diameter of 7 mm, and a minimum wall thickness of 2 mm.

The demonstrator impeller was successfully fabricated using the MEX/M process with 316L filament. As shown in Figure 24a, the green (as-built) sample is positioned on the right, while the sintered part is on the left. The measured shrinkage of the impeller was approximately 13.97% in the X-Y plane and 14.69% along the Z-axis (build direction), which falls within the acceptable limits reported for the MEX/M process [9–13,25].

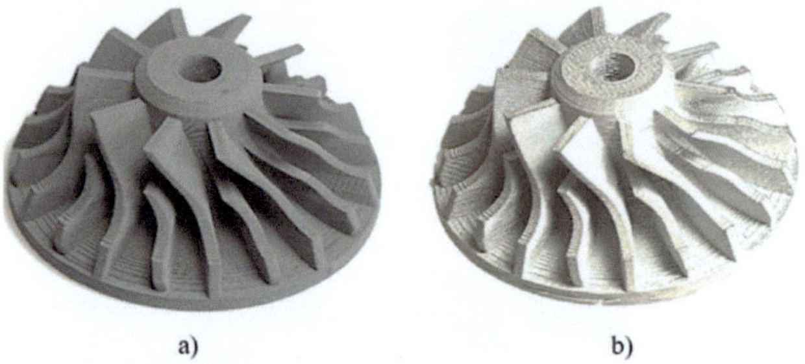

Figure 24. A 316L impeller fabricated using the MEX/M process. (**a**) The green part is on the left; (**b**) the sintered part is on the right.

However, dimensional deviations of up to ±0.2 mm were observed when comparing the sintered impeller with the original CAD model and 3D scan data, analyzed using the VMX500 software (version 2.3.6.74) from KEYENCE GmbH. These deviations are illustrated in Figure 25: (a) shows the 3D scan of the green-state impeller, and (b) compares the sintered component to the scaled CAD model. For accurate comparison, the CAD model was scaled by a factor of 0.853 in the X and Y directions and 0.881 in the Z direction, as determined by the experimentally measured shrinkage rates.

Figure 25. Dimensional deformation of 3D scan impeller. (**a**) Green state; (**b**) sintered state.

Minor discrepancies in the scan data may arise due to the high reflectivity of the metal surface during scanning. Such deviations can be compensated by incorporating appropriate dimensional tolerances in the design phase to ensure proper fit and functionality after the sintering process. The material extrusion of metals (MEX/M) additive manufacturing process demonstrates strong potential for industrial-scale applications, particularly when guided by the design recommendations developed in this study. By applying these principles, MEX/M emerges as a competitive alternative to conventional metal manufacturing methods, enabling advancements in lightweight design, customized component production, and multi-material fabrication.

A comparison with existing design guidelines for the MEX/M process shows differences in the limits on design features when using the same nozzle diameter [15]. For example, horizontal holes with a diameter of 0.68 mm and a maximum deviation of 0.44 mm can already be produced with the design features examined. According to reference [15], a minimum hole size of 1.5 mm is recommended. The existing design recommendations from [15] are material-independent, so they allow for greater tolerance and deviate from the design elements examined in this study.

In conclusion, it can be deduced from the dimensional shrinkage in the various design features that these designs depend on the process parameters of shaping, debinding, and sintering. During the shaping process, the extrusion multiplier, the layer height and the layer thickness which, in turn, depend on the extrusion temperature, influence the bonding of the individual extrusion layers. A leaky bond leads to internal voids, which can cause shrinkage in the Z direction during solidification, as was observed with the different design features. In the debinding process, the debinding temperature, and the temperature gradient have the greatest influence on the final shrinkage. If the debinding temperature is excessively high, the temperature gradient becomes extremely large, leading to anisotropic deformation of the components, causing undesirable deformation in the subsequent sintering process. This observation could be made in the unsupported walls. The heating rate, maximum temperature, and holding times have the greatest influence on dimensional shrinkage in the sintering process. Undesirable deformation of the design features can only be prevented by selecting these process parameters appropriately.

4. Conclusions and Outlook

In this paper, AISI 316L with different design features was manufactured and was evaluated after each process step (shaping, debinding, and sintering). It was shown that defects that occurred in the initial process steps had a negative impact on the later process steps and could not be rectified. With unsupported walls of 1.5 mm thickness, deviations of a maximum of 0.4 mm occurred, corresponding to the nozzle diameter. It could therefore be identified that the nozzle diameter has a significant influence on the accuracy of the different design features. Unsupported walls could already be successfully manufactured with a thickness of 0.2 mm. Cylinders can be produced from a diameter of 0.8 mm, whereby the geometric deviation here may also be 0.42 mm. The hollow cylinders with 2 mm and 4 mm hole diameters can be produced with a wall thickness of 1.6 mm. The maximum deviation here is 0.3 mm. Horizontal overhangs (supported on one and two sides) can be manufactured up to a length of 0.5 mm. Support structures must also be used. Bored holes that are produced lying on the building platform require a minimum diameter of 0.68 mm. Upright holes can have a maximum diameter of 2.53 mm. Larger drill holes must be supported or have a teardrop shape, which can have an angle of 60° to 120°. Overhang angles from 30° can be finished, but deviations of up to 0.45 mm occur in the downskin area.

The design limits were successfully applied to a component so that the deviation after sintering is approximately 0.26 mm. It should be noted that the design guidelines and recommendations for MEX/M were implemented based on 316L tested features with a 0.4 mm nozzle size. The recommended guidelines are specific to the machine setup, process parameters, and filament material specifications with respect to the identified design limits. The use of finer nozzles with a diameter of 0.2 mm can lead to features with smaller deviations, though they may impose additional manufacturing restrictions. However, the printing time in the shaping process is then increased.

As the design guidelines are dependent on material, process parameters, and equipment, these recommendations only apply to the configuration in this paper. If the material, equipment, or process parameters are changed, different specimens may be produced, and different design recommendations may result. Therefore, the methodological approach presented in this paper is generally transferable to other materials and equipment types. Furthermore, the results presented here provide an indication of deviations in design limits.

Author Contributions: Conceptualization, K.A.; methodology, K.A.; software, M.P.R.M.; validation, K.A. and T.R. (Tim Röver); formal analysis, K.A. and C.E.; investigation, T.R. (Titus Rakow); resources, K.A.; data curation, K.A.; writing—original draft preparation, K.A. and M.P.R.M.; writing—review and editing, K.A.; visualization, K.A., T.R. (Titus Rakow) and M.P.R.M.; supervision, K.A. and C.E.; project administration, K.A and C.E.; funding acquisition, K.A. and C.E. All authors have read and agreed to the published version of the manuscript.

Funding: This research received no external funding.

Data Availability Statement: The original data presented in the study are openly available in [22].

Conflicts of Interest: The authors declare no conflict of interest.

Appendix A

Table A1. Optical images of as-built 'green' components.

Unsupported walls (WoSs) 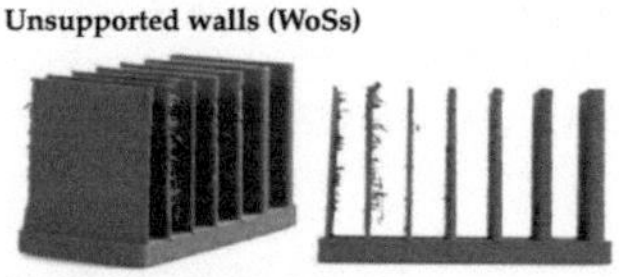 *Thickness (mm): 0.2/0.4/0.6/0.8/1/1.5/2* *Length (mm): 38.5* *Width (mm): 20* *Height (mm): 23*	**Supported horizontal overhangs (ÜZs)**  *Overhang gap (mm): 1/2/3/4/5* *Height (mm): 23* *Width (mm): 3*
Supported walls (WmSs) 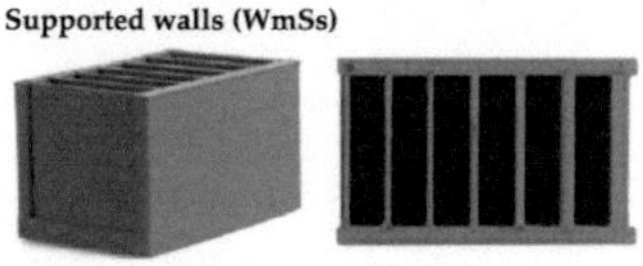 *Thickness (mm): 2/1.5/1/0.8/0.6/0.4/0.2* *Length (mm): 38.5* *Width (mm): 20* *Height (mm): 23*	**Horizontal and vertical boreholes (BH-V, BH-H)** 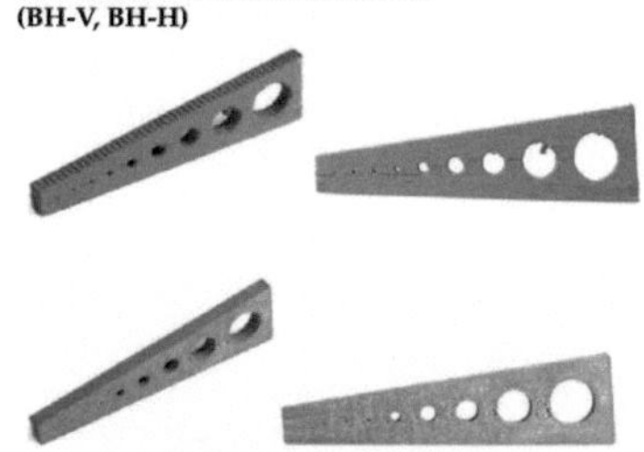 *Bore diameter (mm): 0.4/0.6/0.8/1/2/3/4/6/8*
Vertical cylinder (ZV)  *Bar Ø (mm): 8/6/4/3/2/1/0.8/0.4/0.2* *Height (mm): 20*	**Hollow cylinders (ZH-4)**  *Thickness (mm): 0.4/0.8/1.6/2.4/3.2/4* *Height (mm): 23*

Table A1. *Cont.*

Hollow cylinders (ZH-2, ZH-4)  *Thickness (mm): 0.4/0.8/1.6/2.4/3.2/4* *Height (mm): 23*	**Teardrops features (TDs)**  *Angle (degree): 60/80/100/120* *Diameter (mm): 6*
Horizontal overhangs (HOs)  *Overhang distance (mm): 0.5/1/1.5/2/3/4*	**Inclined walls (DWs)**  *Incline angle (degree): 10/20/30/40/50/60/70/80*

References

1. Lopez, E.; Brueckner, F.; Gruber, S. 21—Industrial applications. In *Fundamentals of Laser Powder Bed Fusion of Metals Additive Manufacturing Materials and Technologies*; Yadroitsev, I., Yadroitsava, I., Plessis, A.D., MacDonald, E., Eds.; Elsevier: Amsterdam, The Netherlands, 2021; pp. 583–595.
2. Sadaf, M.; Bragaglia, M.; Nanni, F. A simple route for additive manufacturing of 316L stainless steel via Fused Filament Fabrication. *J. Manuf. Process.* **2021**, *67*, 141–150. [CrossRef]
3. Suwanpreecha, C.; Manonukul, A. A Review on Material Extrusion Additive Manufacturing of Metal and How It Compares with Metal Injection Moulding. *Metals* **2022**, *12*, 429. [CrossRef]
4. Schatt, W.; Wieters, K.-P.; Kieback, B. *Pulvermetallurgie: Technologien und Werkstoffe*, 2nd ed.; bearbeitete und erweiterte Auflage; Springer: Berlin/Heidelberg, Germany, 2007.
5. Zhong, Y.; Rännar, L.-E.; Liu, L.; Koptyug, A.; Wikman, S.; Olsen, J.; Cui, D.; Shen, Z. Additive manufacturing of 316L stainless steel by electron beam melting for nuclear fusion applications. *J. Nucl. Mater.* **2017**, *486*, 234–245. [CrossRef]
6. García-Rodríguez, S.; Torres, B.; Pulido-González, N.; Otero, E.; Rams, J. Corrosion behavior of 316L stainless steel coatings on ZE41 magnesium alloy in chloride environments. *Surf. Coat. Technol.* **2019**, *378*, 124994. [CrossRef]
7. Gong, H.; Crater, C.; Ordonez, A.; Ward, C.; Waller, M.; Ginn, C. Material Properties and Shrinkage of 3D Printing Parts using Ultrafuse Stainless Steel 316LX Filament. *MATEC Web Conf.* **2018**, *249*, 1001. [CrossRef]
8. Röttger, A.; Boes, J.; Theisen, W.; Thiele, M.; Esen, C.; Edelmann, A.; Hellmann, R. Microstructure and mechanical properties of 316L austenitic stainless steel processed by different SLM devices. *Int. J. Adv. Manuf. Technol.* **2020**, *108*, 769–783. [CrossRef]
9. Martignoni, L.; Vegro, A.; Candidori, S.; Shaikh, M.Q.; Atre, S.V.; Graziosi, S.; Casati, R. Prototyping and characterisation of 316L stainless steel parts and lattice structures printed via metal fused filament fabrication. *Rapid Prototyp. J.* **2024**, *30*, 123–141. [CrossRef]
10. You, S.; Jiang, D.; Wang, F.; Ning, F. Anisotropic sintering shrinkage behavior of stainless steel fabricated by extrusion-based metal additive manufacturing. *J. Manuf. Process.* **2023**, *101*, 1508–1520. [CrossRef]
11. Obadimu, S.O.; Kourousis, K.I. Shrinkage behaviour of material extrusion steel 316L: Influence of primary 3D printing parameters. *Rapid Prototyp. J.* **2022**, *28*, 92–101. [CrossRef]
12. Cerlincă, D.-A.; Tamașag, I.; Beșliu-Băncescu, I.; Severin, T.-L.; Dulucheanu, C. Experimental investigation of FDM manufacturing of 316 l stainless steel. *Int. J. Adv. Manuf. Technol.* **2024**, *135*, 1449–1463. [CrossRef]
13. Alzyod, H.; Ficzere, P. Optimizing fused filament fabrication process parameters for quality enhancement of PA12 parts using numerical modeling and taguchi method. *Heliyon* **2023**, *9*, e14445. [CrossRef] [PubMed]
14. Rebaioli, L.; Fassi, I. A review on benchmark artifacts for evaluating the geometrical performance of additive manufacturing processes. *Int. J. Adv. Manuf. Technol.* **2017**, *93*, 2571–2598. [CrossRef]
15. Desktop Metal. *BMD-Design-Guide: Considerations and Best Practices*; Desktop Metal: Burlington, MA, USA, 2023.
16. Blunk, H.; Seibel, A. Design guidelines for metal binder jetting. *Prog. Addit. Manuf.* **2024**, *9*, 725–732. [CrossRef]
17. Asami, K.; Lozares, J.M.C.; Ullah, A.; Bossen, B.; Clague, L.; Emmelmann, C. Material extrusion of metals: Enabling multi-material alloys in additive manufacturing. *Mater. Today Commun.* **2024**, *38*, 107889. [CrossRef]

18. Asami, M.K.; Herzog, D.; Bossen, B.; Geyer, L.; Klemp, C.; Emmelmann, C. (Eds.) Design Guidelines for Green Parts Manufactured with Stainless Steel. In *The Filament Based Material Extrusion Process for Metals (MEX/M)*; European Powder Metallurgy Association (EPMA): Chantilly, France, 2022. [CrossRef]
19. PT&A, DS-Fil-316L-F-200222-11. Available online: http://www.pt-a.de/filament%20de.htm (accessed on 10 April 2025).
20. Keyence, 3D Scanner—Modellreihe VL | KEYENCE Deutschland. 2025. Available online: https://www.keyence.de/landing/lpc/3d-scanner-modellreihe-vl.jsp?aw=KD_gaBRvlph1437971635824&gad_source=1&gclid=CjwKCAiA74G9BhAEEiwA8kNfpYLN2XsFjCROJXLdUsFD4vQa4FZXNOhQXbI7FvEsnTdpS2k2LvUMvxoChZgQAvD_BwE (accessed on 9 April 2025).
21. Jiang, D.; Ning, F. Additive Manufacturing of 316L Stainless Steel by a Printing-Debinding-Sintering Method: Effects of Microstructure on Fatigue Property. *J. Manuf. Sci. Eng.* **2021**, *143*, 091007. [CrossRef]
22. Asami, M.K.; Medapati, M.P.R.; Rakow, T.; Röver, T.; Emmelmann, C. *Supplementary Data: Design Guidelines for Material Extrusion of Metals (MEX/M)*; TUHH Universitätsbibliothek: Hamburg, Germany, 2025.
23. Asami, K.; Bartsch, K.; Emmelmann, C. Design-Richtlinie für die ressourceneffiziente Gestaltung von Stützstrukturanbindungen im pulverbettbasierten Laserstrahlschmelzen von Ti6Al4V. In Proceedings of the 17th Rapid. Tech 3D Conference, Erfurt, Germany, 22–23 June 2021; Carl Hanser Verlag GmbH & Co. KG: Munich, Germany, 2021.
24. Thingiverse.com, Impeller by Team3187FRC. 2025. Available online: https://www.thingiverse.com/thing:2426666 (accessed on 9 April 2025).
25. Ait-Mansour, I.; Kretzschmar, N.; Chekurov, S.; Salmi, M.; Rech, J. Design-dependent shrinkage compensation modeling and mechanical property targeting of metal FFF. *Prog. Addit. Manuf.* **2020**, *5*, 51–57. [CrossRef]

B.3 Design Optimization of Additively Manufactured Components Using Simulation based Analysis of Infill Structures

The summary of the findings can be found in Chapter 5.3.

Proceedings of the ASME 2023
International Mechanical Engineering Congress and Exposition
IMECE2023
October 29-November 2, 2023, New Orleans, Louisiana

IMECE2023-113065

DESIGN OPTIMIZATION OF ADDITIVELY MANUFACTURED COMPONENTS USING SIMULATION-BASED ANALYSIS OF INFILL STRUCTURES

Karim Asami[1,2], Sebastian Roth[1,2], Michel Krukenberg[1], Claus Emmelmann[1,2]

[1]Technical University Hamburg (TUHH), Hamburg, Germany
[2]iLAS – Institute of Laser and System Technologies, TUHH, Hamburg, Germany

ABSTRACT

Additive manufacturing (AM) has revolutionized the manufacturing industry, allowing the production of complex, lightweight components with specific properties. This has created opportunities for customized parts with optimized performance and cost-efficiency compared to traditional manufacturing.

Infill structures are used to fill the voids inside the component during layer-by-layer manufacturing. These structures, such as grids and honeycombs, optimize mechanical properties, reduce weight, and save production time. However, investigating the mechanical properties of components with different infill structures is typically done experimentally, which is time-consuming and expensive.

This paper proposes a simulation-based approach to investigate the influence of infill structure parameters on the mechanical properties of components made from AISI 316L stainless steel. Two-dimensional grid and honeycomb infill structures are studied, considering infill degree, cell wall thickness and spatial orientation relative to the load direction. The study employs a parametric design approach to generate different infill structures for simulation and validation.

This paper provides a simulation-based methodology for varying and evaluating design parameters for AM processes using grid and honeycomb infill structures. The findings can be used to optimize lightweight component design for specific applications, improving performance and cost-efficiency. The study contributes to the field of Design for Additive Manufacturing (DfAM).

Keywords: Design Analysis, Simulation, Infill, Grid, Honeycomb, 316L, MEX/M, L-PBF, Additive Manufacturing, DfAM

1. INTRODUCTION

Additive Manufacturing (AM), also known as 3D printing, is a revolutionary manufacturing process that has gained significant attention in recent years. Unlike traditional subtractive manufacturing methods, AM builds three-dimensional objects layer-by-layer, allowing for unprecedented design freedom and versatility. With its ability to create complex geometries and functional structures, AM has found applications in various industries, including aerospace, automotive, medical, and consumer products, among others.

The field of AM has evolved rapidly, with advancements in materials, processes, and equipment, leading to improved capabilities and expanding the possibilities of manufacturing. From polymer-based systems to metal and ceramic AM, the technology has opened new paths for producing parts with enhanced mechanical, thermal, and electrical properties. Moreover, the ability to fabricate customized parts with no limitations in the design process has enabled lightweight optimized designs and reduced the material consumption making AM an increasingly attractive option for modern manufacturing.

One approach to generate weight optimized designs is through computer aided topology optimization, where a component is iteratively analyzed for stress distribution and its shape is modified to save material and space. However, this process is computationally intensive and time-consuming, and it results in changes to the external shape of the components, which is not intended in some cases. [1, 2]

Another method is to perform structural optimization by incorporating infill structures in additively manufactured components. Infill structures can be made out of lattice structures, built up in the internal cavities of a component, ideally tailored to the specific load requirements of the component. This approach offers the potential to achieve weight savings without altering the external shape of the components.

Lattice structures are known for their high strength-to-weight ratio. They are typically composed of repeating unit cells with open spaces or voids, which reduces their overall weight while maintaining structural integrity. This makes them ideal for applications where weight reduction is critical, such as aerospace, automotive, and transportation industries. Lattice structures therefore use materials efficiently by optimizing the distribution in the structure regarding the load requirements of the part. [3, 4]

Additionally, lattice structures are highly customizable. The geometry parameters, such as orientation angle, wall thickness and overall filler content can be altered to meet the specific design requirements. This makes them versatile for a wide range of applications and allows the optimization of the infill structure for different use cases. [5] The inner structure of turbine blades for example can be filled with lattice structures. The flow optimized outer shape of the blade remains untouched, while the overall weight is reduced significantly even though the load requirements are still fulfilled.

However, most infill structures are highly anisotropic. [6] Therefore, the exact knowledge of the load capacity in different angles of the lattice structure is required to fulfill the design requirements of the part. In this paper, the focus is on rather simple infill structures, based on honeycombs and grid structures. Those can be manufactured faster in the e.g. the Material extrusion of metals (MEX/M) process due to size, instead of generating more complicated infills structures with e.g. gyroid-cells. [7,8,9] Therefore, this approach can be applied to many already existing use cases without the need of new development cycles.

To solve such a problem, this paper aims to provide a methodology based on simulated infill structures to choose the optimal parameters for a specific lattice structure on a given part with known load requirements. Additionally, angle-dependent properties where infill structures can be altered to tackle dynamic loads or alternating load directions are provided.

2. MATERIALS AND METHODS

2.1 Testing methods

The following testing methods for investigating the load behavior were used: Compression test according to ASTM E9 [10] and 3-point bending test according to ISO 7438 [11].

The compression test according to ASTM E9 is a standardized testing method used to investigate the load behavior of material samples under compression. In this test, a sample is loaded in a testing machine until it fails, refer to [10]. This method allows conclusions to be drawn about important material parameters such as the modulus of elasticity, which describes the stiffness of the material, and the tensile strength, which indicates the maximum load-carrying capacity of the material. The compression test is particularly relevant when components in later application are subjected to combined bending and compressive loads.

The 3-point bending test according to ISO 7438 is another common testing method used to investigate the load behavior of material samples under bending loads. In this test, a sample is supported on two rollers and loaded in the middle until it fails, refer to [11]. This method also allows conclusions to be drawn about material parameters such as the E-modulus and tensile strength. The 3-point bending test is particularly relevant when components in later application are subjected to bending loads, which often act in combination with compressive loads on the component.

In practice, the compression test and the 3-point bending test are commonly used to characterize the load behavior of materials and ensure meeting the requirements of the intended application.

The investigation of the load behavior of infill structures is based on these two testing methods. The two loading conditions are simulated and used for localizing the most heavily loaded areas within the samples and for examining the influences of the parameters of different infill structures. Static test loads are used instead of dynamic loading until sample failure for the investigation of stress distribution within the samples.

2.2 Variation of Infill Structure

In this section, the variation and consequently the parameterization of the infill structures are presented. The starting point of the variations is a solid specimen ($\varphi = 100$ %), which is simulated under compression and 3-point bending. The infill structures are investigated in separate series of tests limited to the degree of infill, the cell wall thickness, and the orientation of the infill relative to the load in order to be able to consider the influence of the individual parameters in isolation. The external dimensions of the specimens correspond to the dimensions of the reference specimens of the test setups: the cylinders have the diameter $d = 40\ mm$ and height $L = 40\ mm$, the beams have a length of $L = 80mm$ and a height/width of $a = 20\ mm$ (see fig. 2).

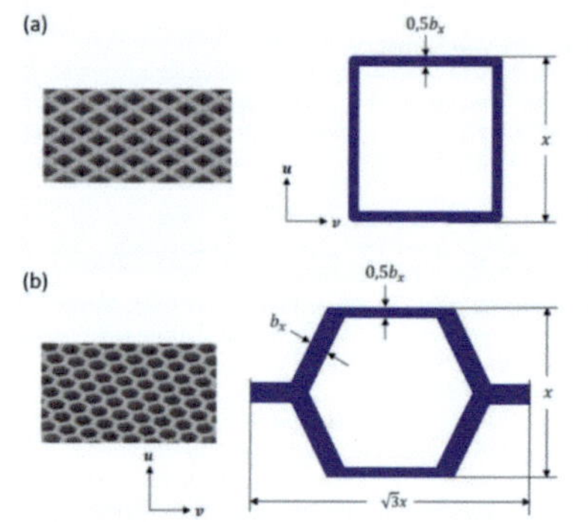

FIGURE 1: Parametrization of the infill structure; (a) parametrization of the grid structure; (b) parametrization of the honeycomb structure.

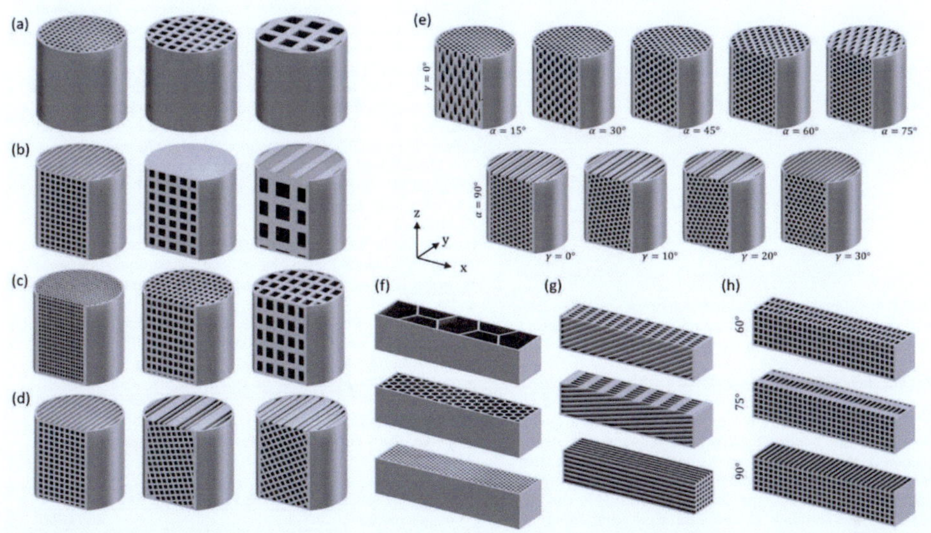

FIGURE 3: Variation of the infill structure; (a)-(e) Variation of the infill structures used in the compression test: Variation of the orientation and angle, variation of the wall thickness, variation of the infill type (grid and honeycomb); (f)-(g) Variation of the infill structures used in the 3-point bending test: variation of the orientation, angle, wall thickness and infill type.

For the parametrization of the infill structures the specimen have to be separated into identical cells that represent the wall of the infill structure as well as the pores. The base area of each of these cells is composed of the proportion of cell wall thickness and pore size. The ratio of the cross-sectional area filled with material to the pore volume represents the infill degree and can be calculated for each cell by equations 1.1 for the grid-based infill structure and with equation 1.2 for the honeycomb infill structure. The wall thickness b_x is varied between 0.5 mm and 10 mm.

$$\varphi_{grid} = \frac{A_{grid} - A_{pore}}{A_{grid}} = \frac{x^2 - (x - b_x)^2}{x^2} \tag{1.1}$$

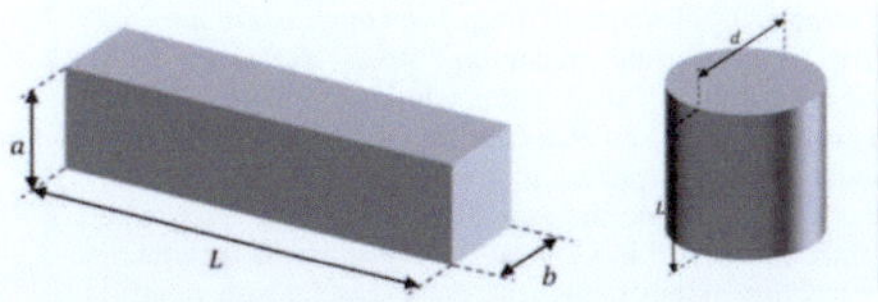

FIGURE 2: Representation of the specimen dimensions.

$$\varphi_{grid} = \frac{A_{hc} - A_{pore}}{A_{hc}} = \frac{\sqrt{3} \cdot x^2 - \sqrt{3} \cdot (x - b_x)^2}{\sqrt{3} \cdot x^2} \tag{1.2}$$

The variation of the infill structure is done in *Autodesk Inventor* by a parametrization of the infill structure according to figure 1. Additionally, the infill structures were oriented horizontally or vertically in the specimens. To tilt the infills, three angle-dependencies were introduced: α rotates the infill structure around the x-axis, β around the y-axis while γ rotates the structure around its longitudinal axis, refer to figure 3 (c)-(e) and (g)-(h). The angles of the compression specimens are varied between α = [0°,15°,...,75°,90°] and γ = [0°,10°,...,30°]. β is not varied separately due to the rotational symmetry of the compression specimens. Because the 3-point bending specimen do not have the same rotational symmetry, more specimen had to be simulated in order to achieve the same angle variations. They were varied between α = [0°,15°,...,75°,90°], β = [0°,15°,...,75°,90°] and γ = [0°,15°,...,45°], which is varied depending on the selected structure with the corresponding angles.

2.3 Simulation setup

In this section the setup of the simulations is presented. The simulation was performed with the FEM software *Altair Hyperworks 2021*. *Hyperworks* provides the benefit, that loads can be directly applied to the infill structure, rather than be absorbed by the shell of the part. A comparison between the

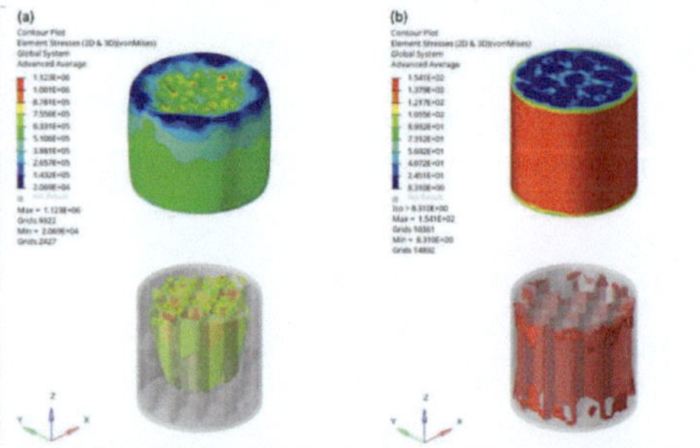

FIGURE 4: Differences in the simulation of the stress distribution with different softwares: (a) Simulation performed with Inventor; (b) Simulation performed with Hyperworks.

results of a simulation that was performed with *Inventor 2022* and a simulation performed with *Hyperworks* is presented in figure 4. The upper pictures are showing the deformed shape of the specimen; the bottom pictures are showing the stress distribution in the infill structure. As one can clearly see, the direct introduction of the load into the shell leads to an incorrect simulation of the loads in the infill structure. This can be neglected by the selection of an ideal rigid connection in Hyperworks and leads to the correct load of the infill structure.

The element size of the used mesh is set to 2 mm and the minimum element size is set to 0.1 mm. Local refinement of the meshing using Proximity and curvature is omitted, because the focus is not on the exact resolution of the stress peaks in small areas of the geometry. The FE model of the cylindrical specimens is constructed from tetra-elements with four nodes each which are well suited for the approximation of complex three-dimensional structures.

For the bending beam, the 3D type of the elements is set to mixed and the 2D type is set to quads. Due to the square bases of the pyramid-shaped elements, an even distribution of the nodes on the specimen surface is achieved, which can be used for the approximation of the support rollers and the compression die. Depending on the internal geometry tetrahedra or pyramids are used to best represent the specimen surface.

To validate the generated mesh, a convergence study was performed. The mesh was implemented for all simulated specimens, regardless of individual size of the analyzed infill structure. 84 cylindrical compression specimen and 96 3-point-bending specimen were simulated.

The simulation is build according to the testing methods presented in section 2.1. In the compression test setup, the mounting of the specimen allows an increase of the contact area between the specimen and the testing machine under the assumption of an ideal frictionless contact. As a result, the stress concentrations in the edge areas of the cylindrical specimen's end faces and bulging of the lateral surface are reduced to maintain the focus on the stresses inside the infill structure. This support is applied to the entire face of the compression specimen.

For the support of a 3-point bending specimen, two support rollers are used [10] to ensure rolling of the beam and prevent an incorrectly high tensile stress at the bottom side of the beam during the bending test. Accordingly, the support is only associated with the specimen surface in the area of these support rollers. The spacing of the support rollers is l=70 mm.

The test load for the compression test according to [9] is set to F_D=50 kN, the test series with grid-shaped infill structure and a variable infill degree is additionally carried out with the loads F_D=100 kN and F_D=200 kN, to verify the assumed linear relationship between test load and simulated stress. A load of F_B=20 kN is chosen for the 3-point bending test simulation, since the stress level of this type of load is higher than that of a pure compression load.

The load analysis is evaluated qualitatively to show differences in stress between configurations. Therefore, the value of the test forces is secondary as long as it is constant across the series of tests to be compared and a linear elastic model is used.

As material, stainless steel 316L was chosen. The material properties that were used in the simulations are presented in table 1.

Table 1: Material properties used for the simulation.

Young's modulus	200 GPa
Poisson's ratio	0,27
Density	7,9 g/cm³

3. RESULTS AND DISCUSSION

The results of FEA strongly depend on the mesh setting. A trade-off between a fine resolution mesh and a reasonable computational effort has to be found in order to obtain reasonable results in an efficient way. Under a load longitudinal to the infill structure, no change of the simulated stress using different mesh sizes occur and does not allow any statement about the quality of the approximation. Under a load transverse to the infill structure, a relationship between stress and finite element size could be identified.

3.1 Compression simulation with the cylindrical shapes

The compression simulation has been performed with cylindrical specimens with grid and honeycomb as infill structures.

Figure 5 shows the maximum stress occurring as a function of the infill degree for cylindrical honeycombs and grid structures with the same wall thickness of 1mm at a compressive force of 50kN. As expected, the maximum stress decreases with increasing infill degree. This is in line with results from different research groups [8, 12] and also confirms the assumption that a higher material volume fraction of infill reduces local and global stress peaks by increasing the area of the absorbing force. The volume fraction of infill has a clear influence on the maximum stresses occurring in the component. For example, with an infill of 10% the maximum stresses are about 215MPa for grid structures and 220MPa for honeycomb structures. In comparison, the stresses with an infill of 30% for grid and honeycomb structures are about 111MPa. This is a reduction in stress of 48% for grid and 30.8% for honeycomb structures. In comparison, the maximum stress in the infill reduces between 60% to 70% from 64.6MPa to 55.6MPa (14%) for the grid and from 63.5MPa to 55.7MPa (12%) for the honeycomb structures. Above 70% infill, the maximum stress is reduced by less than 10% regardless of whether grid or honeycomb structures are applied. Furthermore, in figure 5 it can be observed that both honeycomb and grid structures exhibit similarly large maximum stresses at the same load perpendicular to the infill orientation, since the area filled with infill is of similar size, which is confirmed in other publications [12].

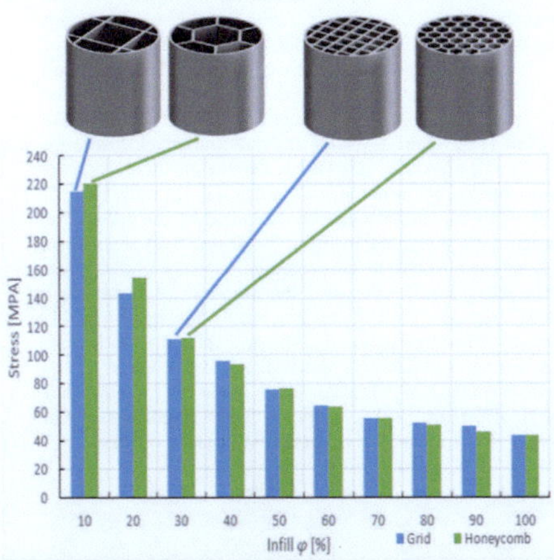

FIGURE 5: Representation of the maximum stress for different infill structures and percentages for cylindrical specimen geometries with a wall thickness of 1mm at a force of 50kN.

Table 2 displays the maximum stresses for grid structures at different infill for 100kN and 200kN force application. A linear relationship between the applied force and the stress can be observed. The stress at 100kN is always approximately half the stress at a force of 200kN, regardless of the infill percentage.

Table 2: Maximum stress depending of force and degree of infill for grid structures.

Infill [%]	Stress at 100kN [MPa]	Stress at 200kN [MPa]
10	415.37	839.52
20	285.69	571.77
30	221.96	445.18
40	179	356.8
50	151.96	304.11
60	126	252.3
70	113.1	226
80	99.56	200.8
90	88.68	177.2
100	87.11	159.6

Figure 6 illustrates the maximum stress as a function of the infill degree and infill orientation for 0° (perpendicular to the load) and 90° (transverse to the load). It is observed that for 90° infill orientation (transverse to the load), the maximum stress appearing also reduces with increasing infill. Furthermore, it can be recognized that at 20% infill the grid structures show a lower stress at 0° orientations with approx. 143 MPa than at 90° orientations with 205 MPa.

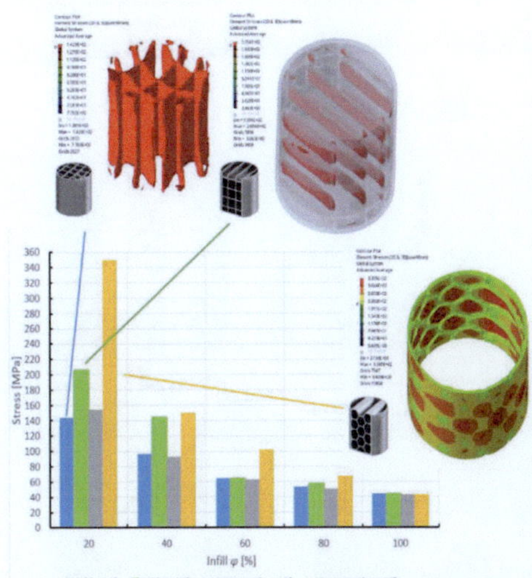

FIGURE 6: Illustration of the maximum stress for different infill structures for cylindrical specimen with a wall thickness 1mm at a force of 50kN at 0° and 90° infill orientation.

The illustrations of the simulations also display the maximum stresses appearing in the infill as well as in the shell structure. It should be mentioned that areas with lower stress levels have been filtered out to visualize more clearly the difference of the highly stressed and thus supporting parts of the specimen. In comparison, the maximum stress for honeycomb structures are higher with 154 MPa for the 0° orientation and approx. 350 MPa for the 90° orientation. In general, it can be observed that the 90°

oriented honeycomb structures have the highest maximum stresses for each percent infill. At 60% infill, the stresses in both orientations of the grid structures and the honeycomb structure at 0° orientations are approximately equal in magnitude of 62-64MPa. Only the honeycomb structure at 90° shows a high stress with approx. 100MPa. If the solid material (100% infill) is considered, the maximum stress is around 44MPa. The difference in the stress less than 25MPa can be observed for the three constellations (see fig. 6). Therefore, for further consideration the infill at 60% is identified as a threshold.

The following figure 7 illustrates the maximum stresses occurring at an infill of 60% with wall thickness of 1mm at different angular orientations α.

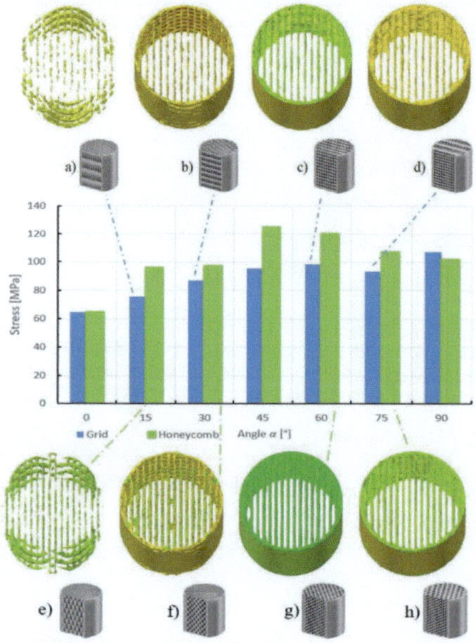

FIGURE 7: Representation of the maximum stress for 60% infill with grid (a-d) and honeycomb (e-h) structures for a wall thickness of 1mm at a force of 50kN for different angles α (the illustrations of the specimen stress distributions are filtered out to the same stress level).

The visualizations of the simulations show maximum stresses occurring in both the infill and the shell for all structures (see fig. 9 a-h). An increase of the stress can be recognized for both infill structures from 0° to 45° orientation from 64MPa to 96MPa (grid) and 125MPa (honeycomb). At 60° orientation, the highest stresses for grid structures occur at the 98MPa stress level. At 45° orientation, the highest stresses occur for honeycomb structures. At 90° orientation, the stresses are higher than at 0° orientation with approximately 102MPa (honeycomb) and 107MPa (grid). In this case, the stress of the honeycomb structures is one time lower than the stress of the grid structures for different angles α.

Varying the angle γ from 0° to 45° for 60% infill, 1mm wall thickness and a force of 50kN shows in the simulations that for the grid structures the maximum stress increases from 107MPa to 127MPa. This can be explained by the occurrence of shear stress in the grid structures due to a different load angle. Varying the angle gamma with the same boundary conditions show no increase in the maximum stress in the simulations for the honeycomb structures, which remain in the range of 97MPa to 102MPa. This is due to the rotational symmetry of the honeycomb, so that shear stress peaks can be avoided when varying the angle γ. Figure 8 shows the maximum stresses for the 0°, 45° and 90° infill orientations at 60% infill as a function of wall thickness. The simulated stress distribution is shown for the 90° orientation honeycomb structures, since the stresses are the highest (maximum 175MPa at 6mm wall thickness) with increasing wall thickness. It is visible that the maximum stresses in this orientation occur in the shell and at the edge structures of the infill. This observation can be explained by the parametric design approach, due to the increase in cell wall thickness allows fewer honeycomb cells to be generated, causing global stress peaks in these areas. Likewise, the stresses for honeycomb structures increase with increasing wall thickness for 45° orientations. The stresses for the 45° grid infill orientation are also increasing dependent to the wall thickness and is in the range from 95MPa to 131MPa. Wall thickness variation has almost no effect for the 0° orientation for honeycomb and grid structures. The stresses vary in these cases around 60MPa.

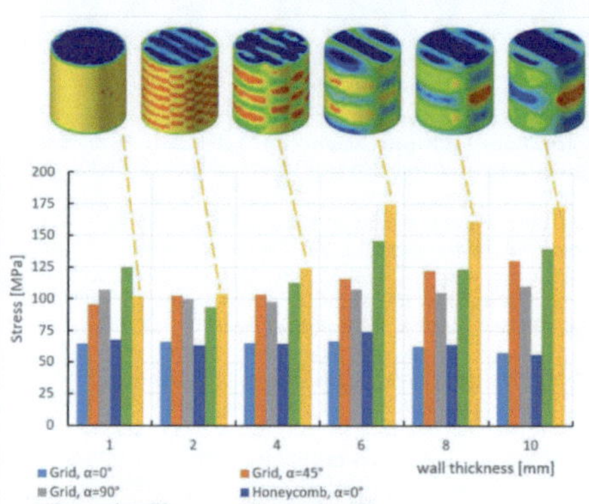

FIGURE 8: Stress distribution for 60% infill at a force of 50kN and 0°, 45° and 90° orientation for different wall thicknesses.

3.2 Bending simulation with beam elements

With the rectangular beams 3-point bending simulations were performed for different grid and honeycomb infill. Figure 9 shows the maximum stresses occurring as a function of the infill degree for rectangular beam elements with honeycombs and grid structures at the same wall thickness of 1mm and a load of 20kN. With increasing infill percentage, the maximum stresses decrease for honeycomb and grid structures. A comparable trend to the compression simulations can be observed, whereby the maximum stresses are consistently higher (see fig.5). Generally, for any percentage of infill, the maximum stresses for honeycomb structures are higher than for grid structures. It is evident that at 10% infill the maximum stress occurring with 1006MPa (grid) and 1121MPa (honeycomb) are significantly higher than the material-dependent tensile and compressive strength respectively of 485MPa [13] and result in component failure. Only above 60% infill the maximum stresses for honeycomb (466MPa) and for grid (400MPa) are below the maximum tensile strength of 316L, thus a threshold can also be identified in this case as well as for the compression simulations.

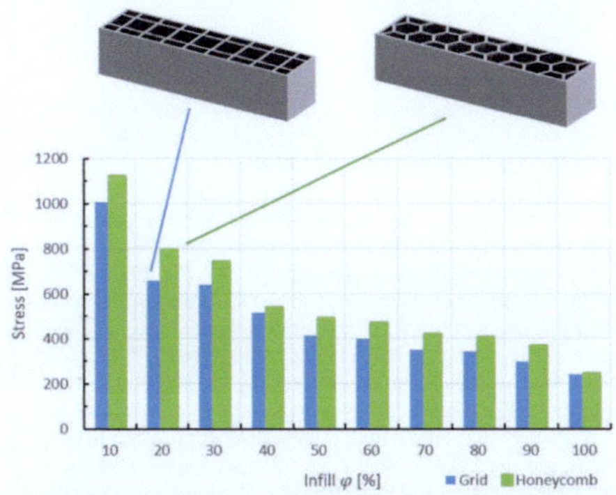

FIGURE 9: Representation of the maximum stress of different infill structures and percentages for rectangular beam specimen geometries with a wall thickness of 1mm and a load of 20kN.

The highest stresses occur in the area of 20kN force application in the specimen surface under the 3-point bending load for a wall thickness of 1mm (see fig. 10). Thus, the different orientations of the infill structures within the beam allow conclusions to be drawn about the stability of the loaded surface. At ($\alpha = 0°$, $\beta = 0°$, $\gamma = 0°$) and ($\alpha = 0°$, $\beta = 90°$, $\gamma = 0°$) the stress on the side walls is small and the transitions between the shell and the cell walls can be localized in the stress distribution on the shell. In addition, the tensile stress on the bottom of the beam under ($\alpha = 0°$, $\beta = 90°$, $\gamma = 0°$) is lower than in the other two configurations, which have similar stress levels on the top and bottom of the specimen. This suggests a high stiffness of the infill structure under ($\alpha = 0°$, $\beta = 90°$, $\gamma = 0°$) since the stress generated by the bending in the shell is lower than the stress due to the load application in the center on the top of the beam. The tensile stress on the bottom side in symmetrically designed bending specimens should theoretically be equal to the compressive stress on the top side. The stress distribution is symmetrically distributed from the neutral axis of the bending specimen to the cell walls above and below and increases with increasing distance from the neutral axis. The cell walls below the loaded surface in the longitudinal direction of the component seem to be the best way to rigidify a component under bending load. Without the cell walls the stresses increase sharply in the remaining vertical walls as well as in the shell of the bending specimen. The cell walls aligned parallel to the loaded surface further reduce stresses in the specimen surface and can be combined with the positive effects of increased shell wall thickness [14].

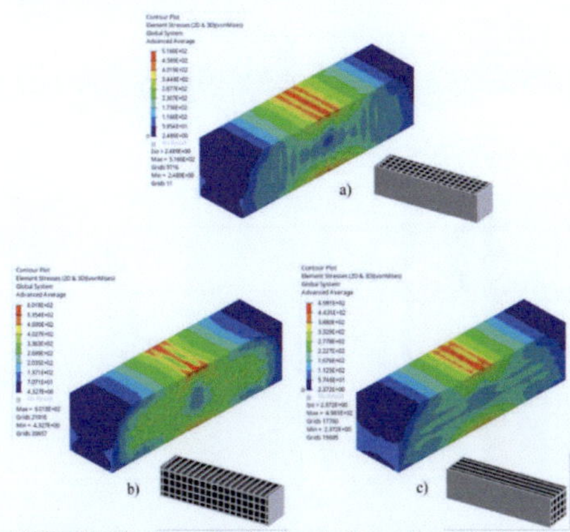

FIGURE 10: Stress distribution in the surface of a bending specimen with grid structure and infill degree of φ=60%, orientation of infill structure: α=0° and β=0° (a), α=90° and β=0° (b), α=0° and β=90° (c).

The diagram in Figure 11 displays the maximum stress as a function of different angles for a wall thickness of 1mm and load of 20kN for grid and honeycomb structures. For the grid structures, the maximum stress decreases steadily from 330MPa (15°) to 302MPa (90°) depending on the angle α. In contrast, the maximum stress for the honeycomb structures increases steadily from 336MPa (15°) to 347MPa (90°) as a function of the angle. Due to the wall orientations grid structures are to be favored for larger angles, as stress peaks could be reduced. For angles β 15° to 60°, the stresses vary between 357MPa and 387MPa for the grid structures and subsequently decrease to 340MPa with greater angle. In contrast, for the honeycomb structures, the stresses decrease from 354MPa (β = 15°) to 300MPa (β = 90°)

and are throughout smaller than for the grid structures at any angle. In the case of the grid structures, as γ is varied, the stresses increase with increasing angle from 411MPa (15°) to 464MPa (45°). A comparison with the honeycomb structures shows that the stresses vary in the range of 441MPa (γ = 30°) and 402MPa (γ = 45°) and tend to decrease at larger angles. The stresses are smaller for 30° and 45° than for the grid structures. A final consideration of the angles leads to the observation that the stresses are the lowest at an angle α for the grid structures. The variation of the angle β has a greater influence on the occurrence of stress peaks for grid and honeycomb structures, which can be better reduced with increasing angle for honeycomb structures. Nevertheless, the stresses are highest at the γ angle for both the grid and honeycomb structures. There seems to be a tendency to reduce the occurrence of stress peaks as the angle increases.

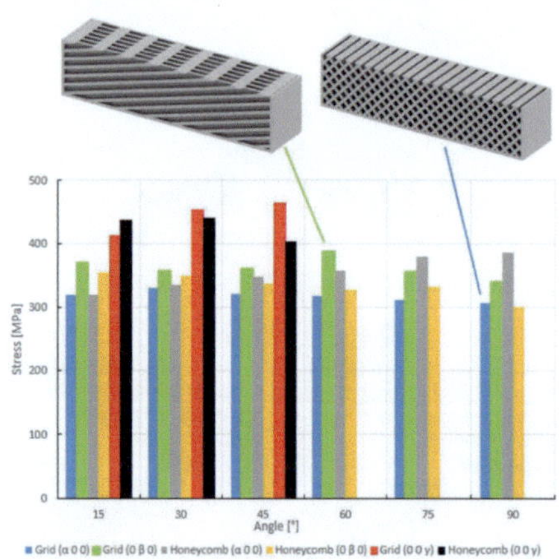

FIGURE 11: Maximum stresses depending on angle (α = 15° to 90°); (β = 15° to 90°); (γ = 15° to 45°) at 1mm wall thickness and load 20kN for grid and honeycomb infill (infill degree φ=60%).

The influence due to a changed cell wall thickness b_x is also investigated under bending load, the considered orientations of the infill structure are (α = 0°, β = 0°, γ = 0°) and can be compared within the test series on the influence of the infill degree at the identical orientations. The infill level of this simulation series is set to φ=60 % (see fig. 12). The maximum stress of the bending specimen frequently occurs in the load application. Increasing the wall thickness for grid structures can lead to a distribution of the stress levels towards the edge regions of the beam elements (fig. 12 a-c). In comparison, the maximum stress level is concentrated in the center of the bending beam, regardless of the wall thickness (fig. 12 d-g). Independent of the infill structure, the stress peaks occur in the areas of the voids of the infill with increasing wall thickness. Likewise, the maximum stress increases with increasing wall thickness for these constellations (see fig.12).

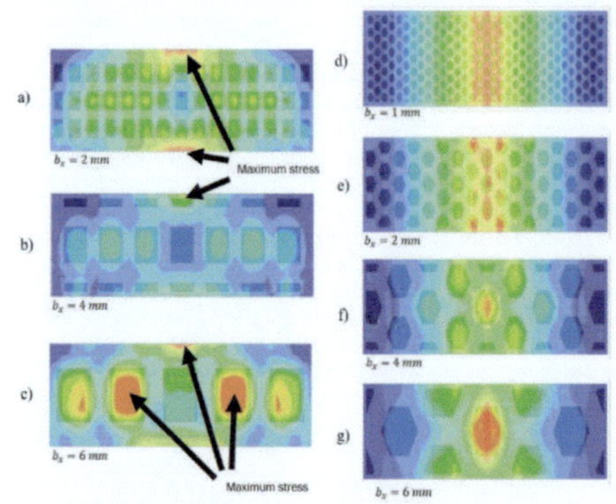

FIGURE 12: Representation of the stress distribution for different wall thicknesses b_x, grid (a-c, side view) and honeycomb (d-g, top view) infill structures.

Figure 13 displays the maximum stresses for the grid and honeycomb structures for different angles and wall thicknesses. It can be observed that the grid structure (α = 0°, β = 90°, γ = 0°) has the lowest stresses with a maximum of 420MPa (b_x =1mm). The wall thickness of 8mm has not been simulated, because the threshold of the stress was already exceeded at 6mm. It appears that the wall thickness has little effect on the stresses in this configuration. The highest stress occurs in the grid structures (α = 90°, β = 0°, γ = 0°) with a wall thickness of 4mm, since in this case there are the fewest walls along the force application and thus the stress is concentrated at 698MPa. Furthermore, it can be recognized an increase of the maximum stress for the honeycomb structures (α = 0°, β = 0°, γ = 0°) angle configuration with increasing wall thickness, due to the fact that the stress peaks concentrate in the voids (compare fig. 12). This finding can also be made for the other honeycomb configurations (α = 90°, β = 0°, γ = 0°) and (α = 0°, β = 90°, γ = 0°) up to a wall thickness of 4mm. At a wall thickness greater than 4mm, the stresses decrease due to the parametric design and less stress peaks appear in the voids. In general, the highest stresses occur at wall thicknesses of 4mm to 8mm, which is due to a smaller number of cells and thus stress peaks are concentrated on a few cells. In the case of high cell wall thicknesses additional cell walls must be inserted in the load application area to ensure uniform load transfer into the infill structure. An alignment of the cell walls along the beam is also advantageous here. Despite the coarser structure an increase in cell size can lead to a relief of the entire component due to the fusion of cell walls and the shell.

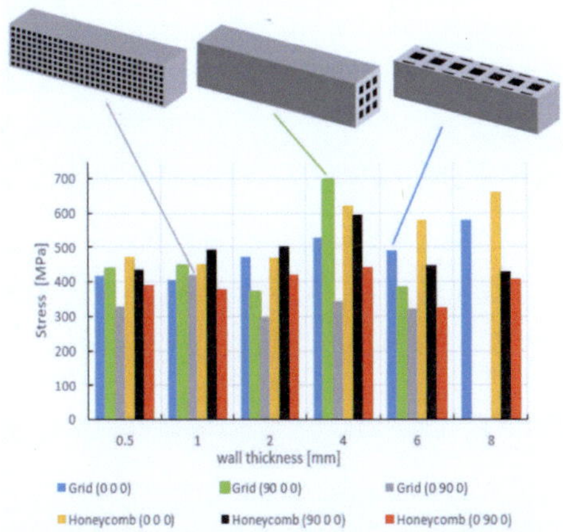

FIGURE 13: Representation of maximum stress for different wall thicknesses, orientations and infill structures (infill degree of φ=60%).

4. CONCLUSION

In this work grid and honeycomb structures were simulatively investigated with compression and bending simulations. It was shown that the parametric design was successfully developed for grid and honeycomb structures. By varying the cell wall thickness, the degree of infill and the orientations, holistic insights into the maximum stresses in the test geometries could be obtained. Both grid and honeycomb structures show superior behavior compared to each other for different configurations and load conditions. With the results one structure cannot be preferred, rather must be used application-dependent. Furthermore, it was found exactly that infill structures have a supporting share in the stresses that occur and 60% infill has comparable maximum stress levels to the solid material and thus a material saving of 40% can be realized for components with infill. With the findings gained from this paper, the production time for components can be significantly reduced where the load situation is estimable and thus component failure prevented. One possible application is for components where the outer structure cannot be changed and yet a material saving is to be achieved, such as in aerospace, automotive, spare parts or toolingmalking. The simulations will be modified for realistic use cases in the future.

Since this study was conducted in a purely simulative manner, it needs to be validated with experimental studies and will be presented in future work. Furthermore, different geometries should be investigated to capture the diversity of 3D-printed components. Likewise, dynamic load conditions in the infill can be supplemented by time-dependent simulations. The compression and 3-point bending tests have been performed to investigate the stresses in the part for the geometries according to the standards. The different configurations can of course be integrated into the other design features of the additively manufactured parts. The simulation results can be successfully applied to extend existing design guidelines and to enable an efficient initial accurate manufacturing. By specific adjustments of the material parameters, these results can be extended not only for the guidelines of the investigated material 316L [15], but also for other design guidelines with materials like IN718 [16] and Ti6Al4V [17]. This opens up new perspectives for the design of infill structures in the context of Design for Additive Manufacturing (DfAM). The ability to apply simulation results to different materials, taking into account specific properties, marks a significant advance in additive manufacturing technology. This enables optimized component manufacturing that simultaneously maximizes material utilization and meets mechanical performance requirements. The results obtained are thus instrumental in expanding the range of applications of DfAM and fully exploiting its potential, making a significant contribution to the development of advanced manufacturing methods.

ACKNOWLEDGEMENTS

The author declare no conflict of interests.

REFERENCES

[1] Sabban, R.; Bahl, S.; Chatterjee, K.; Suwas, S., Globularization using heat treatment in additively manufactured Ti-6Al-4V for high strength and toughness., Acta Mater. **2019**, 162, 239–254.

[2] Ge, J.; Lin, J.; Lei, Y.; Fu, H. Location-related thermal history, microstructure, and mechanical properties of arc additively manufactured 2Cr13 steel using cold metal transfer welding., Mater. Sci. Eng. A 2018, 715, 144–153.

[3] Qian Zhang, Wenwang Wu, Jainlin Liu; „Local Strengthening Design and Compressive Behavior Study of the Triangular Honeycomb Structure" Mdpi Metals 2022, 12(11), 1779; https://doi.org/10.3390/met12111779

[4] Liang-Yu Chen, Shun-Xing Liang, Yujing Liu, Lai-Chang Zhang, Additive manufacturing of metallic lattice structures: Unconstrained design, accurate fabrication, fascinated performances, and challenges, Materials Science and Engineering: R: Reports Volume 146, October 2021, 100648, https://doi.org/10.1016/j.mser.2021.100648

[5] Tomáš Goldmann, Wei-Chin Huang, Sylwia Rzepa, Jan Džugan, Radek Sedláče and Matej Daniel "Additive Manufacturing of Honeycomb Lattice Structure—From Theoretical Models to Polymer and Metal Products" Mdpi Materials 2022, 15(5), 1838; https://doi.org/10.3390/ma15051838

[6] D. Jin, D.J. Tian, J.H. Li and M. Sakane "Low-cycle fatigue of 316L stainless steel under proportional and nonproportional loadings", Fatigue & Fracture of

Engineering Materials & StructuresVolume 39, Issue 7Jul 2016 Pages 791-920

[7] Itthidet Thawon, Thongchai Fongsamootr, Yuttana Monaand Pana Suttaku, Investigation of the Mechanical Properties of Additively Manufactured Metal Parts with Different Relative Densities, Mdpi *Appl. Sci.* 2022, *12*(19), 9915; https://doi.org/10.3390/app12199915

[8] Maitane Gabilondo, Xabier Cearsolo, Mario Arrue and Francisco Castro, Influence of Build Orientation, Chamber Temperature and Infill Pattern on Mechanical Properties of 316L Parts Manufactured by Bound Metal Deposition, Mdpi Materials 2022, 15(3), 1183; https://doi.org/10.3390/ma15031183

[9] Yun Zhou a, Xiaoqing Zuo and Ting Zhang, 316L Stainless Honeycombs Fabricated by Extruding and Sintering, Advanced Materials Research Online: 2010-10-27 ISSN: 1662-8985, Vols. 146-147, pp 1921-1924 doi:10.4028/www.scientific.net/AMR.146-147.1921

[10] ASTM E9-09; Standard Test Methods of Compression Testing of Metallic Materials at Room Temperature, Last Updated: Jan 25, 2018; last access 30.04.23; https://www.astm.org/e0009-09.html

[11] ISO 7438:2020 Metallic materials — Bend test; last access 30.04.2023; https://www.iso.org/standard/72187.html

[12] Pernet, Benoît; Nagel, Jacquelyn Kay; Zhang, Hao; Compressive Strength Assessment of 3D Printing Infill Patterns; Procedia CIRP 105, 2022, 682-687; 10.1016/j.procir.2022.02.114

[13] Material Data Sheet EOS StainlessSteel 316L, https://www.fabb-it.de/files/datenblaetter/edelstahl.pdf, last access 30.04.2023

[14] Lubombo, C., & Huneault, M. (2018). Effect of Infill Patterns on the Mechanical Performance of Lightweight 3D-Printed Cellular PLA Parts. Materials Today Communications. doi:10.1016/j.mtcomm.2018.09.017

[15] Karim Asami et al. Design Guidelines For Green Parts Manufactured With Stainless Steel In The Filament Based Material Extrusion Process For Metals (MEX/M), European Powder Metallurgy Association, Ed., World PM2022 Proceedings, 2022.

[16] D. Herzog, K. Asami, "Design guidelines for laser powder bed fusion in Inconel 718," *Journal of Laser Applications*, vol. 34, no. 1, p. 12015, 2022, doi: 10.2351/7.0000508.

[17] K. Asami, K. Bartsch, and C. Emmelmann, "Design-Richtlinie für die ressourceneffiziente Gestaltung von Stützstrukturanbindungen im pulverbettbasierten Laserstrahlschmelzen von Ti6Al4V: Design rule for efficient support connection point spacing in laser powder bed fusion of Ti6Al4V," in *Proceedings of the 17th Rapid.Tech 3D Conference Erfurt, Germany, 22–23 June 2021*, M. Kynast, G. Witt, and M. Eichmann, Eds., München: Carl Hanser Verlag GmbH & Co. KG, 2021, pp. 80–88.

B.4 Predictive modeling of lattice structure design for 316L stainless steel using machine learning in the L-PBF process

The summary of the findings can be found in Chapter 5.4.

Journal of Laser Applications

RESEARCH ARTICLE | OCTOBER 13 2023

Predictive modeling of lattice structure design for 316L stainless steel using machine learning in the L-PBF process

Special Collection: Proceedings of the International Congress of Applications of Lasers & Electro-Optics (ICALEO 2023)

Karim Asami ; Sebastian Roth ; Michel Krukenberg; Tim Röver ; Dirk Herzog ; Claus Emmelmann

J. Laser Appl. 35, 042046 (2023)
https://doi.org/10.2351/7.0001174

Journal of Laser Applications ARTICLE pubs.aip.org/lia/jla

Predictive modeling of lattice structure design for 316L stainless steel using machine learning in the L-PBF process

Cite as: J. Laser Appl. 35, 042046 (2023); doi: 10.2351/7.0001174
Submitted: 11 July 2023 · Accepted: 27 September 2023 ·
Published Online: 13 October 2023

View Online Export Citation CrossMark

Karim Asami,[1] Sebastian Roth,[1] Michel Krukenberg,[1] Tim Röver,[1] Dirk Herzog,[2] and Claus Emmelmann[1]

AFFILIATIONS

[1]Institute of Laser and System Technologies (iLAS), Hamburg University of Technology (TUHH), Harburger Schloßstraße 28, 21079 Hamburg, Germany
[2]Institute for Industrialization of Smart Materials (ISM), Hamburg University of Technology (TUHH), Eißendorfer Straße 40, 21073 Hamburg, Germany

Note: Paper published as part of the special topic on Proceedings of the International Congress of Applications of Lasers & Electro-Optics 2023.

ABSTRACT

Lattice structures in additive manufacturing of 316L stainless steel have gained increasing attention due to their well-suited mechanical properties and lightweight characteristics. Infill structures such as honeycomb, lattice, and gyroid have shown promise in achieving desirable mechanical properties for various applications. However, the design process of these structures is complex and time-consuming. In this study, we propose a machine learning-based approach to optimize the design of honeycomb, lattice, and gyroid infill structures in 316L stainless steel fabricated using laser powder bed fusion (L-PBF) technology under different loading conditions. A dataset of simulated lattice structures with varying geometries, wall thickness, distance, and angle using a computational model that simulates the mechanical behavior of infill structures under different loading conditions was generated. The dataset was then used to train a machine learning model to predict the mechanical properties of infill structures based on their design parameters. Using the trained machine learning model, we then performed a design exploration to identify the optimal infill structure geometry for a given set of mechanical requirements and loading conditions. Finally, we fabricated the optimized infill structures using L-PBF technology and conducted a series of mechanical tests to validate their performance under different loading conditions. Overall, our study demonstrates the potential of machine learning-based approaches for efficient and effective designing of honeycomb, lattice, and gyroid infill structures in 316L stainless steel fabricated using L-PBF technology under different loading conditions. Furthermore, this approach can be used for dynamic loading studies of infill structures.

Key words: Additive manufacturing, machine learning, DfAM, lattice structures, L-PBF, finite element method, simulation, 316L

I. INTRODUCTION

Additive manufacturing (AM according to ISO/ASTM 52900:2015[1]) has emerged as a promising technology to produce complex components with a variety of benefits and specific attributes, e.g., low thermal conductivity and tailor-made mechanical properties.[2–4] One of the key challenges is to reduce the production time while ensuring the structural integrity of the part. So far, additive manufacturing has primarily been used for prototype manufacturing and small-scale fabrication.[5] Usually, an experimental design setup is required to reduce the time effort and investigation of specific process parameter and the characteristics of the final part.[6,7] Laser powder bed fusion (L-PBF) technology is the most widely used additive manufacturing process due to its high process stability and repeatable manufacturing accuracy.[8]

Journal of Laser Applications ARTICLE pubs.aip.org/lia/jla

However, one solution for the reduction of manufacturing time is the use of infill structures by lattices, which can be applied in areas with lower component utilization and has been already excessively investigated for the L-PBF process.[9–11] These infill structures make it possible to reduce material waste and optimize part weight without compromising the outer contour and material thickness at the stress points.

Infill structures are used in a variety of areas, including mold making, replacement part, and automotive and aerospace applications.[12] For example, in mold making, infill structures can help reduce material waste by filling the cavity inside the part with a lattice structure that has the same outer contour as the original part. This allows efficient use of material while reducing part weight. For instance, in replacement part applications, infill structures can help maintain part performance and function while reducing weight and manufacturing costs.

Owing to the large number of process parameters,[13] machine time, and pre- and postprocess effort, a complete experimental investigation of the large number of different types of infill structures[9] is inefficient. Therefore, simulations based on numerical methods such as finite elements are predominantly used for the investigation of process parameters and characteristic structural properties.[14,15]

Successful manufacturing in the L-PBF process requires experienced designers for the part development with different features, for manufacturing with the process parameters and part orientation, and for postprocessing. In order to avoid multiple trial and error iterations for a product, the design of experiments and simulations, which depend on the computational accuracy and effort, design guidelines for different part features depending on the process and material are developed to ensure successful part production.[16–18]

Machine learning (ML) approaches have been increasingly applied for design aspects in additive manufacturing technologies over recent years to predict the desired component properties as a function of process parameters.[19] For example, ML for the design features has been previously used for roughness prediction of surfaces using regression algorithms,[20] density and melt pool prediction with regression algorithms,[21] and prediction of mechanical properties also applying regression algorithms.[22] Furthermore, a Gaussian-based ML model was used to optimize the process parameters and predict the remelt depth of tracks as a function of the combined laser power and laser scan speed in the L-PBF process.[23,24] For the postprocess, there exists also an optical-based ML model where the data are provided to support vector machine (SVM) classifiers to predict density regions of the part.[25]

There exists an approach that predicts the printability of specific voxel-based infill structure designs using convolutional neural networks (CNNs) as well as neural networks (NNs) for fabrication and, thus, prevents failure during manufacturing.[26] Due to the voxelization size and the hybrid ML architecture, a very high computational time is required, and thus, the efficiency is very limited. This research gap presents the potential for the development of a machine learning model that successfully captures the complex relationships between infill structures and stress with low computational effort, enabling the prediction of the optimal design for infill structures.

With the aid of the ML algorithm, the optimum infill structure with its geometric properties (orientation, wall thickness, etc.) can be determined on the basis of known load conditions, resulting in material and production time savings. Furthermore, this approach can ensure the load-dependent first time right production, as the designs in the algorithm are considered to be manufacturable and can, thus, serve as a complement to existing guidelines. Therefore, this approach contributes one fundamental for the AI-based design for additive manufacturing (DfAM) with unknown design feature combinations that often occur in AM.[16–18]

This investigation explores the potential of machine learning and outlines a possible approach for developing a model that predict stress values based on simulated and experimental performed datasets.

II. MATERIALS AND METHODS

Figure 1 illustrates the methodology used in this paper. For the development of the ML model, parametric designs of infill geometries were initially generated. These infill designs were varied using the parameters displayed in Fig. 1 and simulated with different loading conditions with the aim of determining the maximum stress that occurs in the component. The parameters shown in blue were used as training and test parameters for the ML model. The results of the stress were obtained by compression and 3-point bending tests according to Refs. 27 and 28.

A. Specimen design

The design development of infill structures focuses on grid, gyroid, and honeycomb structures, which were analyzed mainly in the L-PBF process for lightweight application.[29] For the specimen design modeling, cylindrical bars and rectangular beams were considered as basic geometries used in different design applications as particular features. The dimension of cylindrical rods is 40 mm in diameter and height. The rectangular beams have a dimension of

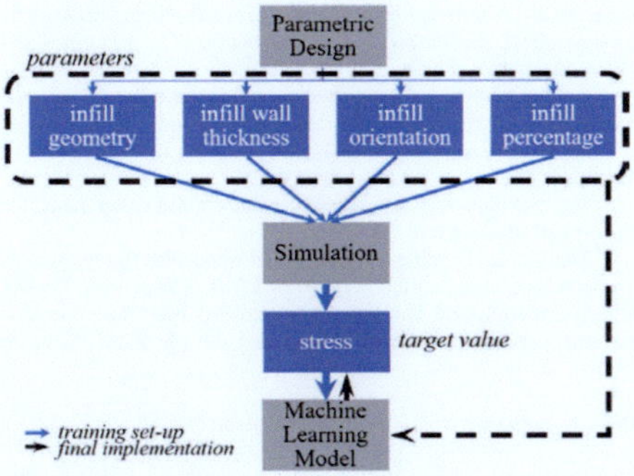

FIG. 1. Methodology for ML model development and stress prediction.

Journal of Laser Applications ARTICLE pubs.aip.org/lia/jla

TABLE I. Specimen load and infill parameter variation.

Parameter	Unit	 Cylindrical rod	 Rectangular beam
Infill type	—	Grid and honeycomb	Grid, honeycomb, and gyroid
Wall thickness	mm	1, 2, 4, 6, 8, and 10	0.5, 1, 2, 4, 6, and 8
Infill orientation	deg	0, 15, 30, 45, 60, 75, and 90	0, 15, 30, 45, 60, 75, and 90
Infill percentage	%	10, 20, … , 100	10, 20, … , 100
Load	kN	50, 100, and 200	10 and 20

80 mm in length and a square edge length of 20 mm. Also the size of the rectangular beams has been varied to a reduced length of 50 mm for manufacturable reasons. The honeycomb and lattice structures were modeled with 3D-CAD software *Autodesk inventor* (Autodesk, San Francisco, USA), and the gyroid structures were designed with *nTopolgy* (nTopology, New York City, USA). Table I presents the different variations in the infill parameters and loads. A total of 167 different samples have been designed with each parameter combination.

For the validation of the simulation, four different beams were manufactured from different variations using the L-PBF method. Two of the samples are gyroid structures with 60% and 80% infill, and one sample each is a lattice and honeycomb structure with 60% infill and 1 mm wall thickness. The specimens were produced on the OCM MPrint (One Click Metal, Tamm, Germany), and the material that was made of is the widely used AISI 316L (1.4404) powder that is also provided by the manufacturer (see Table II), which is used in versatile applications in different fields.[18] Figure 2 shows powder characterization provided by the manufacturer. The particles are mainly spherical with occasional elliptic necks. The particle size distribution is $D_{10} = 19\,\mu$m, $D_{50} = 29\,\mu$m, and $D90 = 43\,\mu$m according to ISO 13322-2.[30]

For the manufacturing of the specimen, the following process parameter was chosen: the laser diameter of 70 μm, the laser power of 200 W, the scan speed of 800 mm/s, and the layer thickness of 40 μm.

Figure 3 illustrates the different manufactured components for the 3-point bending tests.

The 3-point bending tests were performed with the specimens. The machine used is the Zwick/Roell Z010 (Zwick Roell Group, Ulm, Germany), and the force was recorded over time and displacement. All samples were heat treated for 1 h at 900 °C in an Oven ExSO90 by the manufacturer Aim3D (Aim3D GmbH, Rostock, Germany) according to Ref. 31 to reduce thermal stress peaks due to the manufacturing process.

TABLE II. Composition of the elements for 316L powder (Ref. 30).

Element	C	Si	Mn	Cr	Ni	Mo	Fe
wt. %	0.03	1	>2	16–18	11–14	2–3	Basis

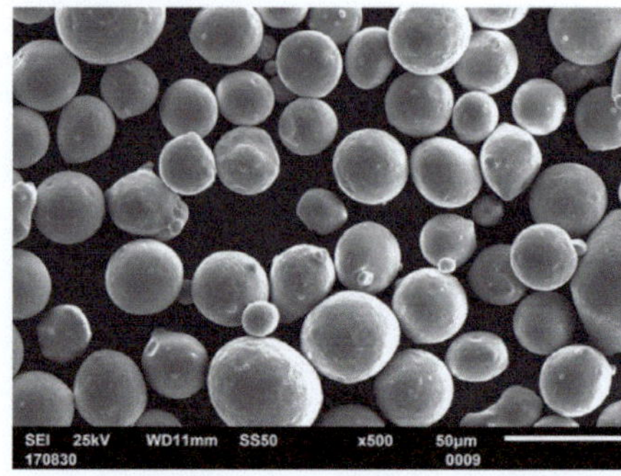

FIG. 2. Illustration of the particle distribution (Ref. 30).

B. Simulation

The pressure and 3-point bending simulations according to Refs. 27 and 28 were performed in *Altair Hyperworks 2021* (Altair, Michigan, USA). A convergence study has been performed in a way that a trade-off between iteration steps and computation time has been selected as the target. The cylindrical bars have been used for compression simulations and the rectangular beams for bending tests. For the cylindrical specimen, tetra-elements with four nodes were used as a mesh. The rectangular beams have been designed for the mesh with tetrahedra and pyramid elements. Simulation was performed for each sample by varying the geometry features and the load condition. The maximum stress for each

FIG. 3. Manufactured specimens of different investigated types of infill: 60% gyroid (top left), 80% gyroid (top right), 60% grid (bottom left), and 60% honeycomb structure (bottom right).

Journal of Laser Applications ARTICLE pubs.aip.org/lia/jla

TABLE III. Assumed material properties.

Element type	Tetra, tetrahedra, and pyramid
Element size	0.01–2 (mm)
Young's modulus	200 (GPa)
Density	7.9 (g/cm^3)
Poisson's ratio	0.27

variation was considered as the target value. The material chosen for the simulation is AISI 316L stainless steel. Table III shows the material properties assumed for the simulations.

In Fig. 4, the simulation approach with different element types is depicted. It is assumed that the force does not impact on a point but on a surface, as in the reality of the tests.

III. MACHINE LEARNING

The machine learning approach for predicting target variables is based on a supervised learning procedure. Regression models are used to model the relationship between the input data and the target variables. In this paper, four specific regression models are considered: decision tree regression (DTR), linear regression (LR), random forest regressor (RFR), and gradient boosting regression (GBR). The 167 simulated compression and 3-point bending test variations of the infill geometries serve as the data basis for the algorithms. The implementation of different models has been performed in PYTHON 3.8. Mainly PANDAS and SCIKIT-learn have been applied for the creation of ML models.[32] The datasets were divided into training and test data for the prediction by different models. Out of the 167 datasets, 70% were allocated to the training data and 30% to the test data. A main criterion for the statement of accuracy for regression-based models is the mean squared error (MSE), which is also applied in this analysis.[21] Another criterion for the accuracy of the models is the consideration of the R^2-score also influenced by the MSE value and examined here.[32] The comparison of the models is done by the normalized values of the variations; thus, the impact of the values occur in sizes from 0 to 1. Finally, predictions are presented for the stresses across the devised loads and different characteristics of the infill structures.

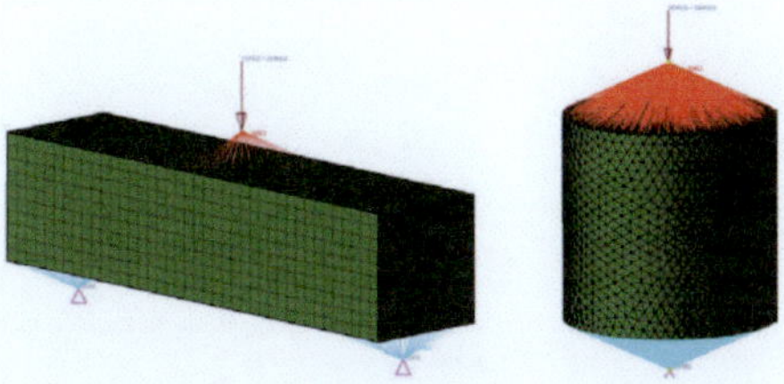

FIG. 4. Compression and 3-point bending simulation with tetra elements (right) and the combination of pyramid and tetrahedra elements (left).

IV. RESULT AND DISCUSSION

All simulations could be executed and the maximum stress could be identified. Depending on the variations, the maximum stresses are in the range from 43.56 MPa (external load of 50 kN, infill of 100%, honeycomb structure, wall thickness of 1 mm, and orientation of 0°) for compression simulation to 1121.35 MPa (external load of 20 kN, infill of 10%, honeycomb structure, wall thickness of 1 mm, and orientation of 0°) for 3-point bending simulation. It is obvious that the 1121.35 MPa maximum stress is higher than the tensile strength of the material with 574 MPa.[30] Nevertheless, these values are considered for qualitative statements about the prediction accuracy of the model. As expected, the maximum stresses increase primarily with increasing load and decreasing percent infill for each orientation and structure. A comparison with other pressure simulations for honeycomb structures shows a comparable stress distribution. It should be mentioned that this comparison is only qualitatively possible, because, on the one hand, unit cells without an external shell were constructed for the simulation and, on the other hand, lower loads were assumed.[10,29] Figure 5 shows the stress

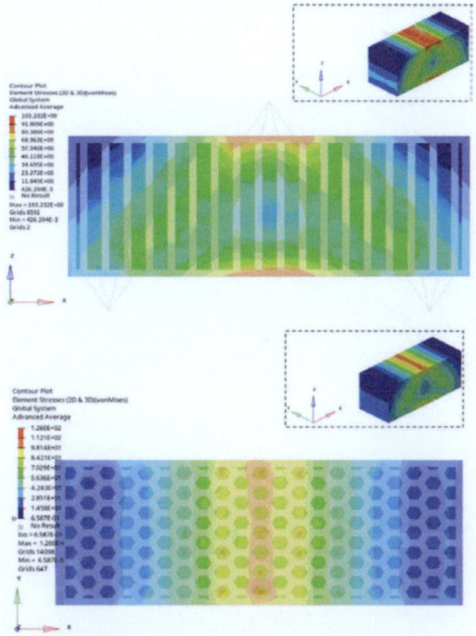

FIG. 5. Stress distribution of the 3-point bending simulation at a load of 10 kN for the bending beams with grid (side view, top image) and honeycomb structure (top view, bottom image).

Journal of Laser Applications ARTICLE pubs.aip.org/lia/jla

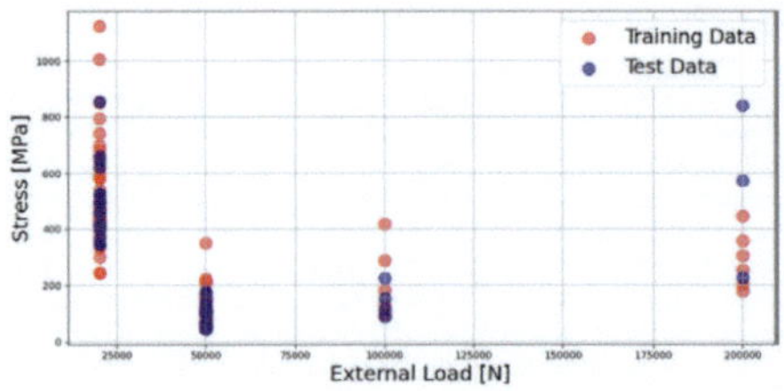

FIG. 6. Distribution of simulated stresses in test and training data.

distribution of the 3-point bending simulation at a load of 10 kN representative for grid (top) and honeycomb (bottom) structures with a length of 50 and 20 mm per edge strand in cross section. It can be observed that the maximum stress in both cases occurs at the point of force application and decreases with increasing distance in all spatial directions. The maximum stress is 103.23 MPa for the grid structure and 125.98 MPa for the honeycomb structure. The prepared samples for the experimental study were performed with the same experimental setup. At a bending load of 10 kN, the lattice specimen has been displaced 1.02 mm in the *z*-direction at the force application point. The simulation shows a displacement of 1.5 mm at the same point. This difference is due to approximation of the simulation and is within the acceptable range. The displacement of the honeycomb structures is 1.6 mm in the simulation and 0.99 mm in the experimental test. The difference is slightly larger here. The displacement of the gyroid infill structure with 80% indicated a displacement with 1.2 mm, a larger plastoelastic deformation. The gyroid infill structure with 60% infill plastically deformed at the load of 10 kN, which could be correctly actuated by the simulation with a higher maximum stress than the tensile strength. The deformation in this case is 4.72 mm.

The maximum stresses of different configurations were divided into training and test data (see Fig. 6). This figure depicts the maximum stress across the external applied load. The training data shown in red have a share of 70% and the test data in blue have a share of 30%. It can be observed that the test data are barely represented at 200 kN. Due to the small dataset and big jumps in the input data, cross-validation of the data has been performed to avoid overfitting using *k*-1 folds ($k = 5$) for the training data and onefold for the test data. The number of repeats is 10.

Figure 7 displays the predictions (*y*-axis) on the simulated data (*x*-axis) for four different models: LR, DTR, GBR, and RFR for the test data. The red lines indicate the point where the predictions coincide with the simulated data. It is obvious that the points

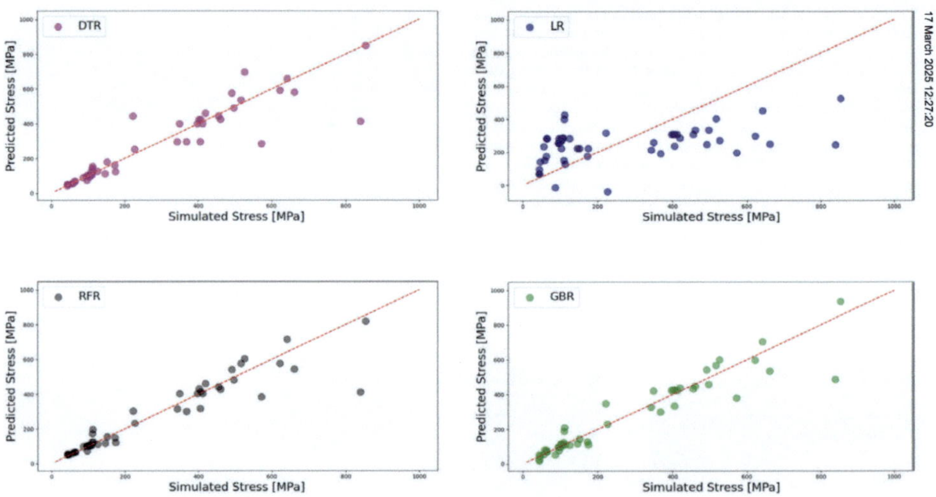

FIG. 7. Stress prediction as a function of simulated stress for different models linear regression (LR, blue, top left), decision tree regression (DTR, magenta, top right), gradient boost regression (GBR, green, bottom left), and random forest regression (RFR, gray, bottom right).

Journal of Laser Applications ARTICLE pubs.aip.org/lia/jla

on the red line, therefore, accurately predict the simulations. It should be mentioned that with increasing distance to the red line, the predictions are less accurate. Since the test data are only available for 20, 50, and 100 kN load cases, a gap between low and high stress regions can be recognized in all diagrams. The LR model makes an accurate prediction in only one case. All other points deviate strongly from the actual values. Therefore, it can be assumed that there is no linear correlation between the input and output parameters. The decision tree model (DTR), in comparison, performs more accurate predictions for lower stresses and underestimates higher stresses. Also, several outliers can be identified in the mid-stress area and just one outlier in the maximum high area. The approximation is acceptable only in the range from 50 to 150 MPa and 400 to 450 MPa. The gradient boost regression (GBR) model also shows a higher approximation accuracy than the LR model but overestimates the stresses for smaller as well as larger loads. Furthermore, it can be observed that in the range of lower stresses, the scatter of the predictions is larger than with the DTR model. In the range above 300 MPa, the deviations of the predictions are smaller compared to the DTR model. Likewise, for large stress, underestimation and overestimation of the simulated stress are apparent. The RFR model shows isolated deviations in the prediction at small and large stresses. In the area of small stresses as in the GBR model, the stresses are increasingly overestimated. This model also has an underestimated prediction for very large stresses (>600 MPa), but it is more accurate than DTR and GBR models. Finally, based on the model comparison, it can be said that LR is unsuitable with the given input parameters, since there is too large scatter independency of the load. The models DTR, GBR, and RFR show a comparably high accuracy in the prediction, whereas the GBR model predicts most frequently.

To examine the trained models for their influencing factors, the data were normalized and the Pearson correlation coefficient was compared. This coefficient is calculated from the variations in the input variables and the output variable.[21] Table IV plots the linear correlations of the input and output variables for different models. It can be recognized that the strongest correlation between load and stress is present for all models. The values vary between 0.716 and 0.723. The second strong dependence is between percent infill and stress at 0.24 for all models. This is plausible since the stress is defined by force per area and, thus, depends directly on both input parameters. The input variables of wall thickness and orientation show comparatively low correlation with stress for all three models, which are in the interval of 0.014 and 0.020, respectively. In this context, the volume percentage of the infill is more decisive than the other parameters. Infill geometry has the lowest correlation to the target stress with values ranging from 0.004 to 0.009 for all models. This is due to the training data, since the maximum stresses of the simulations and the experimental investigations always occurred at the outer contour, and therefore, the infill geometry played an insignificant role in this context.

The evaluation of the trained models is done by considering the root mean square error (RMSE) and the R^2 parameter. RMSE and R^2 are defined as[33]

$$\mathrm{RMSE}(y, \tilde{y}) = \sqrt{\frac{1}{n}\sum_{i=0}^{n-1}(y_i - \tilde{y}_i)^2}, \tag{1}$$

$$R^2 = 1 - \frac{\sum_{i=1}^{n}(y_i - \tilde{y}_i)^2}{\sum_{i=1}^{n}(y_i - \hat{y})^2}. \tag{2}$$

In this case, y_i is the ith prediction, n is the number of simulations, and $\tilde{y}_i$ is the objective quantity and $\hat{y}$ are the average values of y. The R^2 is a variance-based measure of how well the data of a model fit in percentage. In addition, the RMSE has been taken into account to track the extent of difference between the individual predictions.[33] Table V shows the RMSE and R^2 values for the test data of different models. As already observed from Fig. 7, the LR model has a high deviation in the prediction. This is also confirmed by the RMSE of 197.1 MPa, which is unsuitable for robust prediction. The smallest root mean squared error is found in the GBR model. Therefore, it can be assumed that this model can make most accurate predictions. The RFR model has a slightly higher root mean square error than the GBR model, but it is less than 10 MPa. Therefore, this model also indicates comparable prediction accuracy. The DTR model has 86.67 MPa for the test data, a larger error than the RFR and GBR models; however, both errors are in the same order of magnitude (smaller than 100 MPa). The GBR model shows the smallest error and, thus, the highest robustness and flexibility which was previously recognized in Ref. 21. The assumption that the models are overfitted due to a small dataset could not be confirmed although a high deviation in the RMSE would have been an indicator. Attempts to distribute training and test data differently, such as 80% training and 20% test data, resulted in no significantly lower RMSE. The difference in RSME is here for all models less than 8 MPa. The reason for the deviation in the error is apparently due to the large distribution of the variation parameters and, thus, by the small number of simulated data. A larger database and the processing of the available data should significantly reduce the mean squared error. The overfitting can also be excluded by means of the R^2 score because none of the

TABLE IV. Correlation of the input variables of different models, GBR, RFR, and DTR, for stress as the target parameter.

Model	Load	Infill geometry	Wall thickness	Infill orientation	Infill percentage
GBR	0.723	0.004	0.017	0.015	0.24
RFR	0.719	0.008	0.020	0.014	0.24
DTR	0.716	0.009	0.014	0.018	0.24

TABLE V. Mean square error for the test data (MPa) and R^2 (–) score for the investigated models GBR, RFR, DTR, and LR.

Model	Average RMSE (MPa)	R^2
GBR	67.31	0.91
RFR	77.07	0.88
DTR	86.67	0.84
LR	197.1	0.20

Journal of Laser Applications ARTICLE pubs.aip.org/lia/jla

models reaches the value 1.[33] The R^2 value for the LR model displays a value of about 0.20, which by definition is weak. Hence, this is another indication that there is no linear relation between the variations and the target quantity. The models GRB, RFR, and DTR show for R^2 with 0.84– 0.91 a high proportion of independent variables that can be described with the variance of dependent variables. As overfitting of the models could be excluded, all three models seem to be suitable for stress prediction. Ultimately, a larger database may result in a lower squared error and a greater difference in the RMSE of various model accuracies.

Finally, new predictions for previously unknown parameter variations of infill and load have been performed to classify the results. The external load has been assumed to be 10–100 kN in 10 kN steps. A 60% infill volume fraction was assumed for all specimens. The infill structures used are gyroid, grid, and honeycomb, respectively. The wall thickness alternates between 1 and 2 mm and the orientation between 0° and 90°.

Figure 8 below shows the stress predictions of the varied assumed input variables for all four models investigated. It can be recognized that all four models also make a prediction for unknown parameter variations. The LR model deviates significantly in stress prediction from the other models for the all loads, demonstrating again the unsuitability for this approach. The DTR, GBR, and RFR models predict similar stress values for the 50 kN load. These are approximately 60 MPa. The deviations are larger for the prediction at 30 and 10 kN, which are explained in Fig. 7. For smaller loads, the DTR, RFR, and GBR models overestimate and underestimate the simulated stresses, respectively. The predicted stresses for the 10 kN load at GBR are approximately 430 MPa for the honeycomb structure and around 425 MPa for the grid structure. The DTR model predicts 410 MPa for honeycomb and 404 MPa for grid structures. The RFR model predicts 409 MPa for the honeycomb structure and 401 MPa for the grid structure. A comparison with Fig. 7 shows that all models predict a deviation from the simulated stresses that can be explained by the mean square errors, and the data basis is not yet sufficient. Nevertheless, all three models predict qualitatively correctly that the grid structures have a lower maximum stress at a load of 10 kN than the honeycomb structures.

The results of this study demonstrate that various ML algorithms DTR, GRB, and RFR are basically able to make maximum stress predictions for a variety of loads and infill parameters. Furthermore, the investigations could show that some models are more suitable than others for the prediction of stresses depending on the infill parameters and loading situation. Nevertheless, it has also been clearly recognized that the number of data for the development of such an ML algorithm should be significantly larger in order to perform accurate predictions for unknown parameter combination. Nevertheless, the root mean squared error is relatively small at 67.31 MPa for the GBR model. Furthermore, it has been recognized that for the maximum stress in a component with a shell, the volume fraction of the infill is much more dependent than the type and shape of the infill structure. As expected, the GBR model made the most accurate and robust predictions followed by the RFR model. It is obvious that the model prediction is limited to the geometry of the specimens as well as to the use of the AISI 316L material. In order for this model to be more robust in predicting the different infill structures, the stress distribution has to be added to the dataset. Then, areas can be identified in which the different geometric characteristics can be considered more efficiently. Graduated structural combinations are then also predictable.

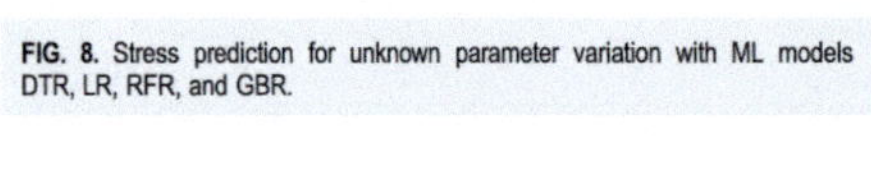

FIG. 8. Stress prediction for unknown parameter variation with ML models DTR, LR, RFR, and GBR.

V. CONCLUSION

In this work, an ML algorithm was developed, which can predict maximum stresses occurring based on the given load and infill parameters of wall thickness, orientation, shape, and volume fraction for the widely used authensitic stainless steel AISI 316L. Bending beams and cylindrical compression specimens with different parameters of the infill structures such as grid, honeycomb, and gyroid were designed, and 167 compression and bending simulations were performed. For experimental validation, four beams with gyroid, honeycomb, and grid structures were fabricated and tested in the L-PBF process. The 167 datasets were then used for testing and training the developed ML algorithm with DTR, GBR, RFR, and LR models. The LR model was found to be unsuitable for this approach. Furthermore, it was shown that the RFR, GBR, and DTR models can make maximum stress predictions with a small relative error. To improve the accuracy, more data are needed for testing and training. It was also concluded that the 70% to 30% split of the training and test data is suitable for this approach. An overfitting of the models could be excluded by the analysis of the R^2 score and the different splitting of the training and test data. In addition, it was validated that a comparison of infill geometries for maximum stresses is only useful if the components have no outer shell. In other cases, the maximum stress occurs directly in the shell in the area of the force application point. In the future, more data will be simulated and validated experimentally in order to improve the accuracy of the ML algorithm and to be able to make quantitative statements for the stress distribution. The stress distribution in the entire component will also be used as a data basis in order to generate a greater influence of various geometric properties of the infill. The aim of this work was to analyze different regression models based on maximum stress, and it was found that one regression model (LR) is unsuitable for stress prediction. In the future, different models such as CNN and deep neural network will be investigated for comparison.[19] These extensions also allow the identification of areas in a component where a certain structure is

Journal of Laser Applications ARTICLE pubs.aip.org/lia/jla

more efficient than another in order to be capable of evaluating graduated structures. In that way, a more comprehensive contribution to the DfAM is provided; in that, specific parameter combinations can be evaluated from the stress distribution even before they have been designed or simulated.

AUTHOR DECLARATIONS

Conflict of Interest

The authors have no conflicts to disclose.

Author Contributions

Karim Asami: Conceptualization (lead); Data curation (lead); Investigation (lead); Methodology (lead); Project administration (lead); Software (lead); Validation (lead); Visualization (lead); Writing – original draft (lead); Writing – review & editing (lead). **Sebastian Roth:** Investigation (supporting). **Michel Krukenberg:** Visualization (supporting). **Tim Röver:** Conceptualization (supporting). **Dirk Herzog:** Methodology (supporting). **Claus Emmelmann:** Supervision (supporting).

REFERENCES

[1]ASTM, ISO/ASTM 52900:2015(E) Standard Terminology for Additive Manufacturing—General Principles—Terminology (2015).

[2]M. Zhao, F. Liu, G. Fu, D. Z. Zhang, T. Zhang, and H. Zhou, "Improved mechanical properties and energy absorption of BCC lattice structures with triply periodic minimal surfaces fabricated by SLM," Materials **11**, 2411 (2018).

[3]Chunze Yan, Liang Hao, Ahmed Hussein, Simon Lawrence Bubb, Philippe Young, and David Raymont, "Evaluation of light-weight AlSi10Mg periodic cellular lattice structures fabricated via direct metal laser sintering," J. Mater. Process. Technol. **214**, 856–864 (2014).

[4]M. Smith, Z. Guan, and W. J. Cantwell, "Finite element modelling of the compressive response of lattice structures manufactured using the selective laser melting technique," Int. J. Mech. Sci. **67**, 28–41 (2013).

[5]T. Wohlers, R. I. Campbell, O. Diegel, R. Huff, and J. Kowen, *Wohlers Report 2019: 3D Printing and Additive Manufacturing State of the Industry* (Wohlers Associates, Fort Collins, CO, 2019).

[6]J. J. S. Dilip *et al.*, "Influence of processing parameters on the evolution of melt pool, porosity, and microstructures in Ti-6Al-4V alloy parts fabricated by selective laser melting," Prog. Addit. Manuf. **2**, 157–167 (2017).

[7]M. Elsayed, M. Ghazy, Y. Youssef, and K. Essa, "Optimization of SLM process parameters for Ti6Al4 V medical implants," Rapid Prototyp. J. **25**, 433–447 (2019).

[8]K. Bartsch, D. Herzog, B. Bossen, and C. Emmelmann, "Material modeling of Ti-6Al-4 V alloy processed by laser powder bed fusion for application in macro-scale process simulation," Mater. Sci. Eng. A **814**, 141237 (2021).

[9]A. Seharing, A. H. Azman, and S. Abdullah, "A review on integration of lightweight gradient lattice structures in additive manufacturing parts," Adv. Mech. Eng. **12**, 168781402091695 (2020).

[10]Q. Zhang, W. Wu, and J. Liu, "Local strengthening design and compressive behavior study of the triangular honeycomb structure," Metals **12**, 1779 (2022).

[11]T. Rosnitschek, A. Seefeldt, B. Alber-Laukant, T. Neumeyer, V. Altstädt, and S. Tremmel, "Correlations of geometry and infill degree of extrusion additively manufactured 316L stainless steel components," Materials **14**, 5173 (2021).

[12]A. J. Pinkerton, "[INVITED] Lasers in additive manufacturing," Opt. Laser Technol. **78**, 25–32 (2016).

[13]T. G. Spears and S. A. Gold, "In-process sensing in selective laser melting (SLM) additive manufacturing," Integr. Mater. Manuf. Innov. **5**, 16–40 (2016).

[14]M. J. Ansari, D.-S. Nguyen, and H. S. Park, "Investigation of SLM process in terms of temperature distribution and melting pool size: Modeling and experimental approaches," Materials **12**, 1272 (2019).

[15]T. Goldmann, W.-C. Huang, S. Rzepa, J. Džugan, R. Sedláček, and M. Daniel, "Additive manufacturing of honeycomb lattice structure-from theoretical models to polymer and metal products," Materials **15**, 1838 (2022).

[16]K. Asami, K. Bartsch, and C. Emmelmann, "Design rule for efficient support connection point spacing in laser powder bed fusion of Ti6Al4V," in *Proceedings of the 17th Rapid Tech 3D Conference, Erfurt, Germany, 22-23 June 2021*, edited by M. Kynast, G. Witt, and M. Eichmann (Carl Hanser Verlag GmbH & Co. KG, München, 2021), pp. 80–88.

[17]D. Herzog *et al.*, "Design guidelines for laser powder bed fusion in Inconel 718," J. Laser Appl. **34**, 012015 (2022).

[18]K. Asami, D. Herzog, B. Bossen, L. Geyer, C. Klemp, and C. Emmelmann, "Design guidelines for green parts manufactured with stainless steel in the filament based material extrusion process for metals (MEX/M)," in *World PM2022 Proceedings, Lyon, France, 9–13 October 2022* (European Powder Metallurgy Association, Chantilly, France, 2022).

[19]J. Jiang, Y. Xiong, Z. Zhang, and D. W. Rosen, "Machine learning integrated design for additive manufacturing," J. Intell. Manuf. **33**, 1073–1086 (2022).

[20]Z. Li, Z. Zhang, J. Shi, and D. Wu, "Prediction of surface roughness in extrusion-based additive manufacturing with machine learning," Robot. Comput. Integr. Manuf. **57**, 488–495 (2019).

[21]M. Kuehne, K. Bartsch, B. Bossen, and C. Emmelmann, "Predicting melt track geometry and part density in laser powder bed fusion of metals using machine learning," Prog. Addit. Manuf. **8**, 47–54 (2023).

[22]I. Baturynska, "Application of machine learning techniques to predict the mechanical properties of polyamide 2200 (PA12) in additive manufacturing," Appl. Sci. **9**, 1060 (2019).

[23]G. Tapia, S. Khairallah, M. Matthews, W. E. King, and A. Elwany, "Gaussian process-based surrogate modeling framework for process planning in laser powder-bed fusion additive manufacturing of 316L stainless steel," Int. J. Adv. Manuf. Technol. **94**, 3591–3603 (2018).

[24]L. Meng and J. Zhang, "Process design of laser powder bed fusion of stainless steel using a Gaussian process-based machine learning model," JOM **72**, 420–428 (2020).

[25]W. Zouhri *et al.*, "Characterization of laser powder bed fusion (L-PBF) process quality: A novel approach based on statistical features extraction and support vector machine," Procedia CIRP **99**, 319–324 (2021).

[26]Y. Zhang, S. Yang, G. Dong, and Y. F. Zhao, "Predictive manufacturability assessment system for laser powder bed fusion based on a hybrid machine learning model," Addit. Manuf. **41**, 101946 (2021).

[27]See https://www.astm.org/e0009-09.html for "Standard test methods of compression testing of metallic materials at room temperature," ASTM E9-09, Jan. 2018 [Online].

[28]See https://www.iso.org/standard/72187.html for "Bend test," ISO 7438:2020 Metallic materials [Online].

[29]Tobias Maconachie *et al.*, "SLM lattice structures: Properties, performance, applications and challenges," Mater. Des. **183**, 108137 (2019).

[30]See https://www.metals4printing.com/ for "Technical data sheet m4p™ 316 l," m4p material solutions GmbH, Austria [Online].

[31]M. Sprengel *et al.*, "Towards the optimization of post-laser powder bed fusion stress-relieve treatments of stainless steel 316L," Metall. Mater. Trans. A **52**, 5342–5356 (2021).

[32]F. Pedregosa *et al.*, "Scikit-learn: Machine learning in Python," J. Mach. Learn. Res. **12**, 2825–2830 (2011).

[33]F. Bodendorf and J. Franke, "A machine learning approach to estimate product costs in the early product design phase: A use case from the automotive industry," Procedia CIRP **100**, 643–648 (2021).

B.5 Application of Machine Learning in Predicting Quality Parameters in Metal Material Extrusion (MEX/M)

The summary of the findings can be found in Chapter 6.

metals

MDPI

Article

Application of Machine Learning in Predicting Quality Parameters in Metal Material Extrusion (MEX/M)

Karim Asami *, Maxim Kuehne, Tim Röver and Claus Emmelmann

Institute of Laser and System Technologies, Hamburg University of Technology, Harburger Schloßstraße 28, 21079 Hamburg, Germany; maxim.kuehne@tuhh.de (M.K.); tim.roever@tuhh.de (T.R.); c.emmelmann@tuhh.de (C.E.)

* Correspondence: karim.asami@tuhh.de

Abstract: Additive manufacturing processes such as the material extrusion of metals (MEX/M) enable the production of complex and functional parts that are not feasible to create through traditional manufacturing methods. However, achieving high-quality MEX/M parts requires significant experimental and financial investments for suitable parameter development. In response, this study explores the application of machine learning (ML) to predict the surface roughness and density in MEX/M components. The various models are trained with experimental data using input parameters such as layer thickness, print velocity, infill, overhang angle, and sinter profile enabling precise predictions of surface roughness and density. The various ML models demonstrate an accuracy of up to 97% after training. In conclusion, this research showcases the potential of ML in enhancing the efficiency in control over component quality during the design phase, addressing challenges in metallic additive manufacturing, and facilitating exact control and optimization of the MEX/M process, especially for complex geometrical structures.

Keywords: additive manufacturing (AM); material extrusion of metals (MEX/M); machine learning (ML); process development; AISI stainless steel 1.4404/316L; design for additive manufacturing (DfAM)

Academic Editors: T. DebRoy, Thomas Niendorf, Dayalan Gunasegaram, Paulo J. Bártolo and Yang Du

Received: 10 March 2025
Revised: 13 April 2025
Accepted: 27 April 2025
Published: 30 April 2025

Citation: Asami, K.; Kuehne, M.; Röver, T.; Emmelmann, C. Application of Machine Learning in Predicting Quality Parameters in Metal Material Extrusion (MEX/M). *Metals* **2025**, *15*, 505. https://doi.org/10.3390/met15050505

1. Introduction

Additive manufacturing (AM) [1] technologies allow, due to layerwise production, the manufacture of complex and intricate geometrical components with reduced material waste and energy consumption. Until now, the laser-beam-based powder bed fusion of metals (PBF-LB/M) has been the most used AM technology in several application industries such as the aerospace, automotive, and medical environments. Therefore, a large amount of data for different PBF-LB/M machines exists, and also approaches to machine learning (ML) prediction promise powerful support in predicting suitable and successful process parameter combinations for different materials [2–6]. The objective of the study in [4] is to utilize ML models to predict the density of AlSi10Mg parts based on various process parameters. To achieve this, 54 experimental data points, as referenced in [7], were used to train and test multilayer perceptron (MLP) models. Among the models evaluated, the MLP was found to be the most accurate and robust prediction model [2]. In addition, two MLP models were trained and tested using density results from simulation data developed in study [4]. However, this did not lead to an improvement in prediction accuracy. Without the inclusion of simulation data, the root mean square error (RMSE) of the MLP model was 0.91%, demonstrating the potential of the MLP algorithm even with a limited amount of experimental data. When incorporating the simulation data, the RMSE increased by 21%, resulting in less accurate predictions. This increase is likely attributable to computational errors associated with the solver used in the simulation. Another study [5] evaluated different ML models for the density prediction of a stainless steel 316L PBF-LB/M specimen with a difference of less than 4% between the experimental results and the ML-based prediction. Different ML models, such as support vector regression (SVR), random forest (RFR), and ridge regression, have been developed to estimate surface roughness in fused deposition modeling (FDM) with high accuracy [8]. Machine learning is also utilized in the design of AM parts. In [9], maximum-stress predictions were made using different regression models, allowing for the efficient use of lattice structures. In this paper, gradient boost regression and random forest regression were the models with the smallest RMSEs of 67.31 MPa and 77.07 MPa for compression and bending specimens, respectively, with R^2 scores of 0.91 and 0.88. These results indicate relatively high performance in terms of predicting mechanical properties and reducing computational simulation time. Additionally, machine learning models are being developed to predict the compressive strength of additively manufactured components, such as PEEK spinal fusion cages, and to improve the geometric accuracy of AM parts through online shape deviation inspection and compensation [10,11]. ML in AM is further used for the prediction of defects in terms of thickness and length. With Gaussian process regression, MLP, RFR, and support vector machine (SVM) models, it was possible to demonstrate significant advantages in processing time and performance as a non-destructive test [12]. These examples show the potential integration of ML in different fields of AM technologies and design processes. It can be used for the prediction of process parameters to achieve successful and desired production, monitor the manufacturing process, and reduce simulation effort for achieving the desired component characteristics. Nevertheless, ML in AM faces several limitations such as data scarcity and quality, variability in product quality, computational challenges, and material and process limitations. The novelty of AM results in a lack of extensive training data, which is crucial for developing robust ML models. Additionally, the quality and standardization of available data are often inadequate, affecting the reliability of predictions [13,14]. AM processes are highly complex and much less established than other manufacturing processes such as machining or casting. Consequently, AM processes are often still lacking in product quality, which poses a significant challenge for ML applications. Such variability complicates the development of predictive models that can generalize well across different conditions [15,16]. The complexity of AM processes requires significant computational resources for ML model training and optimization. This requirement can be a barrier, especially for small-scale operations [13]. The limited library of materials and the presence of processing defects further restrict the applicability of ML in AM. These challenges can hinder the accurate prediction of material performance and process outcomes [17]. Addressing these limitations requires advancements in data collection, standardization, and computational techniques. There are several successful case studies that demonstrate overcoming the limitations of machine learning in additive manufacturing. In [18], machine learning algorithms, such as random forests and support vector machines, are employed to optimize process parameters like laser power and scanning speed. By monitoring the process and predicting the optimal combination of parameters, the variability in printed part quality can be reduced, enhancing reliability in applications such as the aerospace and medical fields. Integrating machine learning with digital twin technology allows for real-time monitoring and control of the AM process. A case study demonstrated the integration's effectiveness in defect detection, achieving high precision and reliability in manufacturing processes [19]. In [20], a hybrid machine learning algorithm was developed to recommend design features during the conceptual phase of AM. This approach was validated through a case study involving the design of R/C car components, proving useful for inexperienced designers by providing feasible design solutions. Furthermore, the synthetic parameters of Ni_3TeO_6 multiferroic materials were experimentally trained and tested with the RFR model using a small dataset of nine data points. The data showed no correlation, and yet a prediction accuracy of 87% was achieved [21]. A final approach of RFR, gradient boosting, and XGBoost ML models was used for the prediction of the corrosion rate of zinc-based alloys with copper, lithium, magnesium, and silver [22]. Here, XGBoost showed the highest prediction accuracy and the highest R^2 value.

The material extrusion of metals (MEX/M, ISO/ASTM 52900) has gained attraction in recent years due to its status as a low-cost AM alternative. The MEX/M process consists of the steps shaping, debinding, and sintering, which are very similar to the metal injection molding process (MIM). In the shaping process, a commercial 3D printer is used for the AM part instead of a mold. The initial intention of this process chain—where the feedstock for the shaping step can be, for example, filament or pellets depending on the extrusion mechanism—was to combine the knowledge of common 3D printing with the MIM process. Nonetheless, the MIM standard of component characteristics is not achievable with MEX/M technology. According to [23], more experimental parameter studies and sintering routines must be developed to achieve suitable material characteristics. In addition, the numerous influencing factors of the multistep process chain make it very difficult and require substantial effort for experimental evaluation. To overcome this hurdle, this paper evaluates an approach using machine learning models with a small batch of experimental data to predict further parameter settings for the MEX/M process. In this way, it is possible to identify suitable process parameters for component design without extensive experimental effort. Up to now, ML has been used in AM technologies for several reasons, either on the experimental side or on the simulation and design side. Furthermore, ML has been applied for quality assurance and predicting process parameter settings. So far, no predictions for the MEX/M process have been made, only for single-stage AM processes, such as the laser-beam-based powder bed fusion (PBF-LB/M) process. Surface roughness and density are key values for meeting the MIM standard of the combined technology and have been set as the target values to be predicted. The choice of the appropriate ML model is challenging, as several ML models are capable of predicting specific parameters, as discussed. Therefore, different models have been evaluated and compared, with a focus on models that can be trained with a small amount of data while achieving sufficiently good prediction.

2. Experimental Approach and ML Methodology

The aim of this study is to develop a predictive model that is able to estimate the surface roughness as well as the sintered density depending on the used process parameters. This provides insights into quality control and process optimization for MEX/M and allows for faster parameter development.

2.1. Specimen Manufacture and Measurements

For the data preparation, different overhang specimens (see Figure 1) have been manufactured on a Renkforce RF2000 (Conrad Electronic SE, Hirschau, Germany) 3D printer with a 0.4 mm nozzle using AISI 316L filament (PT&A GmbH, Dresden, Germany) with a 2.85 mm diameter. The surface roughness of the specimens, which are called green parts after the shaping process, have been measured on the downskin (URz) and upskin (ORz) areas of the overhang parts using a Keyence VHX-S600E (Osaka, Japan). Three areas of each side have been measured three times, and the average of the values has been taken as the average surface roughness of the downskin and upskin areas.

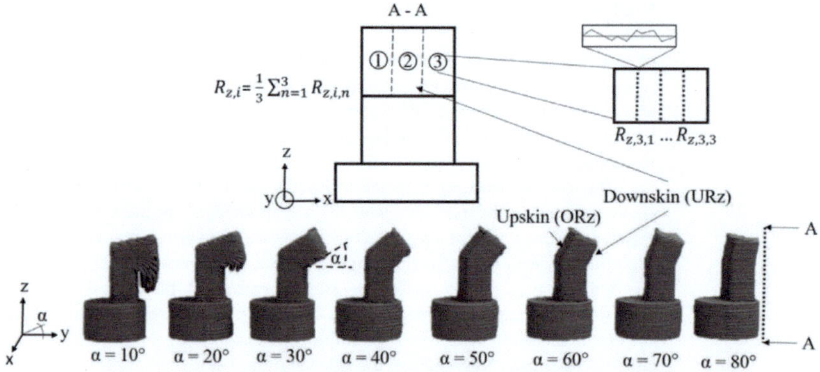

Figure 1. Overhang specimen (green parts) and roughness measuring procedure. Adapted from Ref. [24].

The starting powder composition is shown in [25]. The exact binder composition and particle size distribution of the metal powder are not provided by the manufacturer. The solvent debinding step is performed in 500 mL of acetone for 12 h, heated at 38 °C, as recommended by the filament provider, to achieve a 5% mass loss and open the pores for the thermal debinding step. An SEM image of the debinded 316L filament is provided in [26]. The particle size is in the range of 2 µm to 15 µm. An ExSO90 sinter oven (Aim3D GmbH, Rostock, Germany) was used for the thermal debinding and sintering step of the specimen manufacturing. The thermal debinding and sintering cycles are performed in a combined single step according to the recommendation of the feedstock provider, with 99.9% argon used as the atmosphere for both process steps at a flow rate of 1 L/min. It was found that this setup achieved a sintered density of 97.4% for the 316L specimen, which is higher than the metal injection molding (MIM) standard of 95% density [26]. Figure 2 illustrates the thermal debinding and sintering cycle for the specimens. Two different maximum temperatures were chosen for the printed process parameters. The holding temperatures of 300 °C, 600 °C, and 850 °C were selected to burn out the binder from the samples. The heating rate between each holding temperature is 5 K/min.

The measurements of the different parameter combinations are depicted on the x-axis in Figure 3. On the y-axis, the different roughness measurements of the upskin and downskin surfaces are illustrated. The identification of individual samples was based on the set process parameters. The first number represents the infill percentage [%], the second number denotes the printing speed in millimeters per second [mm/s], the third number corresponds to the layer height in millimeters [mm], and the last number indicates the overhang angle in degrees [°]. For example, the combination 100;2100;0.2;10 signifies that the sample was printed with an infill of 100%, a printing speed of 2100 mm/s, and a layer height of 0.2 mm. The overhang angle of the sample is 10°. As expected, the mean surface roughness for the upskin and downskin areas is higher for lower overhang angles, independent of the printing speed and the percentage infill. A layer height of 0.4 mm causes higher surface roughness for the downskin and upskin areas. The roughness measurements are in the typical range for roughness in the MEX/M process [25].

Furthermore, combinations of additional overhang angles have been printed, debinded, and sintered using the illustrated cycles (Figure 2). Table 1 shows the additional variations of the specimen parameters. The process parameters that were not varied are kept constant according to [25].

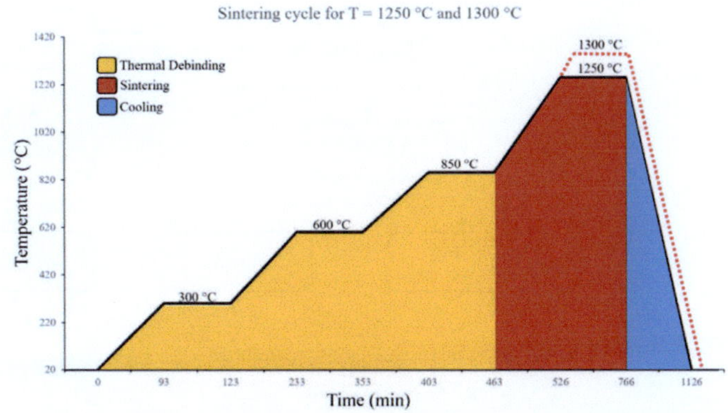

Figure 2. Thermal debinding and sintering cycle for maximum sinter temperature of 1250 °C and 1300 °C. Adapted from Ref. [26].

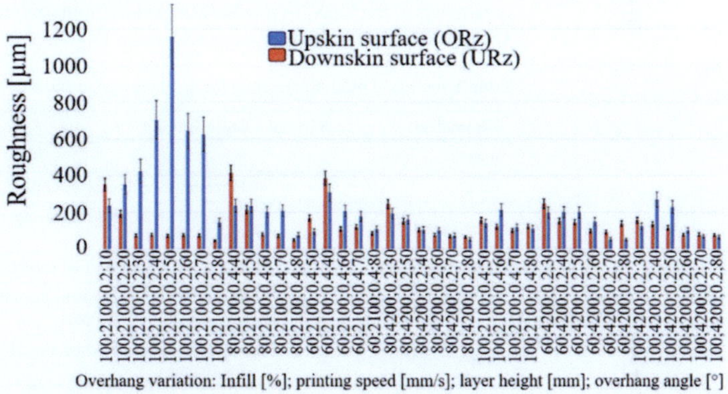

Figure 3. Average surface roughness of the downskin and upskin areas for each process parameter combination of the green parts.

Table 1. Parameter variation of the experimental data.

Variation	Infill [%]	Printing Speed [mm/s]	Layer Height [mm]	Overhang Angle [°]	Sintering Temperature
Green part: 40 combinations	60, 80, 100	2100, 4200	0.2, 0.4	10, 20, ..., 80	-
Sinter part: 48 combinations	50, 100	2100, 4200	0.2, 0.4	10, 20, ..., 80	1250, 1300

2.2. ML Development and Evaluation

The applied methodology and relations between process and parameters are shown in Figure 4. The left side shows the current procedure, where process parameters are selected, a MEX/M process is performed, and the target parameters are investigated. The right side shows the target methodology, where the process parameters are directly fed into the ML model to predict the resulting surface roughness and density. This procedure enables an initial assessment of the part quality without the part actually having to be manufactured.

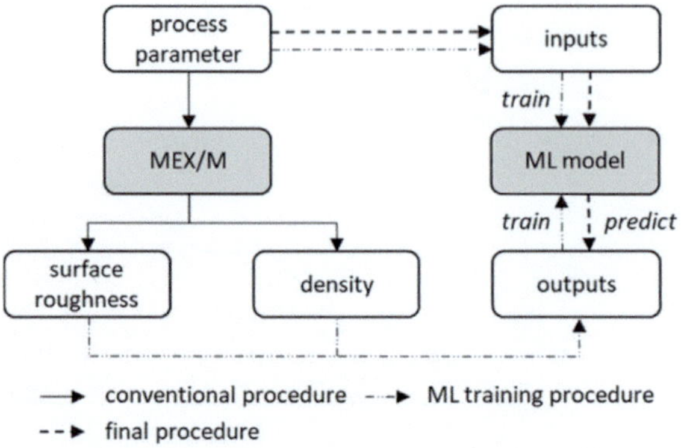

Figure 4. Methodological scheme of the experimental approach and the ML prediction.

The inputs and outputs of the ML model are derived from the process. The process parameters of the MEX/M process are used as inputs and the part properties as outputs. The considered parameters are summarized in Table 2.

Table 2. Input and output parameter for the investigated ML models.

Parameter	Symbol	Description	Unit	Model
Inputs				
Infill	fill	Percentage of the material fill level	%	URz, ORz, density
Print speed	v	Speed of printing head during the process	mm/min	
Layer thickness	l	Height of the individual print layers	mm	
Overhang angle	alpha	Overhang angle of unsupported areas of the printed part	°	
Sinter temperature	st	temperature used when sintering the material	°C	density
Outputs				
Downskin surface roughness	URz	Measured average bottom inclined wall roughness of the component surface	µm	URz
Upskin surface roughness	ORz	Measured average top inclined wall roughness of the component surface	µm	ORz
Density	rho	Part density before sintering	%	density

For data processing and training of the models, Python (v3.8.19) was used with the scikit-learn library (v1.3.0) [27]. These environments were selected to ensure compatibility with the latest machine learning methods as well as robust and efficient processing of the data. The developed suitable ML models and the experimental data are provided in [28].

3. ML Model Setup and Training

The dataset generated for the investigation of surface roughness consists of 40 data points, while the dataset for density comprises 48 data points. The preprocessing of the data consists of four steps: cleaning the data by removing outliers and making sure no values are missing, investigating correlations between features and labels and removing strongly correlated features if necessary, splitting the data into training and test sets, and finally scaling the input data to ensure similar magnitudes of feature values.

When cleaning the data, no missing values and only slight outliers were detected. As the dataset is already very small, it was decided not to remove any further data points as long as the outliers are not attributable to measurement errors. Subsequently, a correlation analysis was performed, including the calculation of the correlation matrix and the Spearman coefficients [29,30]. The plots in Figure 5 show the correlation matrices for the parameters. The calculated coefficients are summarized in Table 3. The results show that URz correlates significantly negatively with angle (−0.63), while ORz also shows a strong negative correlation with print speed (−0.5) and angle (−0.58). Density shows a strong positive correlation with infill (0.87).

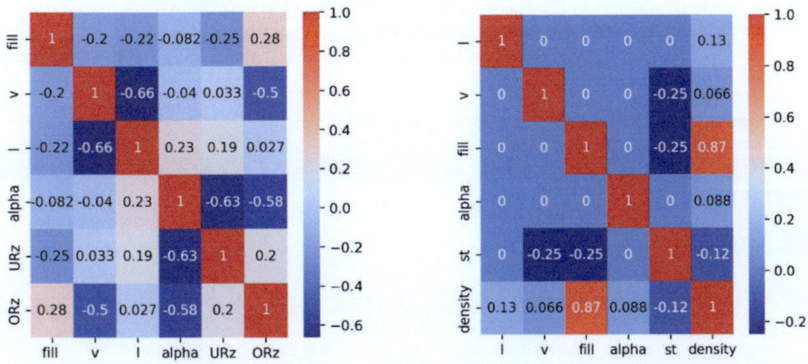

Figure 5. Correlation matrices for the surface roughness (**left**) and density (**right**).

Table 3. Spearmen coefficients for the downskin surface roughness, upskin surface roughness and density.

	URz	ORz	Density
fill	−0.25	0.28	0.87
v	0.03	−0.5	0.07
l	0.19	0.03	0.13
alpha	−0.63	−0.58	0.09
st	-	-	−0.12

For processing in ML algorithms, the data is split into a training set and a test dataset in a ratio of 75% to 25%. A stratified shuffle is applied based on the respective output parameter in order to take the distribution of the target variables into account. This ensures that, despite the small amount of data, results over the entire interval of the target parameter are always included in the training and test datasets. In addition, polynomial features are calculated up to the second degree in order to capture interaction effects [31]. When training the models, it is investigated whether the inclusion of polynomial features offers added value. A feature selection is carried out using SelectKBest. To standardize the feature values, either MinMax or Standard scaling is used to normalize the number spaces to a comparable range, which is important for the accuracy of some ML algorithms [27,32].

The Spearman correlation coefficient is used in the first step to identify nonlinear monotonic correlations between the features and the target variable. Furthermore, SelectKBest is used in the ML pipeline to eliminate irrelevant or redundant features and shorten the computing time. In this method, the features are ranked based on their correlation to the target variable based on a score function f_regression, and the best K features are selected [27]. With GridSearchCV, different values for K are tested.

Permutation Importance is used to quantify the influence of individual features on the model by randomly permuting their values and measuring the change in model performance (see Figure 6). This made it possible to identify l as the most important feature for URz, v as the most important feature for ORz, and fill as the most important feature for density. For ORz and density, there is also a high Spearman correlation (see Figure 5) between the identified features and the target variable. The features are therefore not only statistically significant, but also meaningfully linked to the target variable in terms of content. The relationship between l value and URz is not directly apparent from the Spearman coefficient, which indicates a more complex correlation between the parameters. According to permutation importance, the feature alpha, which is more highly correlated with URz, is the second most important feature for training and therefore just as relevant for model training.

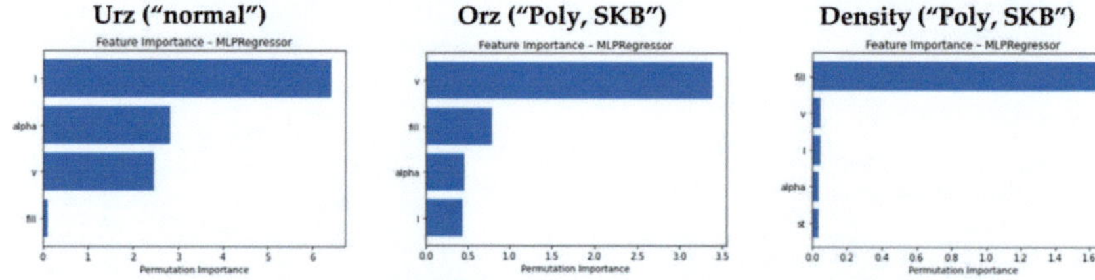

Figure 6. Permutation importance of the features for the ML pipeline.

Pairplots are created to visualize the relationships between each pair of features and are displayed in Figure 7. The absence of a clear linear relationship between the input variables and the target variables suggests the necessity of nonlinear models. The features exhibit only slight visual linear correlations, indicating a degree of independence that is advantageous for many machine learning models. The varying distribution of points highlights the need for normalization, which can be achieved through scaling. Ultimately, the target values indicate specific asperity ranges.

The selection of algorithms for model training was based on the criteria of robustness, prediction accuracy and interpretability. The models considered include linear regression (LR) based on its simplicity and good interpretability for linear relationships [33], random forest regressor (RFR) due to its robustness to outliers and ability to model nonlinear relationships [34], support vector machines (SVM) because the method is suitable for small datasets and offers good generalization capabilities [35], k-nearest-regressor (kNN) based on its simple implementation and adaptation to nonlinear structures [36,37], multilayer perceptron (MLP) for potentially more complex correlations and pattern recognition [38–40], and finally a bagging regressor (Bag) with kNN [41], decision tree regressors, or MLP as base estimators to reduce the variance and increase the stability of the model. The kNN algorithm is based on the fundamental principle of identifying the *k* closest data points based on the input data and a distance metric [42]. The common distance metric is the calculation of the Euclidean distance, which is described in Equation (1) [43].

$$d_E(x,y) = \sqrt{\sum_{i=1}^{k}(x_i - y_i)^2} \quad (1)$$

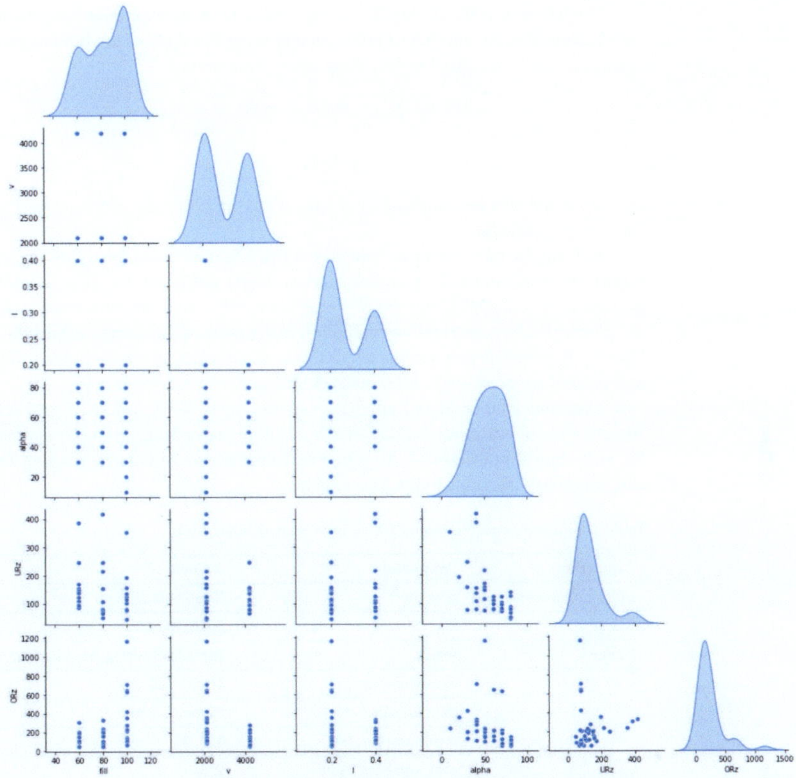

Figure 7. Pairplot displaying feature relationships and distributions.

The algorithm is based on several steps. First, the number of k neighbors is determined. Second, the distances between the object to be classified and all data points are then calculated. All distances are sorted in ascending order to identify the k nearest neighbors. Finally, the classes of the k nearest neighbors are then considered, and based on these classes, the class of the object is determined [43].

RFR is an ensemble algorithm that generates a large number of decision trees during training and combines their outputs to increase the accuracy of the predictions. The main principles are based on ensemble learning, bootstrap aggregation, feature randomness, and prediction [44]. The supervised learning model SVM uses kernel functions to transform nonlinear data into a higher dimension [45], also known as a hyperplane, in order to perform linear classification on the data [43]. MLPs are a type of neural network consisting of several layers of nodes. These include the input layer, one or more hidden layers, and an output layer. Each neuron in one layer is connected to every neuron in the subsequent layer [46]. The MLP output is calculated using a weight function s_j, which is described in Equation (2) as follows [43]:

$$s_j = \sum_{i=1}^{n} (W_{ij} \cdot X_i) - \delta_j, \; j = 1,2,\ldots,t \quad (2)$$

The factor X_i is the ith input, W_{ij} represents the weight amount from the ith node to the jth node, n is the number of nodes, and δ_j is the threshold of the hidden node. For the calculation of the output, the following equations are used:

$$o_p = \sum_{i=1}^{n} (w_{jp} \cdot S_j) - \vartheta_p, \; p = 1,2,\ldots,g \quad (3)$$

$$\text{with } S_j = \frac{1}{1+e^{-s_j}} \quad (4)$$

w_{jp} is the connection weight from the ith to the kth node, and ϑ_p is the threshold of the pth output node.

To train the models, a pipeline was defined that includes the following steps: polynomial features, SelectKBest, scaling, and the respective estimator. To evaluate the model performance, the RMSE and the R^2 value were defined as common error measures [47], supplemented by a qualitative evaluation using plots of the correct versus the predicted values. The RMSE is given in the respective unit of the target value—µm for URz and ORz and percentage points in % for density. RMSE and R^2 are defined in [9].

To optimize the models, hyperparameter tuning is performed using grid search (GridSearchCV) [27] with cross-validation (CV). The corresponding parameter spaces are given for each algorithm in Table 4. All pipelines consider a variation of the scaler, SelectKBest, and polynomial features in the grid search.

Table 4. Selected hyperparameter for all investigated ML models.

Algorithm		Parameter	Values
Polynomial features		activated	True, False
KBest		k	5, 10, 15, 'all'
Scaler		scaler	MinMaxScaler(), StandardScaler()
Estimator	RFR	n_estimators	10, 50, 100
		max_depth	None, 2, 5, 10
		min_samples_leaf	1, 2, 3
		min_samples_split	2, 5
		max_features	None, sqrt
		bootstrap	True, False
	kNN	n_neighbors	3, 5, 7
		weights	uniform, distance
		algorithm	auto, ball_tree, kd,_tree
		p	1, 2
	MLP	hidden_layer_sizes	(10,), (20,), (10, 10,), (25, 25,)
		alpha	0.0001, 0.001, 0.01
		activation	relu, tanh
		solver	lbfgs, adam
		learning_rate	constant, adaptive
	Bag	estimator	kNN, DTR, MLP
		n_estimators	5, 10, 20
		max_samples	0.5, 0.7, 1.0
		bootstrap	True, False
		bootstrap_features	True, False

In addition, to evaluate the generalization capability of the final models, a separate k-fold cross-validation (with k = 5) was performed on the full dataset. The resulting mean performance metrics (e.g., R^2, RMSE) and standard deviations provide insight into the model's stability and expected real-world performance, even under limited data availability, and are displayed in Table 5.

Table 5. Mean performance metric for the output data.

	URz	ORz	Density
RMSE	69.41 ± 44.1	157.29 ± 163.18	2.63 ± 0.84
R^2	0.19 ± 0.32	−1.52 ± 2.33	0.85 ± 0.14

For the training of the ML model, a statistical evaluation was performed, starting with grid-search-based parameter tuning. The search over the respective parameter spaces is performed 10 times each. The best configurations are then selected. The models are then trained 50 times with the selected parameter combinations, and the mean error is determined. This procedure ensures optimal parameter selection and checks the robustness of the model.

4. ML Results

The various ML models LR, RFR, SVM, kNN, MLP, and Bag were trained and evaluated with hyperparameter tuning. Subsequently, the R^2 factor and the RSME for downskin and upskin angles of the experimental data were evaluated (compare Figure 3). Table 6 shows the respective values of each model. The highest R^2 value in the training dataset can be seen in the MLP model for the URz.

Table 6. R^2 value and RMSE of URz and ORz in training and test case for all ML models.

Model	Target Value	Indicators	Training	Test
LR	URz	R^2	0.55	0.51
		RMSE [μm]	57.44	60.21
	ORz	R^2	0.58	0.41
		RMSE [μm]	109.34	237.45
RFR	URz	R^2	0.86	0.70
		RMSE [μm]	32.04	47.51
	ORz	R^2	0.89	0.31
		RMSE [μm]	56.79	255.7
SVM	URz	R^2	0.44	0.39
		RMSE [μm]	63.79	67.60
	ORz	R^2	0.26	0.1
		RMSE [μm]	144.90	293.12
kNN	URz	R^2	0.66	0.65
		RMSE [μm]	49.80	51.27
	ORz	R^2	0.73	0.69
		RMSE [μm]	87.98	170.96
MLP	URz	R^2	0.95	0.26
		RMSE [μm]	18.76	74.66
	ORz	R^2	0.99	0.13
		RMSE [μm]	13.71	288.44
Bag	URz	R^2	0.90	0.53
		RMSE [μm]	27.15	59.10
	ORz	R^2	0.75	0.60
		RMSE [μm]	84.63	194.01

RFR and Bag also show high values of 0.86 and 0.9, respectively. Similarly, these models also have low RMSE values for the training data (MLP 18.75 μm, RFR 32.04 μm and

Bag 27.15 μm). A similar observation can be made in the training dataset for ORz, in which MLP (0.99) also has the highest R^2 value, followed by RFR (0.88) and Bag (0.75). SVM has the lowest R^2 value in the training dataset for URz with 0.44, followed by LR with a value of 0.55. The RMSE value is highest for SVM at 63.79 μm and for LR at 57.44 μm. In the case of ORz, the R^2 factor is also lowest for SVM with 0.26, followed by LR with 0.58. Here, the RMSE values are 144.90 μm (SVM) and 109.34 μm (LR). In the test dataset, the RFR model achieved a maximum coefficient of determination (R^2) of 0.7 for the target variable URz, while the kNN model attained an R^2 value of 0.65. The corresponding RMSE values were 51.27 μm and 47.52 μm, respectively. For the target variable ORz, the kNN model also recorded the highest R^2 value of 0.69, with an RMSE value of 170.96 μm.

Additionally, diagrams of the respective models were generated, illustrating the predictions compared to the experimental data. The red line indicates a perfect prediction. The farther the points are from the line, the less accurate the predictions are. Figure 8 presents the prediction diagrams of the models under examination, facilitating an initial selection of the models in question. The blue dots depict the training data, while the yellow dots represent the test data. For the LR model, it is observed that both the test and training data predictions deviate from the experimental data. This phenomenon is observable for both the downskin angle URz and the upskin angle ORz. Only for small overhang angles do the predictions appear to align more closely with the experimental data, resulting in smaller deviations for both URz and ORz. The SVM model exhibits a behavior similar to that of the LR model's predictions. In this instance as well, the predictions significantly deviate from the actual values. Both the LR and SVM models tend to underestimate the actual surface roughness for URz and ORz. In contrast, the RFR model demonstrates higher accuracy in predicting the training data with larger URz values, a trend that is also observed in the test data. Nonetheless, in both scenarios, the roughness prediction is underestimated.

The kNN model demonstrates that more accurate predictions can be achieved for low URz and ORz values. However, for higher roughness, the predictions deviate from the experimental data. In this context, both the training and test datasets underestimate the actual values for larger roughness values (>200 μm). One reason for this is the low experimentally recorded roughness values. The majority of the data (see Figure 3) show a roughness of less than 250 μm. Because the kNN algorithm makes predictions based on the k nearest neighbors, smaller roughness values are identified as neighbors, resulting in an underestimation of the roughness. The Bag model exhibited a high level of predictive accuracy for URz in the training dataset, including for large roughness values. Nevertheless, for small roughness values, the model showed a tendency to overestimate predictions in the test dataset, whereas for large roughness values, it occasionally underestimated the predictions. In the case of ORz, the training data is even more accurate in its prediction and tends to underestimate small roughness values.

The test data overestimate the experimental data and predict higher values for the roughness. However, for large roughnesses, the predictions in the test dataset tend to be underestimated. In the training dataset, the MLP model shows the smallest deviations in the predicted values for both ORz and URz. Only in the test data, small roughnesses are overestimated and larger roughnesses are underestimated.

Because the training data for MLP and Bag are heavily aligned with the experimental data, while the test data fluctuate more around the exact prediction, the machine learning models appear to be overfitted. For URz and ORz, the MLP model appears to perform the best despite signs of overfitting and is therefore selected for further parameter optimization. The other algorithms are not further investigated, as MLP seems to be the most promising algorithm for this dataset.

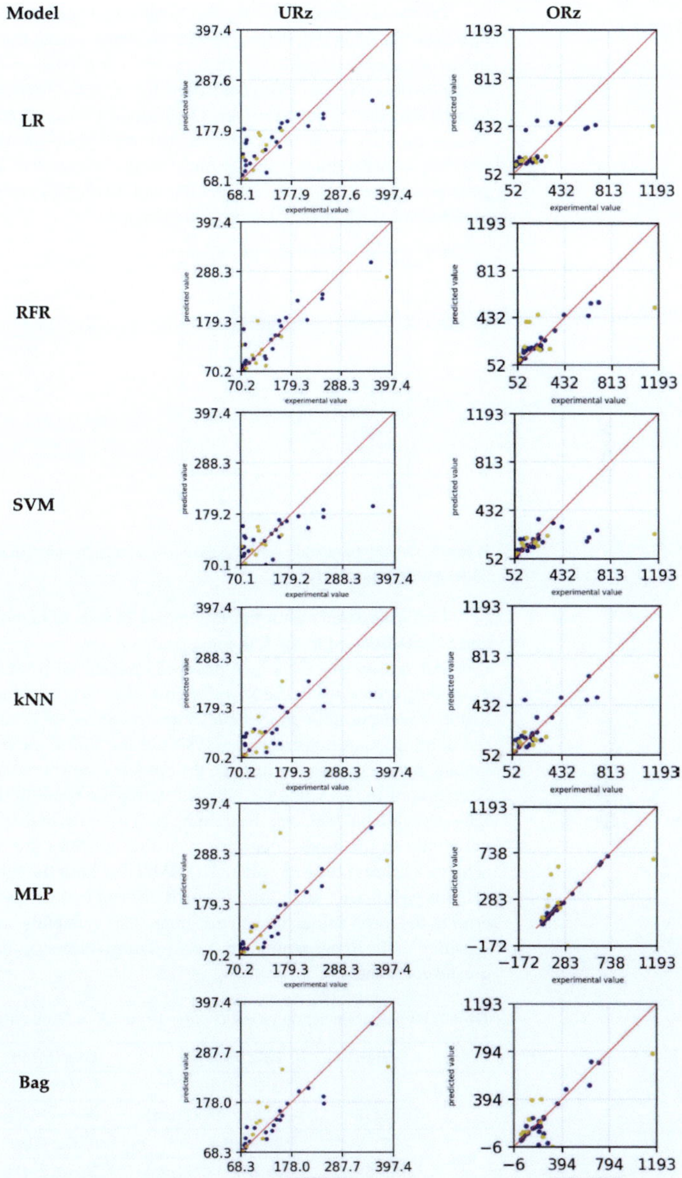

Figure 8. Surface roughness prediction versus experimental measurements in [μm] for the training (blue) and test data (yellow) of all investigated ML models.

The same approach was applied to the target variable density, where, with parameter optimization using polynomial feature importances and the SelectKBest method, the Bag model appears to be the best starting point. The prediction diagram is shown in Figure 9. It can be observed that the predictions for both the training data and the test data are located close to the actual values. Occasionally, the predictions for the training data are underestimated, which may indicate that overfitting is not occurring. Both training and test data oscillate within a similar range around the optimal axis, which means that a good generalization of the model can be assumed. In the following, only the Bag model will be investigated further for the density prediction.

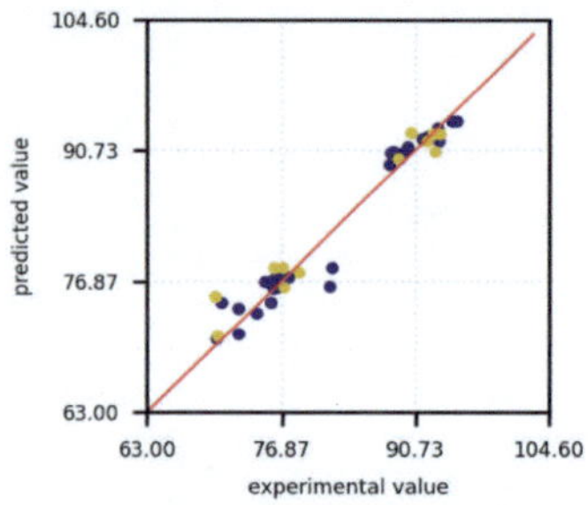

Figure 9. Density prediction versus experimental measurements in percentage points for the training (blue) and test data (yellow).

The hyperparameter optimization over 10 runs using polynomial feature importances and SelectKBest led to the following results in terms of the R^2 factor and RMSE for URz, Orz, and density (see Table 7). A direct comparison with Table 6 for the MLP model shows a significant increase in the R^2 value from 0.25 to approximately 0.71. Additionally, the RMSE was reduced by more than 25 μm to around 49.27 μm. Although the R^2 factor in the training dataset decreased to 0.83 and the RMSE increased to 34.82 μm, the model initially seemed to be overfitted, so the optimization counteracted this trend. The standard deviation of the R^2 value for URz is 0.11 for the training data and 0.3 for the test data. Thus, the variability among the training data is less than that of the test data. This suggests variability in the model's performance and indicates the need for further iterations to achieve a more robust R^2 value. The RMSE for both the training data and test data also exhibits significant variability, making it difficult to draw definitive conclusions about the error of the mean values after 10 iterations. This variability suggests that the MLP model is sensitive to the hyperparameters, necessitating further optimization iterations to achieve greater robustness.

Table 7. Hyperparameter optimization after 10 iterations for the MLP and Bag models.

	Value	**URz (MLP)**	**ORz (MLP)**	**Density (Bag)**
Train	R2	0.83 ± 0.11	0.72 ± 0.12	0.96 ± 0.02
	RMSE	34.82 ± 13.82 [μm]	90.8 ± 34.7 [μm]	1.64 ± 0.34 [%]
Test	R2	0.71 ± 0.3	0.46 ± 0.15	0.91 ± 0.04
	RMSE	49.27 ± 18.95 [μm]	222.92 ± 41.41 [μm]	2.53 ± 0.53 [%]

A similar observation can be made for the ORz MLP model. Here, the R^2 value decreased from 0.99 to 0.72, and the RMSE increased from 13.71 μm to approximately 90.8 μm in the training dataset from the initial iteration step (compare Table 6). The R^2

value for the test data, on the other hand, increased from 0.12 to around 0.46, and the RMSE error was reduced to 222.92 μm. The reduction of the R^2 value in the training data is an indication that overfitting is counteracted, but the high standard deviations of R^2 and RMSE indicate a low robustness of the model.

The Bag model was used for the density forecast. In the training dataset as well as in the test dataset, the R^2 value is already high at 0.96 and 0.91, respectively, so that the independent variables can be well described by the number of dependent variables. The standard deviation is also much smaller than for the other two MLP models, which means that the model can be expected to be more robust. The same applies to the RMSE, which is small at 1.64% (training) and 2.53% (test) and also has a standard deviation of less than 0.55%.

After completing 10 iterations, the optimal hyperparameters were determined by maximizing the coefficient of determination and minimizing the root mean square error. Table 8 shows the values with the optimized hyperparameters. The selected hyperparameters for the MLP model are illustrated in Table 9 and for the Bag model in Table 10.

Table 8. Parameter optimization (best models).

	Value	URz (MLP)	ORz (MLP)	Density (Bag)
Train	R^2	0.96	0.91	0.98
	RMSE	16.04 [μm]	51.29 [μm]	1.12 [%]
Test	R2	0.88	0.91	0.95
	RMSE	32.11 [μm]	91.89 [μm]	1.97 [%]

Table 9. Parameter optimization (selected hyperparameter)—URz, ORz.

Algorithm	Parameter	URz	ORz
Polynomial features	activated	False	True
KBest	K	-	10
Scaler	scaler	MinMaxScaler	MinMaxScaler
Estimator	model	MLP	
Hyperparameters	hidden_layer_sizes	(10,)	(10,)
	alpha	0.01	0.01
	activation	relu	relu
	solver	lbfgs	lbfgs
	learning_rate	adaptive	adaptive

Table 10. Parameter optimization (selected hyperparameter)—density.

Algorithm	Parameter	Density
Polynomial features	activated	True
KBest	K	15
Scaler	scaler	StandardScaler
Bag	estimator	DecisionTreeRegressor
	n_estimators	10
	max_samples	0.7
	bootstrap	false
	bootstrap_features	True

An R^2 of 0.96 shows that the model can explain 96% of the variance of the surface roughness (URz) in the training data. The RMSE is also significantly lower here at 16.04 µm. In the test dataset, the R^2 factor was also increased to 0.88 and also has a low RMSE value of 32.11 µm. A significant increase in the error on the test data indicates that the model has difficulties predicting the test data with the same precision as the training data. There is a possibility of a slight overfitting effect occurring in this case. The R^2 factor of 0.91 does not change for ORz. A constant R^2 value indicates that the model has learned the patterns in the data well and also transfers them to the test data. An RMSE increase from 51.29 µm (training) to 91.89 µm (test) strongly indicates overfitting. Regularization, optimized data processing, and model adjustments can improve the generalization and reduce the test error. For the Bag model for density, the R^2 values of 0.98 (training) and 0.95 (test) are also high. The RMSE is lower in the training (1.12%) than in the test (1.97%). Further optimization iterations should reduce the different errors.

Table 11 indicates the averaged evaluation parameters of 50 iterations with the optimized hyperparameters. The observed discrepancy between training and test data indicates that the model performs better on the training set, which could suggest some degree of overfitting. However, the low standard deviation across both datasets demonstrates a high level of robustness and consistency in the model's predictions. For ORz, the R^2 factor is 0.94 in the training dataset and 0.73 in the test dataset, with a standard deviation of 0.07. Although the performance on the test data is lower, the small standard deviation highlights the stability of the model across samples and its delivery of similar results. Similarly, for density, the R^2 factor is 0.98 on the training set with a standard deviation of 0.003, and 0.92 on the test set with a standard deviation of 0.02. This consistency in performance, coupled with the relatively small deviations, reinforces the robustness of the model, even if the performance on the test data does not reach the same level as that on the training data. Overall, while there is evidence of better training data performance, the low standard deviations suggest that the model's predictions are reliable and not overly sensitive to variations within the datasets. For the RMSE, the error for URz during training is 14.32 with a standard deviation of 3.27, reflecting improvements compared to the results obtained after hyperparameter optimization with 10 iterations. Furthermore, the error in the test set was reduced to 34.93, accompanied by a standard deviation of 6.28. These results indicate that the optimization process enhanced both the accuracy and stability of the predictions for URz.

Table 11. Model training with best parameters (50 runs)—MLP, MLP, and Bagging.

	Value	URz (MLP)	ORz (MLP)	Density (Bag)
Train	R2	0.97 ± 0.01	0.94 ± 0.06	0.98 ± 0.003
	RMSE	14.32 ± 3.27 [µm]	47.67 ± 16.45 [µm]	1.06 ± 0.11 [%]
Test	R2	0.85 ± 0.06	0.73 ± 0.07	0.92 ± 0.02
	RMSE	34.93 ± 6.28 [µm]	155.63 ± 20.47 [µm]	2.51 ± 0.29 [%]

For ORz, the RMSE during training is 47.67 µm with a standard deviation of 16.45 µm. However, the test RMSE is significantly higher at 155.63 µm, with a standard deviation of 20.47 µm. This suggests that, while the model achieved reasonable performance in training, its ability to generalize to test data is more limited for ORz. Additionally, both the roughness and deviations are notably higher for ORz compared to URz.

Importantly, despite the higher error and variability in the predictions for ORz, there was a reduction in the standard deviation for both the training and test datasets when compared to the results obtained after 10 iterations of hyperparameter optimization (as seen in Table 7). This improvement in stability indicates that the optimization process contributed to reducing variability in the predictions, even if the overall performance still shows room for improvement. For density, the RMSE in training is 1.06% with a standard deviation of 0.11%, while in the test dataset, it increases to 2.51% with a deviation of 0.29%. This indicates a noticeable performance drop when transitioning from training to test data. Furthermore, the test RMSE and its deviation are higher compared to the results after 10 iterations of hyperparameter optimization.

Using the selected hyperparameters from Tables 9 and 10 for the MLP and Bag models, the corresponding values for RMSE and R^2 as best results are presented in Table 12. These results offer insights into how the optimized hyperparameters affect model performance in terms of accuracy and explanatory power. For URz, the model demonstrated a strong fit with an R^2 of 0.96 in the training set, indicating that 96% of the variance in the data was explained. However, the test set R^2 decreased to 0.87, and the RMSE increased from 17.38 μm in training to 32.93 μm in testing, suggesting some overfitting. The ORz results showed a similar trend, with a high R^2 of 0.94 in the training set, but a significant drop to 0.79 in the test set. Additionally, the RMSE for ORz was substantially higher, moving from 41.58 μm in training to 138.99 μm in testing, indicating a severe generalization issue. In contrast, the model performed exceptionally well for density, achieving an R^2 of 0.99 in the training set and 0.95 in the test set, with an RMSE of 0.95% and 1.88%, respectively. The minimal difference between the training and test RMSE values for density suggests that the model generalizes well to unseen data. Overall, the results highlight that while the model is highly effective for density, additional improvements in feature selection and regularization are needed to enhance the generalization performance for URz and especially ORz, where overfitting is more pronounced.

Table 12. Model training with best models (best models).

	Value	URz (MLP)	ORz (MLP)	Density (Bag)
Train	R2	0.96	0.94	0.99
	RMSE	17.38 [μm]	41.58 [μm]	0.95 [%]
Test	R2	0.87	0.79	0.95
	RMSE	32.93 [μm]	138.99 [μm]	1.88 [%]

Figure 10 shows the predictions of the experimental values as a further evaluation criterion. In this case too, the blue dots are the training dataset, and the yellow dots are the test data. For URz (left) it can be seen that the trending data for small roughness values deviate from the exact values and make precise predictions with increasing roughness. The scattering for small roughness values can be observed in the test data. With increasing roughness, a deviation from the experimental data can still be observed, but this deviation remains constant with increasing roughness values.

For ORz (Figure 10, center), the model demonstrates superior predictive accuracy for small roughness values (<441 μm), as evidenced by better alignment with the training data and reduced scatter in the predictions. This indicates that the model effectively captures the underlying patterns within this range, likely due to a higher density of data points or less variability in the observed values. However, as roughness increases beyond this threshold, the accuracy of the predictions deteriorates, accompanied by a noticeable increase in scatter. This decline suggests that the model struggles to generalize to larger roughness values, potentially due to a lack of sufficient data or increased complexity in the relationships governing higher roughness levels. Addressing this issue may require additional data for large roughnesses or the inclusion of features that better capture the variability in this range. The situation is the same for the test data of ORz, although the deviation of the predictions is generally significantly larger. In the density prediction (Figure 10, right), the training data are well adapted to the experimental data, and an accurate prediction with few outliers is made for both low and high density values. For the test data, the predictions for low density values tend to be overestimated with a few outliers, while for high density predictions, the values are primarily underestimated.

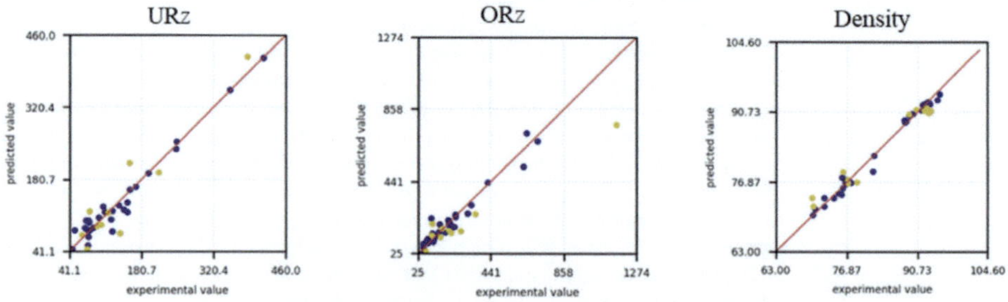

Figure 10. Surface roughness prediction versus experimental measurements for the training (blue) and test data (yellow) of all investigated ML models for URZ (**left**) in µm, ORz (**center**) in µm, and density (**right**) percentage points.

5. Discussion

The initial attempt to utilize and optimize a single ML model for predicting all three parameters —URz, ORz, and density—was unsuccessful. The primary limitation was the small size of the measurement dataset, coupled with the substantial variability in the parameter values. These factors hindered the model's ability to capture consistent patterns across all targets, ultimately preventing it from providing reliable and usable predictions. Furthermore, the small dataset restricts the model's ability to generalize to unseen data, and with the limited samples, the risk of overfitting increases, reducing predictive reliability for new process conditions. Additionally, the small datasets may not fully capture the variability in process parameters, leading to biased or unstable model performance. This observation underscores the need for either an increased dataset size, feature-specific models, or tailored preprocessing strategies to better account for the distinct characteristics and variances of each parameter in order to improve model robustness and generalizability. As a result, it became necessary to train and test the models individually for each parameter. The evaluated models LR, RFR, SVM, kNN, MLP, and Bag demonstrated varying levels of performance and robustness during hyperparameter optimization. As anticipated, the LR model failed to achieve a high R^2 value and a low RMSE due to the complexity of the relationships between input and output variables. The predictions from the LR model exhibited significant deviations from the actual experimental data, indicating that this approach is unsuitable for accurately modeling the investigated parameters. According to [9], this is an indication that conventional regression models cannot solve the problem and that an ML approach is required. This can be attributed to the inherent limitations of LR in capturing complex, nonlinear relationships within the data. The MEX/M process involves multiple interacting parameters, such as material properties, process conditions, and geometric features, which exhibit nonlinear dependencies. Because LR assumes a linear relationship between the input features and the target variables, it was unable to adequately model the intricate patterns present in the dataset. Consequently, the LR model showed poor generalization performance and was deemed unsuitable for making accurate predictions in this context.

After 10 optimization iterations, the RFR model showed worse results than the MLP and Bag models. A negative value of R^2 of -1.97 led to the exclusion of the model from further investigations. A negative R^2 value means that the model performs worse than a trivial model that only uses the mean value of the target variable as a prediction. Specifically, the model not only does not explain any variance in the data, but also provides predictions that increase the error variance, which is also observed in [48]. For RFR, the negative R^2 values may be attributed to over-complexity in the model, such as an excessive number of trees or insufficient regularization during hyperparameter tuning. This caused the model to overfit to subtle patterns or noise in the training dataset, which were not representative of the overall data distribution. In both cases, negative R^2 values highlight that the models performed worse than a simple baseline prediction (e.g., predicting the mean of the target variable). This underscores the importance of carefully balancing model complexity and regularization during hyperparameter optimization to ensure reliable predictions and robust generalization. Moreover, with a maximum of 48 data points, the dataset is relatively small and noisy in this case, which prevents the RFR model from learning meaningful patterns and can also result in a negative R^2 value. In addition, because some of the input features correlate weakly with the target values, the model struggles to make accurate predictions [44].

The SVM model appears to have difficulties in accurately capturing the variability of the URz and ORz values. The data could be nonlinearly separable or have a highly nonlinear pattern that the SVM model does not capture effectively. With a small number of data points, SVM may have difficulty making robust predictions, as it is highly dependent on the number and quality of support vectors, also observed in [49]. SVM is sensitive to unscaled input data. If the data have not been correctly normalized or standardized, this can affect performance. A similar observation was made in the roughness study of SVM in [50] where the accuracy of 0.56 was also the worst for roughness predictions. This poor performance can be attributed to several factors inherent to the nature of the dataset and the characteristics of the SVM model. Firstly, SVM models, especially when used with nonlinear kernels, can struggle with small datasets that have high variability or noise. The limited number of experimental data points in this study likely led to overfitting during training, where the model attempted to fit to every detail in the data, including noise. This overfitting reduced the model's ability to generalize to unseen data, resulting in poor test performance. Secondly, the high dimensionality or complex feature interactions in the dataset may have exacerbated the model's inability to find an optimal decision boundary. SVM requires careful tuning of hyperparameters such as the regularization parameter C and the kernel-specific parameters (e.g., gamma for the RBF kernel). Without sufficient data to guide this tuning process effectively, the model's performance likely deteriorated further. Lastly, SVMs are sensitive to feature scaling and may fail to perform well if input features are not properly normalized. Given these challenges, SVM was unable to capture the intricate relationships in the dataset, resulting in poor predictive accuracy and high error rates, making it the least effective model for this application.

For the kNN model, a negative R^2 factor also appeared in the test for URz after 10 training iterations. In addition, a value of 1 and an RMSE of 0 occurred in the training for the R^2 value of ORz. A negative R^2 value indicates that the kNN model's predictions are worse than a simple baseline model. This suggests that the model fails to capture the patterns in the test data, which may be due to overfitting to the training data. An R^2 value of 1 and an RMSE of 0 indicate that the kNN model memorized the training data perfectly for ORz. While this may seem ideal, it often signals overfitting, where the model performs exceptionally well on the training data but struggles to generalize to unseen test data. This issue arises because the kNN algorithm stores the training dataset and bases predictions on the closest neighbors. If k is too small (e.g., k = 1), the model essentially memorizes the training points, making it sensitive to noise and unable to generalize well to unseen data. This behavior indicates that the models were overfitting to the training data, leading to a significant loss of generalization ability. For kNN, this issue likely arose from an inappropriate selection of the number of neighbors (k). Therefore, the kNN model was not considered further for parameter optimization.

The experimentally measured surface roughnesses are significantly higher than the roughness values found in the literature due to the overhang angle [51–53]. As shown in [51], the process parameters have a significant influence on the surface roughness of the components. A comparison with the recorded measurement data shows that the geometry of the inclined walls has a much greater influence (factor 10) than the parameter variations. The measured values, which act as a dataset for the ML models, also vary considerably. For URz, the variance in the measurement data is notably lower compared to ORz, which facilitates more effective training and hyperparameter optimization. The reduced variability allows the model to capture the underlying patterns more reliably, leading to improved predictive accuracy. After 50 iterations of hyperparameter optimization, the predictions for URz exhibit higher precision and better alignment with experimental values compared to ORz. This suggests that the lower variance in the measurement data for URz not only enhances model training but also contributes to improved generalization performance. In contrast, the higher variance observed in the ORz data likely introduces additional challenges in modeling and optimization, resulting in less accurate predictions.

The average accuracy of the predictions is defined as:

$$Average\ accuracy = 1 - \frac{RMSE}{\sum_1^n R_{z_n}} \quad (5)$$

The accuracy in the test dataset for URz is 75.71%, which makes a prediction possible. For ORz, the average accuracy is only 39.26% and an exact prediction is not generally feasible. The MLP model performed more accurately for the prediction of URz than for ORz based on the evaluation criteria examined and the predictions of the test and training data. The representativeness and size of the training dataset are critical for the model's ability to generalize to new data [54]. Because the ORz data are highly variable and there is only a small amount of data, an accurate prediction is difficult and requires the input of more data. In addition, even after 50 training sessions, the standard deviation of 20.47 µm in the test case is a power of 10 higher than for URz, and a prediction is inadmissible. Larger roughness values may be associated with increased complexity or variability in the underlying physical processes such as the staircase effect [55,56], making them harder to predict accurately.

The density of the sample geometries was experimentally determined to range between 70% and 95% using the Archimedes method. While this does not meet the typical standards for metal injection molding (MIM), it falls within the empirically established density range for the material extrusion for metals (MEX/M) process, as reported in the literature [23]. This suggests that the density values observed are consistent with the inherent characteristics of the MEX/M process, which is known to result in slightly lower densities compared to MIM due to differences in processing parameters and material behavior. The average accuracy of the density predictions in the test dataset using the Bag model, calculated according to Equation (5), is 97.44%. This indicates that the model provides highly accurate predictions within the same order of magnitude as the experimental data. The deviation of only 1.55% compared to the accuracy of [4] for the density prediction underscores the reliability of the model in capturing the underlying patterns of the dataset. Such a small deviation highlights the model's robustness and suitability for predicting density in similar experimental setups.

The insights gained from the performance of the suitable ML models make it possible to identify process parameter combinations for the MEX/M processes for the desired surface roughness in a component with overhang features. On the one hand, this can estimate and minimize the post-processing effort. On the other hand, the surface roughness can be determined by prediction for surfaces where post-processing is not possible. As a result, the cost of manufacturing a component using the MEX/M process can also be optimized in the wider sense, and cost estimation in AM processes can be better determined by knowing the more precise post-processing costs [57]. Density prediction in the MEX/M process helps to significantly reduce the experimental effort regarding process parameters in the shaping, debinding, and sintering steps, and in this way it helps to find suitable process parameter combinations that have the highest achievable density. The findings regarding density can contribute to design guidelines in MEX/M green part manufacturing [25] and thus produce functional prototypes. Because there is a strong correlation of the infill percentage with the density in the Bag model, it is now possible to evaluate porous components in the context of a solid material using the density prediction, which has only been investigated experimentally in isolated cases to date [23]. The methodology is basically transferable to other AM processes, so that the density prediction in the PBF-LB/M process can be applied in principle, and thus the material characteristics can be added to the predictions of the maximum stress in [9].

6. Conclusions

For the MEX/M process, the upskin and downskin surface roughness values, along with the density, were empirically determined as experimental values for various overhang angles. These measurements provide critical insights into the relationship between geometric features and resulting material properties, which are essential for optimizing process parameters and achieving desired part characteristics in MEX/M applications. The experimental data served as the foundation for predicting the upskin and downskin surface roughnesses, as well as the density of MEX/M samples. To achieve this, machine learning models, including LR, RFR, SVM, kNN, MLP, and Bag, were trained on the dataset. The performance of these models was evaluated using the R^2 coefficient and RMSE as metrics. Additionally, prediction plots were analyzed to assess the accuracy and reliability of the models in capturing the relationships between experimental variables and target properties. The following findings were obtained from the investigations.

The LR model proved to be unsuitable for predicting the upskin and downskin roughnesses, as well as the density, due to its low R^2 values and high RMSE in both training and testing phases.

The SVM model exhibited the lowest R^2 value and the highest RMSE among all evaluated models, making it even less suitable than the LR model for predicting the upskin and downskin roughnesses of MEX/M samples.

The kNN and RFR models demonstrated greater robustness compared to LR, achieving relatively higher R^2 values and lower RMSE during initial evaluations. However, after several iterations of hyperparameter optimization, both models exhibited negative R^2 values for certain target variables, particularly in the test phase.

The MLP model was identified as a strong candidate for predicting the surface roughness (upskin and downskin) in MEX/M samples. Its suitability stems from its ability to handle complex, nonlinear relationships within the data, which are characteristic of the MEX/M process. The MLP model's architecture, consisting of interconnected layers of neurons, enables it to capture intricate patterns and interactions among the input features, making it well-suited for modeling the variability in surface roughness. Moreover, through proper hyperparameter optimization, including the adjustment of the number of hidden layers, neurons, activation functions, and learning rates, the MLP achieved high R^2 values and low RMSE, demonstrating its robustness and accuracy in both the training and test datasets for URz. Furthermore, an average accuracy of only 39% is too imprecise, so no predictions can be made for ORz with the current dataset. In the case of iterative learners such as MLP, early stopping could be employed based on validation loss to further avoid overfitting to the training data.

The Bag model was identified as a suitable approach for predicting the density of MEX/M samples, achieving an average prediction accuracy of 97.44%. This high level of accuracy can be attributed to the ensemble nature of the Bag method, which combines multiple base models to reduce variance and improve robustness.

An attempt to predict both surface roughness and density within a single model for different stages of the process chain was not feasible. This limitation can be attributed to the fundamentally different natures of the target variables and the complex interactions between process parameters at various stages.

Overall, the failure of regression models underscores the necessity of employing more experimental data and eventually advanced machine learning approaches, such as ensemble methods or neural networks, which are better equipped to handle the nonlinear relationships and complex data structures inherent to additive manufacturing processes.

The results demonstrate that the choice of model and its hyperparameter optimization significantly influence the predictive accuracy for MEX/M sample properties, providing a robust framework for future process optimization and design improvements. With the aid of ML models, it will be possible in future to reduce intensive experimental test series and efficiently find suitable material properties.

Author Contributions: Conceptualization, K.A.; Methodology, K.A. and M.K.; Software, K.A. and M.K.; Validation, K.A., T.R. and C.E.; Formal analysis, K.A. and T.R.; Investigation, K.A.; Resources, K.A. and C.E.; Data curation, K.A.; Writing—original draft, K.A. and M.K.; Writing—review & editing, K.A.; Visualization, K.A.; Supervision, K.A. and C.E.; Project administration, K.A.; Funding acquisition, K.A. All authors have read and agreed to the published version of the manuscript.

Funding: Publishing fees supported by Funding Programme Open Access Publishing of Hamburg University of Technology (TUHH).

Data Availability Statement: The original contributions presented in this study are included in the article. Further inquiries can be directed to the corresponding author.

Conflicts of Interest: The authors declare no conflict of interest.

References

1. *ISO/ASTM 52900:2023*; Additive Manufacturing—General Principles—Part Positioning, Coordinates and Orientation. ISO: Geneva, Switzerland, 2023.
2. Kuehne, M.; Bartsch, K.; Bossen, B.; Emmelmann, C. Predicting melt track geometry and part density in laser powder bed fusion of metals using machine learning. *Prog. Addit. Manuf.* **2023**, *8*, 47–54. [CrossRef]
3. Bartsch, K. *Digitalization of Design for Support Structures in Laser Powder Bed Fusion of Metals*; Light Engineering für die Praxis; Springer Nature: Cham, Switzerland, 2023; ISBN 978-3-031-22955-8.
4. Bossen, B.; Kuehne, M.; Kristanovski, O.; Emmelmann, C. Data-driven density prediction of AlSi10Mg parts produced by laser powder bed fusion using machine learning and finite element simulation. *J. Laser Appl.* **2023**, *35*, 042023. [CrossRef]
5. Gor, M.; Dobriyal, A.; Wankhede, V.; Sahlot, P.; Grzelak, K.; Kluczyński, J.; Łuszczek, J. Density Prediction in Powder Bed Fusion Additive Manufacturing: Machine Learning-Based Techniques. *Appl. Sci.* **2022**, *12*, 7271. [CrossRef]
6. Kumar, S.; Gopi, T.; Harikeerthana, N.; Gupta, M.K.; Gaur, V.; Krolczyk, G.M.; Wu, C. Machine learning techniques in additive manufacturing: A state of the art review on design, processes and production control. *J. Intell. Manuf.* **2023**, *34*, 21–55. [CrossRef]
7. Wischeropp, T.M.; Tarhini, H.; Emmelmann, C. Influence of laser beam profile on the selective laser melting process of AlSi10Mg. *J. Laser Appl.* **2020**, *32*, 022059. [CrossRef]

8. Wu, D.; Wei, Y.; Terpenny, J. Predictive modelling of surface roughness in fused deposition modelling using data fusion. *Int. J. Prod. Res.* **2019**, *57*, 3992–4006. [CrossRef]
9. Asami, K.; Roth, S.; Krukenberg, M.; Röver, T.; Herzog, D.; Emmelmann, C. Predictive modeling of lattice structure design for 316L stainless steel using machine learning in the L-PBF process. *J. Laser Appl.* **2023**, *35*, 042046. [CrossRef]
10. Sivakumar, N.K.; Palaniyappan, S.; Bodaghi, M.; Azeem, P.M.; Nandhakumar, G.S.; Basavarajappa, S.; Pandiaraj, S.; Hashem, M.I. Predictive modeling of compressive strength for additively manufactured PEEK spinal fusion cages using machine learning techniques. *Mater. Today Commun.* **2024**, *38*, 108307. [CrossRef]
11. Wang, H.; Al Shraida, H.; Jin, Y. Predictive modeling for online in-plane shape deviation inspection and compensation of additive manufacturing. *Rapid Prototyp. J.* **2024**, *30*, 350–363. [CrossRef]
12. Rodríguez-Martín, M.; Fueyo, J.G.; Gonzalez-Aguilera, D.; Madruga, F.J.; García-Martín, R.; Muñóz, Á.L.; Pisonero, J. Predictive Models for the Characterization of Internal Defects in Additive Materials from Active Thermography Sequences Supported by Machine Learning Methods. *Sensors* **2020**, *20*, 3982. [CrossRef]
13. Sarkon, G.K.; Safaei, B.; Kenevisi, M.S.; Arman, S.; Zeeshan, Q. State-of-the-Art Review of Machine Learning Applications in Additive Manufacturing; from Design to Manufacturing and Property Control. *Arch. Comput. Methods Eng.* **2022**, *29*, 5663–5721. [CrossRef]
14. Zhang, Y.; Safdar, M.; Xie, J.; Li, J.; Sage, M.; Zhao, Y.F. A systematic review on data of additive manufacturing for machine learning applications: The data quality, type, preprocessing, and management. *J. Intell. Manuf.* **2023**, *34*, 3305–3340. [CrossRef]
15. Mahmoud, D.; Magolon, M.; Boer, J.; Elbestawi, M.A.; Mohammadi, M.G. Applications of Machine Learning in Process Monitoring and Controls of L-PBF Additive Manufacturing: A Review. *Appl. Sci.* **2021**, *11*, 11910. [CrossRef]
16. Razvi, S.S.; Feng, S.; Narayanan, A.; Lee, Y.-T.T.; Witherell, P. A Review of Machine Learning Applications in Additive Manufacturing. In Proceedings of the 39th Computers and Information in Engineering Conference, Anaheim, CA, USA, 18–21 August 2019; American Society of Mechanical Engineers: Houston, TX, USA, 2019; Volume 1, p. V001T02A040.
17. Babu, S.S.; Mourad, A.-H.I.; Harib, K.H.; Vijayavenkataraman, S. Recent developments in the application of machine-learning towards accelerated predictive multiscale design and additive manufacturing. *Virtual Phys. Prototyp.* **2023**, *18*, e2141653. [CrossRef]
18. Gaikwad, M.U.; Gaikwad, P.U.; Ambhore, N.; Sharma, A.; Bhosale, S.S. Powder Bed Additive Manufacturing Using Machine Learning Algorithms for Multidisciplinary Applications: A Review and Outlook. *Recent. Pat. Mech. Eng.* **2025**, *18*, 12–25. [CrossRef]
19. Jyeniskhan, N.; Keutayeva, A.; Kazbek, G.; Ali, M.H.; Shehab, E. Integrating Machine Learning Model and Digital Twin System for Additive Manufacturing. *IEEE Access* **2023**, *11*, 71113–71126. [CrossRef]
20. Yao, X.; Moon, S.K.; Bi, G. A hybrid machine learning approach for additive manufacturing design feature recommendation. *Rapid Prototyp. J.* **2017**, *23*, 983–997. [CrossRef]
21. Botella, R.; Fernández-Catalá, J.; Cao, W. Experimental Ni3TeO6 synthesis condition exploration accelerated by active learning. *Mater. Lett.* **2023**, *352*, 135070. [CrossRef]
22. Davletshin, A.; Korznikova, E.A.; Kistanov, A.A. Machine Learning Prediction of the Corrosion Rate of Zinc-Based Alloys Containing Copper, Lithium, Magnesium, and Silver. *J. Phys. Chem. Lett.* **2025**, *16*, 114–122. [CrossRef]
23. Suwanpreecha, C.; Manonukul, A. A Review on Material Extrusion Additive Manufacturing of Metal and How It Compares with Metal Injection Moulding. *Metals* **2022**, *12*, 429. [CrossRef]
24. Herzog, D.; Asami, K.; Scholl, C.; Ohle, C.; Emmelmann, C.; Sharma, A.; Markovic, N.; Harris, A. Design guidelines for laser powder bed fusion in Inconel 718. *J. Laser Appl.* **2022**, *34*, 012015. [CrossRef]
25. Asami, M.K.; Herzog, D.; Bossen, B.; Geyer, L.; Klemp, C.; Emmelmann, C. Design Guidelines for Green Parts Manufactured with Stainless Steel in the Filament Based Material Extrusion Process for Metals (MEX/M). In Proceedings of the World PM2022 Congress & Exhibition, Lyon, France, 9–13 October 2022. [CrossRef]
26. Asami, K.; Crego Lozares, J.M.; Ullah, A.; Bossen, B.; Clague, L.; Emmelmann, C. Material extrusion of metals: Enabling multi-material alloys in additive manufacturing. *Mater. Today Commun.* **2024**, *38*, 107889. [CrossRef]
27. Pedregosa, F.; Varoquaux, G.; Gramfort, A.; Michel, V.; Thirion, B.; Grisel, O.; Blondel, M.; Prettenhofer, P.; Weiss, R.; Dubourg, V.; et al. Scikit-learn: Machine Learning in Python. *J. Mach. Learn. Res.* **2011**, *12*, 2825–2830.
28. Asami, M.K.; Kuehne, M. Machine Learning Models and Data for the Application of Machine Learning in Predicting Quality Parameters in Metal Material Extrusion (MEX/M). Available online: https://tore.tuhh.de/entities/product/c0c59885-cf8e-4191-acd1-4d0097b5a57c (accessed on 26 April 2025).
29. Othman, W.; Hamoud, B.; Kashevnik, A.; Shilov, N.; Ali, A. A Machine Learning-Based Correlation Analysis between Driver Behaviour and Vital Signs: Approach and Case Study. *Sensors* **2023**, *23*, 7387. [CrossRef]

30. Huang, J.; Wei, Y. Data-driven analysis of stroke-related factors and diagnostic prediction. In Proceedings of the 2024 IEEE 6th Advanced Information Management, Communicates, Electronic and Automation Control Conference (IMCEC), Chongqing, China, 24–26 May 2024; IEEE: Piscataway, NJ, USA, 2024; pp. 60–65.
31. Rezazadeh, A. Toe-Heal-Air-Injection Thermal Recovery Production Prediction and Modelling Using Quadratic Poisson Polynomial Regression. *arXiv* **2020**, arXiv:2012.02262.
32. Géron, A. *Hands-on Machine Learning with Scikit-Learn, Keras, and TensorFlow: Concepts, Tools, and Techniques to Build Intelligent Systems*, 2nd ed.; O'Reilly: Beijing, China; Boston, MA, USA; Farnham, UK; Sebastopol, CA, USA; Tokyo, Japan, 2019; ISBN 978-1-4920-3264-9.
33. Xiao, W.; Wang, Y. The Application of Partially Functional Linear Regression Model in Health Science. *Sci. Discov.* **2020**, *8*, 134. [CrossRef]
34. Roy, M.-H.; Larocque, D. Robustness of random forests for regression. *J. Nonparametric Stat.* **2012**, *24*, 993–1006. [CrossRef]
35. Adankon, M.M.; Cheriet, M. Support Vector Machine. In *Encyclopedia of Biometrics*; Li, S.Z., Jain, A.K., Eds.; Springer: Boston, MA, USA, 2015; pp. 1504–1511. ISBN 978-1-4899-7487-7.
36. Rachdi, M.; Laksaci, A.; Kaid, Z.; Benchiha, A.; Al-Awadhi, F.A. *k* -Nearest neighbors local linear regression for functional and missing data at random. *Stat. Neerl.* **2021**, *75*, 42–65. [CrossRef]
37. Kramer, O. Dimensionality Reduction by Unsupervised K-Nearest Neighbor Regression. In Proceedings of the 2011 10th International Conference on Machine Learning and Applications and Workshops, Honolulu, HI, USA, 18–21 December 2011; IEEE: Piscataway, NJ, USA, 2011; pp. 275–278.
38. Multilayer Perceptron. In *Pattern Recognition and Image Preprocessing*; Bow, S.-T., Ed.; Signal Processing and Communications; CRC Press: Boca Raton, FL, USA, 2002; pp. 201–224. ISBN 978-0-8247-0659-3.
39. Taud, H.; Mas, J.F. Multilayer Perceptron (MLP). In *Geomatic Approaches for Modeling Land Change Scenarios*; Camacho Olmedo, M.T., Paegelow, M., Mas, J.-F., Escobar, F., Eds.; Lecture Notes in Geoinformation and Cartography; Springer International Publishing: Cham, Switzerland, 2018; pp. 451–455. ISBN 978-3-319-60800-6.
40. Hassan, H.A.; Abd Ghani, M.N.A.; Zabidi, A.; Md Adnan, W.N.W.; Rahman, J.A.; Rusni, I.M. Pattern Classification in Recognising Idgham Maal Ghunnah Pronunciation Using Multilayer Perceptrons. In Proceedings of the 2024 IEEE 6th Symposium on Computers & Informatics (ISCI), Kuala Lumpur, Malaysia, 10 August 2024; IEEE: Piscataway, NJ, USA, 2024; pp. 333–338.
41. Meng, L.; McWilliams, B.; Jarosinski, W.; Park, H.-Y.; Jung, Y.-G.; Lee, J.; Zhang, J. Machine Learning in Additive Manufacturing: A Review. *JOM* **2020**, *72*, 2363–2377. [CrossRef]
42. Altman, N.S. An Introduction to Kernel and Nearest-Neighbor Nonparametric Regression. *Am. Stat.* **1992**, *46*, 175–185. [CrossRef]
43. Le, T.M.; Vo, T.M.; Pham, T.N.; Dao, S.V.T. A Novel Wrapper–Based Feature Selection for Early Diabetes Prediction Enhanced With a Metaheuristic. *IEEE Access* **2021**, *9*, 7869–7884. [CrossRef]
44. Breiman, L. Random Forests. *Mach. Learn.* **2001**, *45*, 5–32. [CrossRef]
45. Vapnik, V.N. An overview of statistical learning theory. *IEEE Trans. Neural Netw.* **1999**, *10*, 988–999. [CrossRef]
46. Bisong, E. The Multilayer Perceptron (MLP). In *Building Machine Learning and Deep Learning Models on Google Cloud Platform*; Apress: Berkeley, CA, USA, 2019; pp. 401–405. ISBN 978-1-4842-4469-2.
47. Chicco, D.; Warrens, M.J.; Jurman, G. The coefficient of determination R-squared is more informative than SMAPE, MAE, MAPE, MSE and RMSE in regression analysis evaluation. *PeerJ Comput. Sci.* **2021**, *7*, e623. [CrossRef] [PubMed]
48. Klusowski, J.M. Sharp Analysis of a Simple Model for Random Forests. *arXiv* **2020**, arXiv:1805.02587. [CrossRef]
49. Battineni, G.; Chintalapudi, N.; Amenta, F. Machine learning in medicine: Performance calculation of dementia prediction by support vector machines (SVM). *Inform. Med. Unlocked* **2019**, *16*, 100200. [CrossRef]
50. Toorandaz, S.; Taherkhani, K.; Liravi, F.; Toyserkani, E. A novel machine learning-based approach for in-situ surface roughness prediction in laser powder-bed fusion. *Addit. Manuf.* **2024**, *91*, 104354. [CrossRef]
51. Caminero, M.Á.; Romero Gutiérrez, A.; Chacón, J.M.; García-Plaza, E.; Núñez, P.J. Effects of fused filament fabrication parameters on the manufacturing of 316L stainless-steel components: Geometric and mechanical properties. *Rapid Prototyp. J.* **2022**, *28*, 2004–2026. [CrossRef]
52. Gloeckle, C.; Konkol, T.; Jacobs, O.; Limberg, W.; Ebel, T.; Handge, U.A. Processing of Highly Filled Polymer–Metal Feedstocks for Fused Filament Fabrication and the Production of Metallic Implants. *Materials* **2020**, *13*, 4413. [CrossRef]
53. Damon, J.; Dietrich, S.; Gorantla, S.; Popp, U.; Okolo, B.; Schulze, V. Process porosity and mechanical performance of fused filament fabricated 316L stainless steel. *Rapid Prototyp. J.* **2019**, *25*, 1319–1327. [CrossRef]
54. Siamidoudaran, M.; İşçioğlu, E. Injury Severity Prediction of Traffic Collision by Applying a Series of Neural Networks: The City of London Case Study. *PROMET—TrafficTransportation* **2019**, *31*, 643–654. [CrossRef]
55. Eyercioğlu, Ö.; Aladağ, M. Non-Planar Toolpath for Large Scale Additive Manufacturing. *Int. J. 3D Print. Technol. Digit. Ind.* **2021**, *5*, 477–487. [CrossRef]

56. Rajput, A.S.; Babu, P.; Das, M.; Kapil, S. Influence of toolpath strategies during laser polishing on additively manufactured biomaterials. *Surf. Eng.* **2024**, *40*, 967–982. [CrossRef]
57. Asami, K.; Herzog, D.; Deutschmann, T.; Röver, T.; Kelbassa, I.; Emmelmann, C. Methodology for Cost Estimation Using Characteristic Factors in Additive Manufacturing. *J. Jpn. Soc. Powder Powder Metall.* **2025**, *72*, S75–S82. [CrossRef]

B.6 Metallic Bipolar Plate Production Through Additive Manufacturing: Contrasting MEX/M and PBF-LB/M Approaches

The summary of the findings can be found in Chapter 7.1.

Journal of
Experimental and
Theoretical Analyses

Article

Metallic Bipolar Plate Production Through Additive Manufacturing: Contrasting MEX/M and PBF-LB/M Approaches

Karim Asami, Sebastian Roth, Jan Hünting, Tim Röver and Claus Emmelmann

Special Issue
Featured Papers for *Journal of Experimental and Theoretical Analyses (JETA)*

Edited by
Prof. Dr. Marco Rossi

MDPI

https://doi.org/10.3390/jeta3020012

Journal of
Experimental and Theoretical Analyses

MDPI

Article

Metallic Bipolar Plate Production Through Additive Manufacturing: Contrasting MEX/M and PBF-LB/M Approaches

Karim Asami [1,*], Sebastian Roth [2], Jan Hünting [1], Tim Röver [1] and Claus Emmelmann [1]

[1] Institute of Laser and System Technologies (iLAS), Hamburg University of Technology (TUHH), Harburger Schloßstraße 28, 21079 Hamburg, Germany; jan.huenting@tuhh.de (J.H.); tim.roever@tuhh.de (T.R.); c.emmelmann@tuhh.de (C.E.)
[2] Quest One GmbH, Viktoriaring 23, 22143 Hamburg, Germany; sebastian.roth@questone.com
* Correspondence: karim.asami@tuhh.de

Abstract: Additive manufacturing (AM) technologies have witnessed remarkable advancements, offering opportunities to produce complex components across various industries. This paper explores the potential of AM for fabricating bipolar plates (BPPs) in fuel cell or electrolysis cell applications. BPPs play a critical role in the performance and efficiency of such cells, and conventional manufacturing methods often face limitations, particularly concerning the complexity and customization of geometries. The focus here lies in two specific AM methods: the laser powder bed fusion of metals (PBF-LB/M) and material extrusion of metals (MEX/M). PBF-LB/M, tailored for high-performance applications, enables the creation of highly complex geometries, albeit at increased costs. On the other hand, MEX/M excels in rapid prototyping, facilitating the swift production of diverse geometries for real-world testing. This approach can facilitate the evaluation of geometries suitable for mass production via sinter-based manufacturing processes. The geometric deviations of different BPPs were identified by evaluating 3D scans. The PBF-LB/M method is more suitable for small features, while the MEX/M method has lower deviations for geometrically less complex BPPs. Through this investigation, the limits of the capabilities of these AM methods became clear, knowledge that can potentially enhance the design and production of BPPs, revolutionizing the energy conversion and storage landscape and contributing to the design of additive manufacturing technologies.

Keywords: additive manufacturing (AM); bipolar plates; fuel cell (PEMFC); electrolysis cell (PEMEC); material extrusion of metals (MEX/M); laser powder bed fusion of metals (PBF-LB/M); prototyping; design for AM (DfAM)

Academic Editor: Marco Rossi

Received: 10 March 2025
Revised: 5 April 2025
Accepted: 9 April 2025
Published: 14 April 2025

Citation: Asami, K.; Roth, S.; Hünting, J.; Röver, T.; Emmelmann, C. Metallic Bipolar Plate Production Through Additive Manufacturing: Contrasting MEX/M and PBF-LB/M Approaches. *J. Exp. Theor. Anal.* **2025**, 3, 12. https://doi.org/10.3390/jeta3020012

1. Introduction

In the imminent shift towards a renewable-based economy, hydrogen is emerging as a pivotal energy carrier due to its high energy density and environmentally friendly production possibilities [1–3]. However, the predominant method of hydrogen production, relying on fossil fuels like petroleum and coal, raises concerns about escalating greenhouse gas emissions [4–6]. As an efficient and promising alternative, water electrolysis stands out, with proton exchange membrane electrolyzer/fuel cells (PEMECs/PEMFCs, hereafter referred to as PEMFCs) offering numerous advantages, such as high-purity hydrogen, the capability to produce compressed gases for direct use or storage and a swift response to current fluctuations [7–9]. Despite garnering significant attention, PEMFCs face several hurdles hindering their widespread commercialization, including cost, performance and durability concerns [10–12].

Central to PEMFCs are bipolar plates (BPPs), essential for distributing reactants and products as well as conducting the current. These plates constitute a substantial portion (from 40% to up to 51%) of the overall PEMFC cost [13]. Furthermore, the distribution of reactants and the electric contact resistance between BPPs and current distributors highly influence the cell performance [14]. Designing and manufacturing BPPs with intricate flow channels poses challenges, particularly in managing mass transport. Addressing these challenges, additive manufacturing (AM) is emerging as a transformative technology capable of rapidly producing complex structures like lattices and triply periodic minimal surface (TPMS) structures from digital 3D computer-aided design (CAD) models. AM offers significant advantages for several innovative industries like aerospace, energy and healthcare [15–17]. Specialized components can be created with AM that result in high-performance, lightweight and cost-effective PEMFCs for future applications [13].

Stainless steel (SS) has emerged as a viable alternative material for BPPs, offering moderate mechanical strength at a lower cost compared to titanium and graphite. However, in harsh environments, such as on the anode side of PEMFCs, SS may require corrosion-resistant coatings like Au, Au/Ni, TiN, Pt/Fe, Ti/Pt, SnO_2 or CrN [14,18]. These coatings are applied through various techniques, such as the promising active screen plasma nitriding technique [19]. Conversely, the cathode side experiences less intense corrosion, making untreated SS suitable for this application [14]. Studies have been conducted on the properties and manufacturability of different BPP designs manufactured using PBF-LB/M, such as wineglass-shaped channels and vertical honeycomb geometries. However, a research gap remains regarding the investigation of more complex BPP channel geometries [20,21].

The laser powder bed fusion (PBF-LB/M) and material extrusion of metals (MEX/M) [22] are applied to manufacture BPPs and current distributors as a single AISI SS 316L part with various flow field designs. The experimental findings are especially important in the validation process of flow field simulations that can quickly identify the optimal geometry for the given operating conditions of a PEMFC.

Therefore, this paper aims to analyze the suitability of the two AM processes for the production of metallic BPPs and recognize their advantages and disadvantages. For this, BPPs with six different flow field designs were manufactured and geometrically compared to each other by microscopy and 3D scans.

2. Manufacturing Methods

2.1. MEX/M

The material extrusion of metals is an innovative AM technique revolutionizing metal component production. At its core lies screw-, plunger- or filament-based material extrusion, in which a feedstock composed of metal powder dispersed in a plastic matrix serves as the raw material. The granulate, bar or filament feedstock is extruded layer by layer, akin to in the fused deposition modeling (FDM) process, to construct the desired geometry. This additive approach allows for precise control over material deposition, facilitating the creation of intricate designs [23,24].

MEX/M follows a structured, three-step process: printing, debinding and sintering. First, the desired geometry is printed layer by layer by melting the feedstock. Following printing, the green parts formed by the deposited layers undergo debinding, where the plastic binder within is removed, leaving a porous structure of metal powder particles, also called the backbone. Subsequently, the debinded green parts, also called brown parts, are thermally debinded and sintered at high temperatures. This process involves heating the brown parts below their melting point in order to fuse the metal powder particles [23,25,26]. This results in metal parts with the desired mechanical properties and relative densities exceeding 97% [23,27].

Hot isostatic pressing (HIP) is a post-processing strategy that can significantly improve MEX/M parts by increasing their density to almost 100% and thus increasing their mechanical strength. However, due to its high price, HIP is only used when necessary. HIP and other, more conventional post-processing technologies like surface treatment, coating, subtractive machining and heat treatment allow for the fine-tuning of the material properties and surface finishes, enabling customized solutions for various industrial applications [23].

The disadvantages of the MEX/M process include a resolution that is limited by the size of the nozzle, inferior surface finishes and the high complexity of the process because of the need for debinding and sintering [23,28]. The shrinkage of MEX/M components during the sintering process can be more than 20% and makes it difficult to achieve the required dimensional accuracies [29]. Additionally, intricate components are often in need of support structures for both printing and sintering and add another layer of complexity to the MEX/M process [30].

MEX/M offers several advantages over traditional manufacturing methods. Firstly, it provides a high degree of design flexibility, enabling the fabrication of complex geometries without the restrictions of conventional machining or casting processes. Moreover, MEX/M eliminates the need for molds, which are mandatory for the metal injection molding process, which has process similarities to the MEX/M process chain, as well as reducing tooling costs and lead times. Additionally, the most common MEX/M approach (filament-based) eliminates the necessity of depowdering the final part after sintering, streamlining post-processing workflows compared to powder-based AM processes. Lastly, the widespread availability of MEX/M feedstocks made from various materials contributes to the accessibility and scalability of this AM process [31].

2.2. *PBF-LB/M*

The laser powder bed fusion of metals is a widely recognized AM technology that has gained prominence due to its capability to fabricate complex, dense and robust metal parts with an exceptional geometrical resolution. This technology is particularly valued for its versatility in processing a diverse array of materials, including high-strength aluminum alloys, high-melting-point elements and high-entropy alloys, which makes it suitable for various demanding applications across industries such as aerospace, automotives and healthcare [15,32,33].

One of the significant advantages of PBF-LB/M is its ability to produce intricate and lightweight components, enabling the design and manufacture of complex geometries that would be challenging or impossible to achieve with traditional manufacturing methods. However, to ensure the quality and integrity of these complex parts, the use of support structures is essential. Properly designed support structures prevent defects and warping during the build process due to thermally induced residual stresses. The development of advanced software tools for support generation and optimization has become crucial in this context [34].

Despite its strengths, PBF-LB/M has certain limitations, particularly concerning the surface quality and feature size. These challenges can be mitigated through hybrid manufacturing approaches, where PBF-LB/M is combined with subtractive techniques for post-processing. This integration enhances the overall functionality of the parts, enabling the production of components that meet stringent performance criteria [35]. Post-processing techniques such as hot isostatic pressing and solution annealing are also vital in the PBF-LB/M process. These methods are employed to improve the microstructure of the printed parts, reduce porosity and enhance the mechanical properties, ensuring that the final components meet the required specifications for critical applications [36].

Another constraint of the PBF-LB/M process is the need to get rid of semi-sintered or loose powder which is trapped in the pores of the solidified material or in the cavities of the component's geometry after build completion. The remaining powder can be a health hazard to humans, block intricate cavities like cooling channels or fully sinter in thermal post-processes like HIP or heat treatment [37]. Typically, powder removal is achieved by loosening the powder first and then flushing it out with the help of a fluid [38]. Since these powder removal methods have their own restrictions, it is important to address this issue early in the design stage of PBF-LB/M components by avoiding closed cavities or undercuts.

In addition, the development of new materials for PBF-LB/M can be advanced through in situ alloying, where elemental powders are mixed during the printing process. While this approach opens up new possibilities for creating custom alloys, achieving a homogeneous mixture remains a challenge that requires further research and optimization [39,40].

Finally, advancements in beam-shaping techniques have shown promise in addressing challenges related to laser–material interaction. By tailoring the shape of the laser beam, it is possible to improve both the quality and efficiency of the PBF-LB/M process, paving the way for the more reliable and cost-efficient production of high-performance metal parts [41].

It is clear that both technologies have different boundary conditions, such as their minimum possible layer height and different resolution rates. The initial intention was to identify their respective potential. The PBF-LB/M process is suitable for manufacturing complex structures that cannot be produced conventionally with a density of more than 99%. The MEX/M process, on the other hand, which includes sintering and debinding steps rather than metal injection molding (MIM), can act as a complementary solution to the MIM process and enable the mass production of functional products [23].

3. Methodology

The different flow fields of the BPPs were designed in *Autodesk Inventor 2023* (Autodesk, San Rafael, CA, USA) based on the state of the art [42,43]. The designs included a classic serpentine flow field structure as a starting point for further investigations; see Figure 1a. This serpentine structure was altered with various features that could positively influence the performance of a fuel cell in different ways. The dimensions of the square BPPs were 33 mm in width and length and 3 mm in thickness. For the investigation and evaluation, only the width and length were used, as it was evident that the MEX/M process would be less accurate regarding the thickness due to the greater layer heights. Unless otherwise specified, the channels of all examined geometries were 1 mm in width and height. To see all the standard tessellation language (STL) files of the BPP designs, scanning data and the measurement point locations of the scan analysis, the reader is kindly referred to the supplementary dataset of this work [44]. The bars that separated the serpentine flow channel were altered with grooves (0.3 mm high and 0.4 mm wide) that connected the neighboring channels with each other; see Figure 1b. This had a homogenizing effect on the reactant distribution at the interface and thus ensured a homogenized operating state.

The designs in Figure 1c–e were primarily aimed at avoiding or reducing water accumulation in the channels. The wave structure in Figure 1c varied the width of the channel from 1.1 mm to 1.4 mm and prevented the accumulation of water in rectangular geometries. In Figure 1d, it can be seen that 2 mm long and 0.7 mm high pyramids were integrated into the channels, through which local turbulence was introduced into the flow of reactants, which also prevented the accumulation of water and locally increased the flow velocities. A similar, albeit homogeneous, effect was achieved by the wavy-shaped bottom of the channel, which is shown in Figure 1e. Because of this particular design, any water that formed could not accumulate. The iterative change in the channel height from 0.477 mm to 1.0 mm led to a repeatedly locally influenced flow velocity, which resulted in an overall more homogeneous flow. The serpentine [narrowing] design from Figure 1f exhibited a steady narrowing of the channel in the direction of the outlet from 1.7 mm to 0.3 mm in width. This ensured increasing pressure in the rear area of the channel and thus led to a more homogeneous pressure distribution within the BPP.

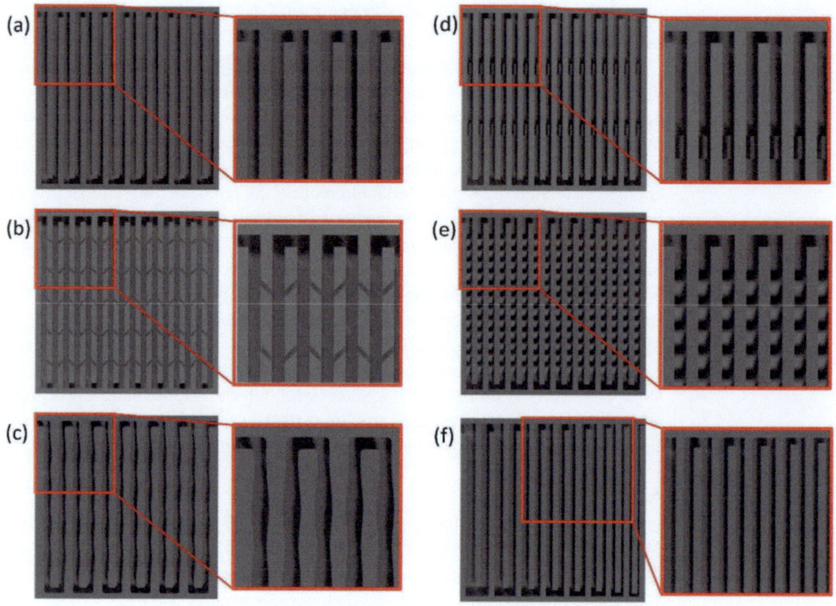

Figure 1. Flow field design of the manufactured bipolar plates; structures (**a**–**f**): serpentine, rib grooves, wave structure, triangular baffles, wavy bottom, serpentine [narrowing].

Changing a flow field design using grooves has already been investigated experimentally in [45,46]. The channel width and height were set to 1 mm, as in this work, with a groove cross-section of 0.2 mm. Compared to a simple serpentine structure, the result was an increase in the current density of 16% and an increase in the power density of 3%. Conventionally manufactured wave structures were produced from a graphite material with varying widths of 1 mm to 1.25 mm, 1 mm to 1.5 mm and 1 mm to 1.75 mm [47]. The results of the experimental investigation with different cell temperatures and air and hydrogen flow rates showed that the wave structures with a width of 1 mm to 1.25 mm had a higher performance than the other wave structures and achieved a 20.15% increase in the maximum output power with a hydrogen and air flow compared to conventional serpentine structures [47].

A simulative study of pyramid structures was conducted in [48], and the results showed that the reactant gas distribution in the gas diffusion layer could be increased compared to that of the conventional structure.

Wavy bottom channels were investigated simulatively in [49,50] and an increase in the pressure drop and an increase in the mass transfer of the reactant gas were revealed. The current density could be increased further by 21.8% with this design feature compared to a simple serpentine structure. In this case, the channel sizes were also 1 mm in the cross-sectional area, and the wave bottom structure had a height of 0.5 mm [49].

Narrower serpentine channels were simulated in [51], where the channels were continuously reduced from a width of 1.5 mm to 1.9 mm to a width of 0.5 mm to 0.1 mm. It was shown that a reduction in the channel width generally provided an advantage regarding the uneven flow distribution at high current densities compared to conventional serpentine structures. At the largest reduction in the channel width, the current density limit was 1.35 times higher than in the conventional design [51].

The BPPs were manufactured using the MEX/M and PBF-LB/M methods to compare both manufacturing processes regarding their suitability in the production of BPPs with complex geometric features and high-dimensional accuracy.

The manufacturing of the BPPs using the MEX/M process was performed on a *Renkforce RF 2000* (Conrad Electronic SE, Hirschau, Germany) 3D printer with an extruder that had been optimized for constant metal filament feeding. A *V6* nozzle (E3D, Chalgrove, UK), made of hardened steel and with a diameter of 0.4 mm, was used. The slicing of the CAD models was performed with *Netfabb Ultimate 2023* (Autodesk, San Rafael, CA, USA) software for all PBF-LB/M models, and for the MEX/M models, a *Simplify3D* slicer (Cincinnati, OH, USA) was used. Available MEX/M design guidelines were adopted and are applied in this paper, including in regard to the process parameters (Table 1) and the powder composition of the 316L SS [52]. The sinter process and debinding of the BPPs was performed according to [27]. For the chemical debinding, each BPP was submerged for 12 h at 40 °C in 500 mL acetone as a solvent. The thermal debinding and sintering were performed in an *Aim3D ExSO90* (New AIM3D GmbH, Rostock, Germany) sinter oven. Argon (99.9%) was used for the sintering atmosphere, and the sintering cycle used, with a maximum temperature of 1286 °C, is shown in Figure 2.

Table 1. Parameters of the MEX/M manufacturing process.

AM Machine	Renkforce RF 2000
Extrusion temperature	135 °C
Heat bed temperature	40 °C
Nozzle diameter	0.4 mm
Printing speed	2100 mm/s
Extrusion multiplier	1.35
Skirt layer	2
Retraction distance	4.5 mm
Layer height	200 µm
Material	316L stainless steel

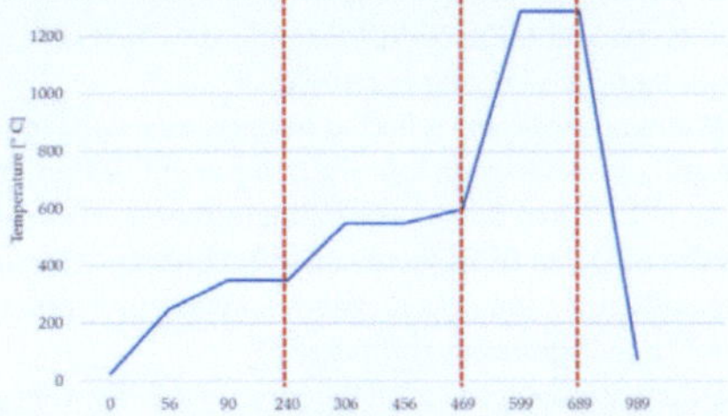

Figure 2. Representation of the sintering cycle for all BPPs produced in the MEX/M process.

The PBF-LB/M BPPs were produced using an *MPrint* 3D printer (One Click Metal GmbH, Tamm, Germany), and the process parameters shown in Table 2 were used. As in the MEX/M process, the PBF-LB/M BPPs were printed using a 316L feedstock. The spherical particle size distribution of the PBF-LB/M material was D_{10} = 19 µm, D_{50} = 29 µm and D_{90} = 43 µm according to ISO 13322-2, which was provided by the manufacturer, *Material 4 Print GmbH & Co. KG* (Löhne, Germany) [53,54].

Table 2. Parameters of the PBF-LB/M manufacturing process.

AM Machine	One Click Metal MPrint
Laser power	200 W
Scan speed	800 mm/s
Hatch distance	0.1 mm
Layer thickness	40 µm
Laser diameter	70 µm
Material	316L stainless steel

The MEX/M filament from *PT+A GmbH* (Dresden, Germany) also had spherical metal particles, where the sizes were in the range of 2 µm–15 µm [27,55]. Based on a direct comparison, the MEX/M feedstock appeared to use finer metal powder than the PBF-LB/M feedstock.

For the dimensional accuracy measurement of the manufactured BPPs, a *Keyence VL-500* (Keyence, Osaka, Japan) coordinate measuring machine 3D scanning system was used. It had a geometric resolution of ±10 µm. The associated scanning software, *VL-500 application series* (Keyence, Osaka, Japan), enabled the analysis of the scanning data by combining scans from different angles and performing 3D CAD comparisons that visualized geometric deviations between the printed BPPs and the original CAD data as false-color pictures. A *VHX-S600E* microscope from *Keyence* (Keyence, Osaka, Japan) was used to create images of small BPP features.

4. Results and Discussion

When measuring the dimensions of the MEX/M BPPs, it was observed that the channels were narrower and the ribs wider than those designed in CAD software. This was due to a combination of over-extrusion during printing and the shrinkage of the component in the sintering step. To measure these inaccuracies, three cubes with dimensions of $10 \times 10 \times 10$ mm^3 were fabricated and measured both as green parts and as sintered components in the same process chain. The shrinkage from the green body to the sintered part in the XY direction was 0.870 mm (8.70%), and in the Z direction, it was 0.896 mm (8.96%). The shrinkage was greater in the Z direction because of the higher thermal gradient, the effect of the layer direction and gravity [23,56,57]. The shrinkage of the measured cubes was below the typical MPIF 35 MIM standard, which is in the range of 12–20%, and is an indication of suitable sintering parameters [23].

This shrinkage was also reflected in the dimensions of the BPPs. By incorporating the shrinkage and over-extrusion into the component design, this effect can be compensated for in future BPPs. Therefore, in the following comparison of dimensions, the digital CAD models of the MEX/M BPPs were scaled according to the measured shrinkage values to enable a meaningful comparison. This step was not necessary for the PBF-LB/M BPPs as they did not show significant shrinkage.

Three-dimensional scans of all the produced BPPs were generated by using the *Keyence VL-500* and is shown, e.g., in Figure 3. Subsequently, the scanned data were overlaid onto the original CAD geometry to generate 3D CAD comparisons using the provided software.

The color mapping of the structures represents the deviations of the scans (from +0.2 mm to −0.2 mm) from the reference geometry of the BPPs. In the color mapping, blue areas indicate deviations towards the inside and red areas indicate deviations towards the outside of the designed BPPs. Green areas represent little to no deviations. From the images, only qualitative statements regarding printing accuracy can be made. This is due to the fact that artifacts and inaccuracies arise during data processing, making precise quantitative analysis impossible.

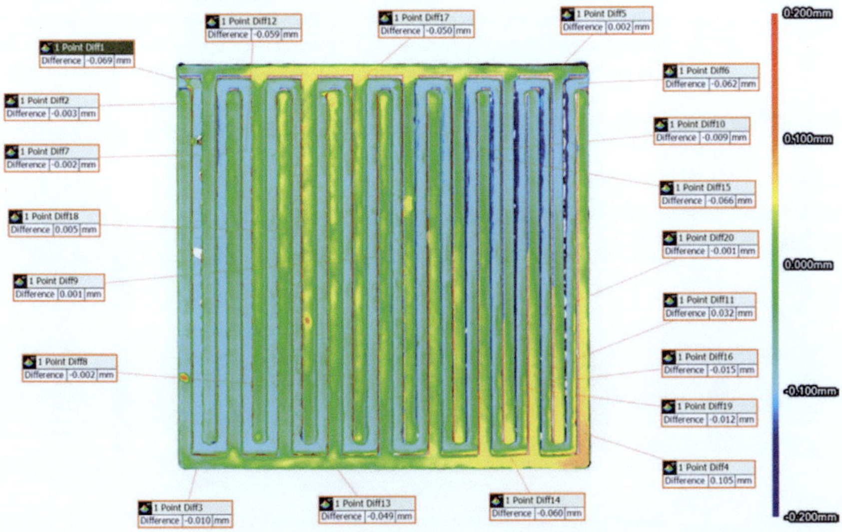

Figure 3. Locations of all 20 measurement points used for the geometric analysis of the scanned BPPs, shown for an example of the serpentine structure produced in the PBF-LB/M process.

To enable a comparison between the two manufacturing methods, 20 measurement points on the surface of all the scanned BPPs were created and examined. The location of these points was selected in such a way as to obtain results that were as holistic as possible, choosing geometrically interesting and highly deformed areas (the gas inlet and outlet, outer and inner walls, the bottom of the gas channel and geometric features if present in the BPP geometry). The location of all the measurement points is shown in Figure 3 using the example of the serpentine structure produced in the PBF-LB/M process.

The mean deviations from the CAD geometries, as well as the standard deviations among the measurement points, were calculated and analyzed. A graph showing the mean and standard deviations of all the produced BPPs is presented in Figure 4, allowing for a comparative analysis of how the two manufacturing processes performed in terms of accuracy and consistency.

The MEX/M process produced smaller mean deviations from the original CAD models for the serpentine, rib groove and serpentine (narrowing) geometries compared to the PBF-LB/M process. For instance, the serpentine geometry had the lowest mean deviation of all the MEX/M BPPs with a deviation of 0.031 mm, whereas this was more than twice as high at 0.065 mm for PBF-LB/M. Similarly, the rib groove geometry showed a lower mean deviation of 0.041 mm with MEX/M compared to 0.053 mm in the PBF-LB/M process. Notably, rib grooves exhibited the second-lowest mean deviations of all geometries in both processes. Also, the serpentine (narrowing) BPPs could be produced with higher accuracy using the MEX/M process, demonstrating a significantly lower mean deviation of 0.042 mm compared to those produced using PBF-LB/M with 0.059 mm. On the other hand, both geometries with small features (triangular baffles and wavy bottom) and the wave structure performed better in the PBF-LB/M process. The wave structure exhibited a slightly lower mean deviation in the PBF-LB/M (0.060 mm) process compared to MEX/M (0.067 mm). Triangular baffles and wavy bottom geometries followed a similar trend, with those produced using MEX/M showing significantly higher mean deviations (0.077 mm and 0.081 mm, respectively) than in the PBF-LB/M process (0.047 mm and 0.056 mm, respectively). These discrepancies likely stemmed from the small features in these two geometries. A more detailed description of this will be given further down in this section.

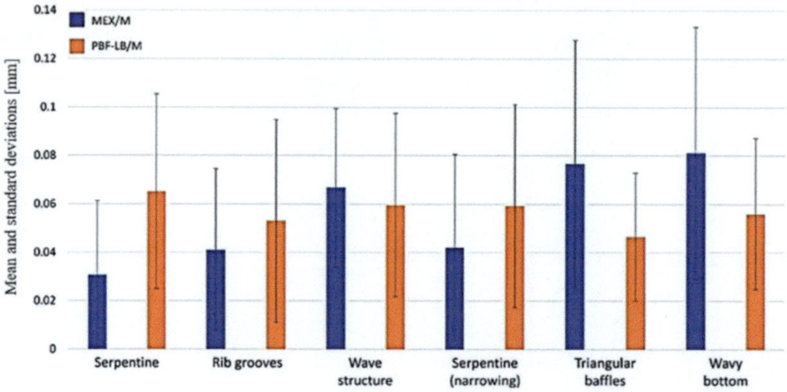

Figure 4. Mean and standard deviations of the produced BPPs from their original CAD geometries.

Notably, the triangular baffle geometry showed the lowest mean deviations in the PBF-LB/M process but the second highest in the MEX/M process. Generally, the mean deviations across all geometries were more consistent when manufactured using the PBF-LB/M process and more geometry-dependent when manufactured using MEX/M.

The standard deviations indicated the measurement consistency, and both processes exhibited varying degrees of variability across the examined geometries (see Figure 4). Serpentine, rib groove and wave structures showed standard deviations in the MEX/M process in a small range from 0.031 mm to 0.033 mm and thus less variability than in the PBF-LB/M process, with values between 0.038 mm and 0.042 mm for the same geometries. The serpentine (narrowing) geometry also exhibited a slightly lower standard deviation with MEX/M manufacturing, at 0.038 mm, compared to 0.042 mm with PBF-LB/M manufacturing. As seen in the mean deviations, MEX/M struggled with the small features in the triangular baffles and wavy bottom geometry, having standard deviations of 0.051 mm and 0.052 mm, respectively. In contrast, the PBF-LB/M process produced significantly more consistent standard deviation values of 0.026 mm and 0.031 mm, respectively. Overall, the PBF-LB/M process tended to have more consistent results for the assessed geometries with small and intricate features, as indicated by generally lower standard deviations compared to those produced using MEX/M. For all other geometries, the MEX/M process exhibited lower standard deviations and thus more consistent results.

The BPPs generally exhibited high standard deviation values. For some structures, such as the narrowing serpentine structures, triangular baffles, and wavy bottom structures, the ranges fluctuated by more than 50% relative to the mean deviation. Possible reasons for this were, on the one hand, that the points were set manually in the evaluation of the structures, and therefore it was not always possible to evaluate the same point on the BPPs. Therefore, there was generally a measurement inaccuracy in the evaluation, which was unavoidable due to technical limitations. On the other hand, white areas of the BPPs with standard deviations of more than 50% can be recognized in Figures 5 and 6. These areas contained the smallest design features and were not fully captured by the scanning system, meaning that there was also a fundamental source of measurement error. In order to reduce the high fluctuations in the measurements, more measuring points must be set, and the scan areas will have to be divided into several smaller segments.

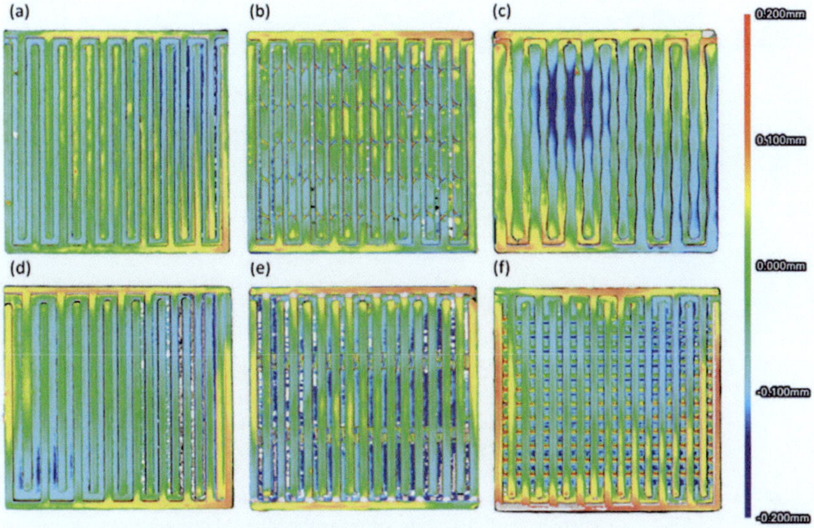

Figure 5. Three-dimensional CAD comparison of the BPPs produced in the MEX/M process; structures (**a–f**): serpentine, ribgrooves, wave structure, serpentine [narrowing], triangular baffles, wavy bottom.

A qualitative examination of the 3D CAD comparison images shown in Figures 5 and 6 revealed that the PBF-LB/M process had a tendency to produce BPPs with corners that bent upwards, a characteristic which was less pronounced in the MEX/M BPPs.

A small study on stacked BPPs evaluated the possibility of manufacturing several BPPs in one pass and simultaneously reducing the warpage of the outer corners. Therefore, three serpentine BPPs were oriented in a standing position, connected by 2 mm wide extrusions and manufactured using MEX/M from 316L SS. Subsequently, the individual BPPs were separated with a metal saw and polished. The results were BPPs with significantly reduced warpages at the corners, of which an example is shown in Figure 7. The red areas in Figure 7b are due to inaccuracies in the separation process. A disadvantage of this procedure was the overhangs that were created by orienting the BPPs in a standing position. Further research should explore this promising approach.

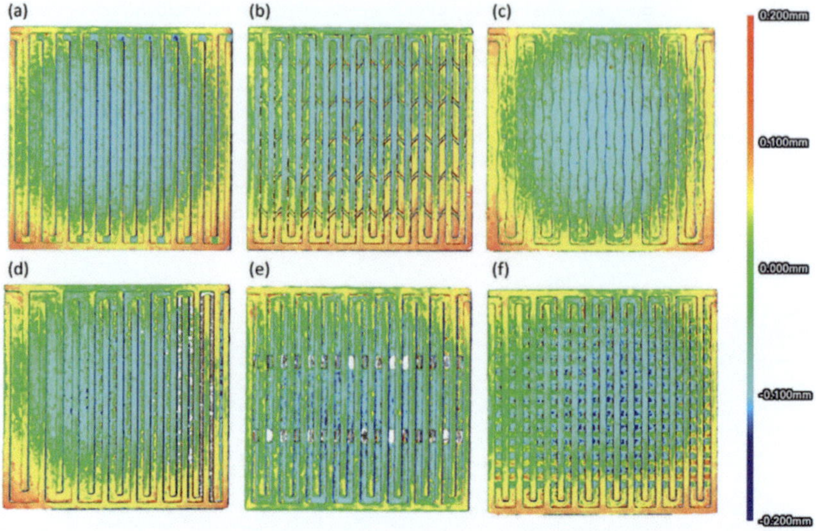

Figure 6. Three-dimensional CAD comparison of the BPPs produced in the PBF-LB/M process; structures (**a–f**): serpentine, rib grooves, wave structure, serpentine [narrowing], triangular baffles, wavy bottom.

It can be stated that the PBF-LB/M process exhibited more consistent mean deviations across all BPP geometries, whereas the mean deviations of the MEX/M process were more geometry-dependent. The BPPs with no intricate features showed a lower mean deviation when produced using MEX/M than using PBF-LB/M. In contrast, the PBF-LB/M process showed significantly lower mean and standard deviations when it came to small and complex features, as can be seen in the triangular baffles and wavy bottom geometries.

Generally, the BPPs produced using the MEX/M process became less accurate as the structures became smaller. For example, when comparing the triangular baffles manufactured using both processes (see Figure 8), the difference quickly became apparent: the pyramid-shaped structures produced using the MEX/M process, particularly at the tip, could not be manufactured true to the CAD. This was mainly due to the selection of a nozzle that was too wide, but it could also have been influenced by other process settings. Previous test series have shown that even finer structures can be produced accurately and consistently with the CAD model if the settings are adjusted and a finer nozzle is used [58]. Additionally, the tendency to produce corners bent upwards was significantly less pronounced in the MEX/M process.

The choice of a manufacturing process depends heavily on various parameters like the complexity of the model and the tendency of the design to warp at the outer corners. According to the findings of this paper, the MEX/M process shows great promise for the fabrication of BPPs in the near future, especially when the problems with small and complex features are overcome. For this to happen, subsequent research in this field is needed.

J. Exp. Theor. Anal. **2025**, 3, 12

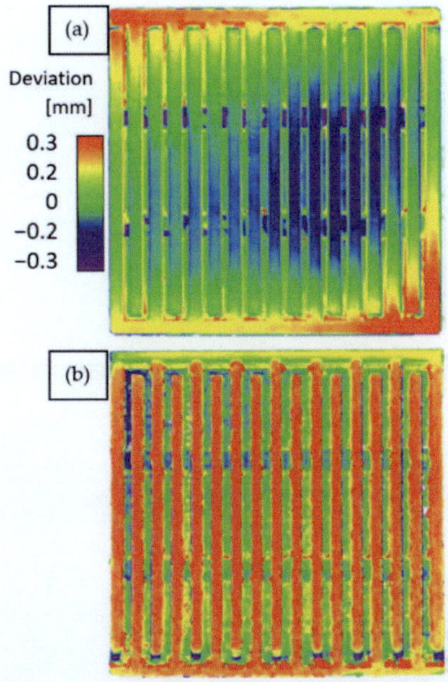

Figure 7. Three-dimensional CAD comparison of serpentine BPPs made from 316L using MEX/M. Manufactured (**a**) alone and horizontally and (**b**) stacked in a vertical, standing-up orientation.

The economic feasibility of both manufacturing methods was assessed using the cost tool developed in [59], which contains the system and material prices used in this work and roughly estimates the costs as a function of the number of units. It should be noted that the costs are approximated based on the component volume using a simplified approach. It is assumed that the component is a cuboid fully composed of the material, thereby estimating the maximum costs. The cost tool is freely available at [60] and can be supplemented with additional -machine setups. Based on the BPP dimensions and volume, excluding post-processing, the component costs for a single BPP were EUR 121 in the MEX/M process and EUR 115 in the PBF-LB/M process. It is interesting to note that the MEX/M process chain for the production of a BPP generally assumed to be cost-effective was more expensive than the PBF-LB/M process chain. One possible reason for this is the cost of the multi-stage process chain, which requires more energy, as well as the significantly higher material costs of the filament, which were more than 4.5 times higher than the cost of the 316L powder for the PBF-LB/M process [59]. At EUR 270, the production costs for eight BPPs in the MEX/M process were exactly twice as high as the costs for production in the PBF-LB/M process. Other factors, such as several cost-intensive sintering runs, played a role due to the limited chamber volume of the sinter oven used. Furthermore, the stacking possibility was not considered in the cost estimation, meaning that the costs were overestimated and the maximum capacity of the chamber was not considered. Therefore, selecting the PBF-LB/M process for the manufacturing of more complex components appears to be both more precise and cost-effective.

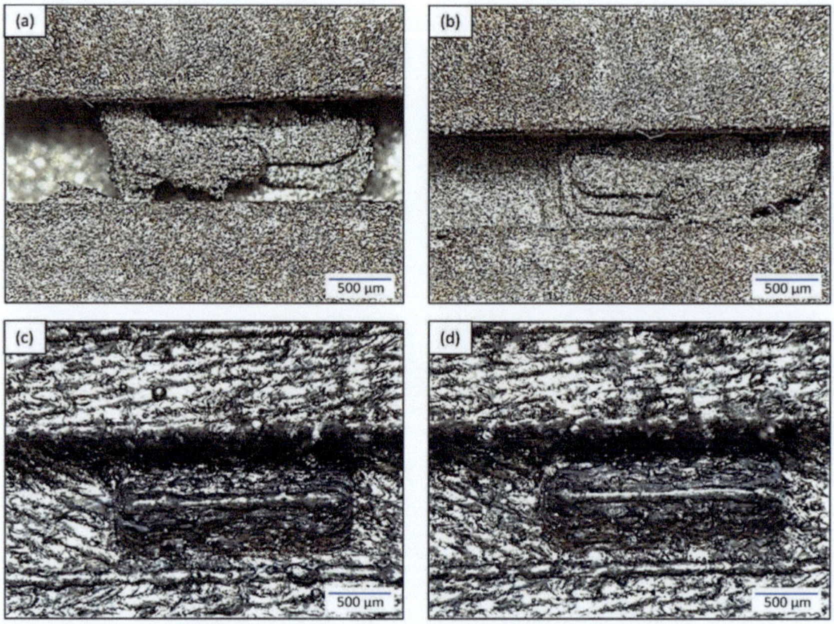

Figure 8. Comparison of (**a**,**b**) MEX/M- and (**c**,**d**) PBF-LB/M-produced pyramids inside the channels of the triangular baffle geometry.

5. Conclusions

In this paper, the manufacturability in the MEX/M and PBF-LB/M processes of six different bipolar plate geometries with 316L material was investigated. These technologies basically showcase the possible production methods for different bipolar plate geometries. In the MEX/M process, the limits of the technology were identified in the production of small features. For example, the pyramids in the triangular baffle structures were not produced with dimensional accuracy and shape fidelity, thereby compromising the intended function of the pyramids. One way to produce more precise features in this process is to reduce the extrusion multiplier, which has a significant influence on the extrusion process in green part production [61]. Furthermore, a smaller nozzle can also ensure a higher level of detail.

Nevertheless, the MEX/M process also had some manufacturing advantages, especially compared to the PBF-LB/M process. For example, serpentines, rib grooves and narrowing serpentines could be produced with a lower standard deviation, which was in the range of 0.03 mm to 0.04 mm. The warpage of the thin bipolar plates could be avoided by stacking several plates or by printing a complete PEMFC. The PBF-LB/M method was better suited than the MEX/M method for producing complex, irregular structures such as wavy bottom structures or wave structures, with mean deviations of 0.06 mm and 0.058 mm, respectively.

To summarize the potential of the two AM methods, it can be said that the PBF-LB/M method can be used for more complex structures with detailed geometric features. The MEX/M method offers a good trade-off between sufficient accuracy for simpler geometric bipolar plates and the necessary precision. Therefore, possible applications of the PBF-LB/M method include prototyping geometrically complex bipolar plates and performing experimental evaluations to compare with simulation results. The MEX/M process can also be used for prototype production, but here the experimental evaluation should focus on preparation for mass production using the MIM process.

Author Contributions: Conceptualization, K.A.; methodology, K.A.; software, J.H.; validation, K.A. and T.R.; formal analysis, K.A. and J.H.; investigation, K.A. and S.R.; resources, S.R. and J.H; data curation, K.A. and J.H.; writing—original draft preparation, K.A. and J.H.; writing—review and editing, K.A.; visualization, K.A. and J.H.; supervision, K.A. and C.E.; project administration, K.A.; funding acquisition, K.A. and C.E. All authors have read and agreed to the published version of the manuscript.

Funding: This research received no external funding.

Institutional Review Board Statement: Not applicable.

Informed Consent Statement: Not applicable.

Data Availability Statement: All data are published in: Asami, M.K.; Roth, S.; Hünting, J.; Röver, T.; Emmelmann, C. Metallic Bipolar Plate Production Through Additive Manufacturing: Contrasting MEX/M and PBF-LB/M Approaches; TUHH Universitätsbibliothek: Hamburg, Germany, 2025. https://doi.org/10.15480/882.14567 [44].

Conflicts of Interest: Author Sebastian Roth was employed by the company: Quest One GmbH. The remaining authors declare that the research was conducted in the absence of any commercial or financial relationships that could be construed as a potential conflict of interest.

Abbreviations

3D	Three-dimensional
AM	Additive manufacturing
BPPs	Bipolar plates
CAD	Computer-aided design
FDM	Fused deposition modeling
HIP	Hot isostatic pressing
MEX/M	Material extrusion of metals
MIM	Metal injection molding
MPIF 35	Materials Standard 35 for MIM Parts
PBF-LB/M	Laser powder bed fusion of metals
PEMEC/PEMFC	Proton exchange membrane electrolyzer/fuel cell
SS	Stainless steel
STL	Standard tessellation language
TPMS	Triply periodic minimal surface

References

1. Agyekum, E.B.; Nutakor, C.; Agwa, A.M.; Kamel, S. A Critical Review of Renewable Hydrogen Production Methods: Factors Affecting Their Scale-Up and Its Role in Future Energy Generation. *Membranes* **2022**, *12*, 173. [CrossRef] [PubMed]
2. Züttel, A.; Remhof, A.; Borgschulte, A.; Friedrichs, O. Hydrogen: The future energy carrier. *Philos. Transact. A Math. Phys. Eng. Sci.* **2010**, *368*, 3329–3342. [CrossRef] [PubMed]
3. Sarmah, M.K.; Singh, T.P.; Kalita, P.; Dewan, A. Sustainable hydrogen generation and storage—A review. *RSC Adv.* **2023**, *13*, 25253–25275. [CrossRef]
4. Budsberg, E.; Crawford, J.T.; Morgan, H.; Chin, W.S.; Bura, R.; Gustafson, R. Hydrocarbon bio-jet fuel from bioconversion of poplar biomass: Life cycle assessment. *Biotechnol. Biofuels* **2016**, *9*, 170. [CrossRef]
5. Howarth, R. Methane emissions and climatic warming risk from hydraulic fracturing and shale gas development: Implications for policy. *Energy Emiss. Control Technol.* **2015**, *2015*, 45–54. [CrossRef]

6. Melnyk, O.; Shumylo, O.; Onishchenko, O.; Mykhailova, I.; Obniavko, T.; Korobko, T. Concept and Prospects for the Use of Hydrogen Fuel in Maritime Transport. *Collect. Sci. Works Ukr. State Univ. Railw. Transp.* **2023**, *203*, 96–105. [CrossRef]
7. Bessarabov, D.; Wang, H.; Li, H.; Zhao, N. (Eds.) *PEM Electrolysis for Hydrogen Production: Principles and Applications*, 1st ed.; CRC Press: Boca Raton, FL, USA, 2016. [CrossRef]
8. Sun, X.; Xu, K.; Fleischer, C.; Liu, X.; Grandcolas, M.; Strandbakke, R.; Bjørheim, T.S.; Norby, T.; Chatzitakis, A. Earth-Abundant Electrocatalysts in Proton Exchange Membrane Electrolyzers. *Catalysts* **2018**, *8*, 657. [CrossRef]
9. Esposito, D.V.; Yim, K.; Fraga Alvarez, D.V.; Cohen, L.; Jin, J.; West, A.C.; Weimer, M.S.; Moulton, S.; Dameron, A.A.; Ayers, K.E.; et al. (Invited) Proton Exchange Membrane Electrolyzers Based on Sub-Micron Thick Membranes. *ECS Meet. Abstr.* **2023**, *MA2023-01*, 2028. [CrossRef]
10. Kassim, Z.; Shaffee, S.N.A.; Aminuddin, F.A.; Krishnan, R.K.; Ghapar, N.A.A.; Russell, L.; Perez, V.G.; Sarawagi, A. Improving Green Hydrogen Production through Proton Exchange Membrane Electrolyzer Simulation Study. In Proceedings of the SPE Gas & Oil Technology Showcase and Conference, Dubai, United Arab Emirates, 8 May 2024; SPE: Dubai, United Arab Emirates, 2024; p. D021S018R002. [CrossRef]
11. Motupally, S.; Mishra, L.; Subramanian, V.R. (Digital Presentation) Modeling Water Transport in Proton Exchange Membrane Electrolyzers through First Principles. *ECS Meet. Abstr.* **2023**, *MA2023-02*, 2158. [CrossRef]
12. Zeng, Z.; Bliznakov, S.; Bonville, L.J.; Maric, R. Advanced Porous Transport Layers for PEM Water Electrolyzers: Impact of the Interfacial and Bulk Properties of the PTLs on the Electrolyzers Performance. *ECS Meet. Abstr.* **2023**, *MA2023-01*, 2056. [CrossRef]
13. Bliznakov, S.; Zeng, Z.; Ouimet, R.; Niedzwiecki, A.; Capuano, C.; Ayers, K.E.; Bonville, L.J.; Maric, R. (Invited) Innovative Membrane Electrode Assemblies for the Next Generation Proton Exchange Membrane Water Electrolyzers. *ECS Meet. Abstr.* **2023**, *MA2023-01*, 1992. [CrossRef]
14. Lettenmeier, P.; Wang, R.; Abouatallah, R.; Burggraf, F.; Gago, A.S.; Friedrich, K.A. Coated Stainless Steel Bipolar Plates for Proton Exchange Membrane Electrolyzers. *J. Electrochem. Soc.* **2016**, *163*, F3119–F3124. [CrossRef]
15. Sun, C.; Wang, Y.; McMurtrey, M.D.; Jerred, N.D.; Liou, F.; Li, J. Additive manufacturing for energy: A review. *Appl. Energy* **2021**, *282*, 116041. [CrossRef]
16. Kamal, M.; Rizza, G. Design for metal additive manufacturing for aerospace applications. In *Additive Manufacturing for the Aerospace Industry*; Elsevier: Amsterdam, The Netherlands, 2019; pp. 67–86. [CrossRef]
17. Sheoran, A.J.; Kumar, H.; Arora, P.K.; Moona, G. Bio-Medical applications of Additive Manufacturing: A Review. *Procedia Manuf.* **2020**, *51*, 663–670. [CrossRef]
18. Lin, K.; Li, X.; Dong, H.; Du, S.; Lu, Y.; Ji, X.; Gu, D. Surface modification of 316 stainless steel with platinum for the application of bipolar plates in high performance proton exchange membrane fuel cells. *Int. J. Hydrogen Energy* **2017**, *42*, 2338–2348. [CrossRef]
19. Lin, K.; Qiao, J.; Gu, D.; Wang, H.; Shi, B.; Zhang, W.; Shan, J.; Xu, Y.; Tian, L. Active screen plasma nitriding of laser powder bed fusion processed 316L stainless steel for the application of fuel cell bipolar plates. *Virtual Phys. Prototyp.* **2023**, *18*, e2225490. [CrossRef]
20. Lin, K.; Xu, Y.; Gu, D.; Shan, J.; Shi, K.; Zhang, W. Laser Powder Bed Fusion of Multifunctional Bio-inspired Vertical Honeycomb Sandwich Structures: For the Application of Lightweight Bipolar Plates of Proton Exchange Membrane Fuel Cells. *Chin. J. Mech. Eng.* **2024**, *37*, 102. [CrossRef]
21. Lin, K.; Qiao, J.; Shi, K.; Dong, W.; Gu, D. Laser powder bed fusion of micro-channels for the application of proton exchange membrane fuel cell bipolar plates. *CIRP J. Manuf. Sci. Technol.* **2023**, *43*, 193–204. [CrossRef]
22. *ISO/ASTM 52900:2021*; Additive Fertigung—Grundlagen—Terminologie. ISO: Geneva, Switzerland, 2022.
23. Suwanpreecha, C.; Manonukul, A. A Review on Material Extrusion Additive Manufacturing of Metal and How It Compares with Metal Injection Moulding. *Metals* **2022**, *12*, 429. [CrossRef]
24. Meng, F.; Beretta, M.; Selema, A.; Sergeant, P.; Vleugels, J.; Desplentere, F.; Ferraris, E. Production and characterisation of filament-based Material Extrusion (MEX) additively manufactured copper parts. *Procedia CIRP* **2024**, *121*, 234–239. [CrossRef]
25. Santos, C.; Gatões, D.; Cerejo, F.; Vieira, M.T. Influence of Metallic Powder Characteristics on Extruded Feedstock Performance for Indirect Additive Manufacturing. *Materials* **2021**, *14*, 7136. [CrossRef]
26. Galantucci, L.M.; Pellegrini, A.; Guerra, M.G.; Lavecchia, F. 3D Printing of parts using metal extrusion: An overview of shaping debinding and sintering technology. *Adv. Technol. Mater.* **2022**, *47*, 25–32. [CrossRef]
27. Asami, K.; Crego Lozares, J.M.; Ullah, A.; Bossen, B.; Clague, L.; Emmelmann, C. Material extrusion of metals: Enabling multi-material alloys in additive manufacturing. *Mater. Today Commun.* **2024**, *38*, 107889. [CrossRef]
28. Caminero, M.Á.; Romero Gutiérrez, A.; Chacón, J.M.; García-Plaza, E.; Núñez, P.J. Effects of fused filament fabrication parameters on the manufacturing of 316L stainless-steel components: Geometric and mechanical properties. *Rapid Prototyp. J.* **2022**, *28*, 2004–2026. [CrossRef]
29. Obadimu, S.O.; Kourousis, K.I. Shrinkage behaviour of material extrusion steel 316L: Influence of primary 3D printing parameters. *Rapid Prototyp. J.* **2022**, *28*, 92–101. [CrossRef]

30. Ahn, J.; Doh, J.; Kim, S.; Park, S. Knowledge-Based Design Algorithm for Support Reduction in Material Extrusion Additive Manufacturing. *Micromachines* **2022**, *13*, 1672. [CrossRef]
31. Gonzalez-Gutierrez, J.; Cano, S.; Schuschnigg, S.; Kukla, C.; Sapkota, J.; Holzer, C. Additive Manufacturing of Metallic and Ceramic Components by the Material Extrusion of Highly-Filled Polymers: A Review and Future Perspectives. *Materials* **2018**, *11*, 840. [CrossRef]
32. Martucci, A.; Aversa, A.; Lombardi, M. Ongoing Challenges of Laser-Based Powder Bed Fusion Processing of Al Alloys and Potential Solutions from the Literature—A Review. *Materials* **2023**, *16*, 1084. [CrossRef]
33. Salmi, M. Additive Manufacturing Processes in Medical Applications. *Materials* **2021**, *14*, 191. [CrossRef]
34. Dimopoulos, A.; Chryssinas, G.; Mavroforaki, D.; Gan, T.-H.; Chatzakos, P. An Interactive Web-Based Platform for Support Generation and Optimisation for Metal Laser Powder Bed Fusion. *Materials* **2024**, *17*, 1639. [CrossRef]
35. Holder, D.; Henn, M.; Buser, M.; Hagenlocher, C.; Onuseit, V.; Graf, T. Bridging additive and subtractive manufacturing: Exceeding the limits of conventional manufacturing with combined continuous-wave and ultrafast lasers. *PhotonicsViews* **2024**, *21*, 50–53. [CrossRef]
36. Kimm, J.; Hanke, S.; Weber, S.; Lentz, J. Processing of High Interstitial Austenitic Steel with Powder Bed Fusion-Laser Beam/Metal: Evolution of Chemical Inhomogeneity and Microstructural Features during Postprocessing. *Adv. Eng. Mater.* **2024**, *26*, 2301902. [CrossRef]
37. Hunter, L.W.; Brackett, D.; Brierley, N.; Yang, J.; Attallah, M.M. Assessment of trapped powder removal and inspection strategies for powder bed fusion techniques. *Int. J. Adv. Manuf. Technol.* **2020**, *106*, 4521–4532. [CrossRef]
38. Sing, S.L.; An, J.; Yeong, W.Y.; Wiria, F.E. Laser and electron-beam powder-bed additive manufacturing of metallic implants: A review on processes, materials and designs. *J. Orthop. Res.* **2016**, *34*, 369–385. [CrossRef]
39. Huber, F.; Rasch, M.; Schmidt, M. Laser Powder Bed Fusion (PBF-LB/M) Process Strategies for In-Situ Alloy Formation with High-Melting Elements. *Metals* **2021**, *11*, 336. [CrossRef]
40. Venkatesh Kumaran, S.; Torralba, J.M. Laser Powder Bed Fusion Processing of Low Cost CoCrFeNiMoxNby High Entropy Alloys with Promising High-Temperature Properties via In Situ Alloying Commercial Powders. *Metals* **2024**, *14*, 500. [CrossRef]
41. Paillier, G.; Prätzsch, N. Beam shaping to tackle laser powder bed fusion challenges: Multi-plane light conversion to increase the printing speed and reduce hot-cracking. *PhotonicsViews* **2024**, *21*, 30–35. [CrossRef]
42. Yin, W.; Guo, P.; Wei, J.; Peng, J.; Bian, N.; Xing, H. Development Status of Flow Field Design for Bipolar Plates in Hydrogen Fuel Cells. *J. Eng. Res. Rep.* **2024**, *26*, 250–261. [CrossRef]
43. Balamurugan, C.; Jeno, S.; Adhikesavan, C.; Praveen, A. *Design and Analysis of Bipolar Plates in Proton Exchange Membrane Fuel Cells Using Computational Fluid Dynamics Analysis*; SAE International: Columbus, OH, USA, 2024; p. 2024-01-4324. [CrossRef]
44. Asami, M.K.; Roth, S.; Hünting, J.; Röver, T.; Emmelmann, C. *Metallic Bipolar Plate Production Through Additive Manufacturing: Contrasting MEX/M and PBF-LB/M Approaches*; TUHH Universitätsbibliothek: Hamburg, Germany, 2025. [CrossRef]
45. Marappan, M.; Palaniswamy, K.; Velumani, T.; Chul, K.B.; Velayutham, R.; Shivakumar, P.; Sundaram, S. Performance Studies of Proton Exchange Membrane Fuel Cells with Different Flow Field Designs—Review. *Chem. Rec.* **2021**, *21*, 663–714. [CrossRef]
46. Koresawa, R.; Utaka, Y. Water control by employing microgrooves inside gas channel for performance improvement in polymer electrolyte fuel cells. *Int. J. Hydrogen Energy* **2015**, *40*, 8172–8181. [CrossRef]
47. Seyhan, M.; Akansu, Y.E.; Murat, M.; Korkmaz, Y.; Akansu, S.O. Performance prediction of PEM fuel cell with wavy serpentine flow channel by using artificial neural network. *Int. J. Hydrogen Energy* **2017**, *42*, 25619–25629. [CrossRef]
48. Barati, S.; Khoshandam, B.; Ghazi, M.M. An investigation of channel blockage effects on hydrogen mass transfer in a proton exchange membrane fuel cell with various geometries and optimization by response surface methodology. *Int. J. Hydrogen Energy* **2018**, *43*, 21928–21939. [CrossRef]
49. Sun, F.; Su, D.; Yin, Y.; Pang, B.; Guo, J. Effects of Combined Baffles on the Proton Exchange Membrane Fuel Cell Performance. *Int. J. Electrochem. Sci.* **2022**, *17*, 221134. [CrossRef]
50. Liu, Y.; Liu, P.; Ren, J.; Jin, Z.; Han, X. Effect of flow channel shapes of proton exchange membrane fuel Cell on its performances. *Int. J. Electrochem. Sci.* **2022**, *17*, 220319. [CrossRef]
51. Wang, C.-T.; Ou, Y.-T.; Wu, B.-X.; Thangavel, S.; Hong, S.-W.; Chung, W.-T.; Yan, W.-M. A modified serpentine flow slab for in Proton Exchange Membrane Fuel Cells (PEMFCs). *Energy Procedia* **2017**, *142*, 667–673. [CrossRef]
52. Asami, K.; Herzog, D.; Bossen, B.; Klemp, C.; Geyer, L.; Emmelmann, C. Design Guidelines For Green Parts Manufactured From Stainless Steel In The Filament Based Material Extrusion Process For Metals (MEX | M). In *World PM2022 Proceedings*; EPMA: Lyon, France, 2022. [CrossRef]
53. Asami, K.; Roth, S.; Krukenberg, M.; Röver, T.; Herzog, D.; Emmelmann, C. Predictive modeling of lattice structure design for 316L stainless steel using machine learning in the L-PBF process. *J. Laser Appl.* **2023**, *35*, 042046. [CrossRef]
54. Technical Data Sheet m4p^{TM} 316 l. Feistritz i.R. Available online: https://www.metals4printing.com/path/app/?qs_page=F2E685075C7378F5FE3FA2BE952763874D72F446&qs_productId=31D00E5751FA4C775C33275302EACA5B918A58C5# (accessed on 5 February 2025).

55. Metall Filamente für FFF/FDM. Available online: http://www.pt-a.de/filament%20en.htm (accessed on 5 February 2025).
56. Quarto, M.; Carminati, M.; D'Urso, G. Density and shrinkage evaluation of AISI 316L parts printed via FDM process. *Mater. Manuf. Process.* **2021**, *36*, 1535–1543. [CrossRef]
57. Shaikh, M.Q.; Singh, P.; Kate, K.H.; Freese, M.; Atre, S.V. Finite Element-Based Simulation of Metal Fused Filament Fabrication Process: Distortion Prediction and Experimental Verification. *J. Mater. Eng. Perform.* **2021**, *30*, 5135–5149. [CrossRef]
58. Singh, P.; Balla, V.K.; Atre, S.V.; German, R.M.; Kate, K.H. Factors affecting properties of Ti-6Al-4V alloy additive manufactured by metal fused filament fabrication. *Powder Technol.* **2021**, *386*, 9–19. [CrossRef]
59. Asami, K.; Herzog, D.; Deutschmann, T.; Röver, T.; Kelbassa, I.; Emmelmann, C. Methodology for Cost Estimation Using Characteristic Factors in Additive Manufacturing. *J. Jpn. Soc. Powder Powder Metall.* **2025**, *72*, S75–S82. [CrossRef]
60. Asami, M.K.; Herzog, D.; Deutschmann, T.; Röver, T.; Kelbassa, I.; Emmelmann, C. *Cost Application for the Publication: Methodology for Cost Estimation Using Characteristic Factors in Additive Manufacturing*; TUHH Universitätsbibliothek: Hamburg, Germany, 2024. [CrossRef]
61. Ghorbani, J.; Koirala, P.; Shen, Y.-L.; Tehrani, M. Eliminating voids and reducing mechanical anisotropy in fused filament fabrication parts by adjusting the filament extrusion rate. *J. Manuf. Process.* **2022**, *80*, 651–658. [CrossRef]

B.7 Material extrusion of metals: Enabling multi-material alloys in additive manufacturing

The summary of the findings can be found in Chapter 8.

Materials Today Communications 38 (2024) 107889

Contents lists available at ScienceDirect

Materials Today Communications

journal homepage: www.elsevier.com/locate/mtcomm

Material extrusion of metals: Enabling multi-material alloys in additive manufacturing

Karim Asami [a,*], José M. Crego Lozares [a,b], Abid Ullah [a,b,*], Bastian Bossen [a], Leighton Clague [c], Claus Emmelmann [a]

[a] *Institute of Laser and System Technologies, Hamburg University of Technology, Harburger Schloßstraße 28, Hamburg 21079, Germany*
[b] *Centre for Additive Manufacturing, School of Engineering, RMIT University, Melbourne, VIC 3000, Australia*
[c] *XERION Berlin Laboratories GmbH, Groß-Berliner Damm 84A, Berlin 12487, Germany*

ARTICLE INFO

Keywords:
Additive Manufacturing
Multi-Material Alloys
MEX/M Technology
316 L
17–4PH
Sintering
Filament Feedstock

ABSTRACT

This article discusses the use of Material Extrusion of Metals (MEX/M - defined in ISO/ASTM 52900) technology for producing multi-material components, which offers a cost-effective and efficient method without requiring major adaptations to existing equipment. The article presents the successful manufacture of 316L/17–4PH multi-alloy steel cubes for four different sintering temperatures using commercially available equipment and feedstock. The feasibility study of the multi-material samples was supported by analyzing the shrinkage, optical density, and microstructure as output parameters. Unlike other works, the multi-material produced parts showed regular shrinkage behaviour and no major deformation, indicating the possibility of creating functional parts without the need for post-processing. Interestingly, relevant findings were made on the different shrinkage values obtained for the constellations considered. Further in-depth investigation of each parameter involved in all the stages of the chain value (extrusion, debinding, and sintering) will be key to support the present results. Additional research into the mechanical and magnetic properties of the samples produced could lead this combination of materials to applications in various industries including automotive and aerospace.

1. Introduction

The advent of additive manufacturing (AM) has furthered many areas of industry with the ability to create novel geometries that were not possible with long-established manufacturing processes, such as milling and casting [1,2]. By far the most widespread AM process is the Material Extrusion of Polymers (MEX/P), due to its low cost and fast production of parts [3,4]. A recent development within the AM space is the filament-based material extrusion of metals (MEX/M - defined in ISO/ASTM 52900), also known as Metal Fused-Filament Fabrication (FFF) as per ISO 52900, whereby a metallic powder held within a polymer matrix is extruded using MEX/P style printers [5,6]. Similar to Metal Injection Molding (MIM), the part is then debound chemically and/or thermally [7,8], and then placed in a sintering furnace until it densifies. The result is a high-density metallic part with densities in the range of 95–98% [9]. Due to its similarity to both MEX/P and MIM, MEX/M has a solid foundation in the literature [10–13]. Thus, a wide range of materials is readily available in different forms (rods, pellets, cartridges) due to the usage of feedstock similar to that used in MIM [14,15].

One interesting use case of MIM is Multi-material (MM) parts. Previously referred to as two-component MIM (2 C-MIM) different materials are formed together during injection molding and are subsequently debound and co-sintered [16]. This allows not only for a reduction in the amount of post-processing required but also increases the functionality of parts [17]. A range of different material combinations have already been tested, such as steel-steel [18,19], other metal-metal [20–22] metal-ceramic [17,23] as well as ceramic-ceramic [24,25]. Previous MM studies using 2 C-MIM have generally necessitated the use of adjusted powder composition fraction to yield successful results [17,19]. Parallel to MEX/M, there is also existing research regarding MM parts produced using MEX/P, where the discrete deposition of multiple materials using separate nozzles, as well as the mixing of two materials within a single nozzle head has also been investigated [26,27]. Late investigations have also focused on optimizing the interlayer strength of the polymer parts, especially at the interface between two materials [28–31]. Since the

* Corresponding authors at: Institute of Laser and System Technologies, Hamburg University of Technology, Harburger Schloßstraße 28, Hamburg 21079, Germany.
E-mail addresses: karim.asami@tuhh.de (K. Asami), abid.ullah@tuhh.de (A. Ullah).

https://doi.org/10.1016/j.mtcomm.2023.107889
Received 10 September 2023; Received in revised form 12 December 2023; Accepted 15 December 2023
Available online 19 December 2023

MEX/P and MEX/M share the same hardware, the previous study could be covered for metal MM parts.

MM deposition using MEX/M is therefore well-posed, due to the fact that equipment and research exist on the MEX/P side, while the feedstock material combinations have been investigated from the MIM side. Conversely, MEX/M offers some advantages with respect to MEX/P and MIM. Namely, the feedstock of different MEX/M materials has a similar, or custom, polymer matrix. This allows for uniform printing parameters during the printing stage, which leads to better adhesion between different layers. With respect to MIM, MEX/M can create parts that aren't fully dense, i.e. parts with an infill pattern as typically seen in MEX/P, which can help speed the debinding process [32]. Additionally, the geometric freedom enabled by MEX/M that is not accessible through MIM could be used to fine-tune the interface of two materials, to increase the chances of successful sintering.

To date, there has been a very limited amount of research regarding MM parts produced using MEX/M. Mousapour et al. [33] investigated the use of 316L stainless steel and high-carbon iron MM cubes. Mixed, coupled, and functionally graded samples were produced, using commercially available filament with mixed results. By optimizing the sintering conditions concerning temperature, holding time, and heating rate, densities of 91–92% were achieved. All samples showed extensive deformation, indicating the need for post-processing to obtain usable components. Abel et al. [34] experimented with Zirconia/17–4PH stainless steel components, with self-prepared feedstock. Samples were produced without major defects, showing good adhesion at the interface on a macroscopic level. To that end, existing literature lacks a thorough investigation of manufacturing MM components using MEX/M, despite the literature that exists from MIM already, as well as MEX/M being well suited to MMAM concerning other AM processes, especially in terms of cost-effectiveness [35].

Depending on the material combination, there is a wide range of applications for MM MEX/M components. The work of Imgrund et al. [19] details the manufacture of magnetic-nonmagnetic parts using a combination of 17–4PH and 316L stainless steel, albeit with micrometal injection molding. Two demonstrators were produced, one being a magnetic positioning encoder incorporated with a hall sensor, the other being a miniature tachometer. Zirconia/17–4PH components, as produced by Abel et al. [34] can be used to create heating elements, with the electrical conducting stainless steel sitting within an isolating zirconia matrix. 17–4PH/316L was identified as an interesting material combination for MM MEX/M, due to existing research from MIM, the availability of feedstock, as well as the similar sintering behaviors of the two materials [19]. In this paper, the feasibility of MM MEX/M is studied. For this, the bonding between the two materials will be assessed by analyzing the shrinkage, optical density, and microstructure for four proposed sintering temperatures. The aim of this research is to prepare the foundations of the process towards achieving a successful manufacturing process that enables quick changes in design and enables a higher degree of complexity than 2 C-MIM for MM parts.

2. Methods and materials

The test samples are manufactured with AISI 316L (EN 1.4404) and AISI 630 / 17–4PH (EN 1.4542) filament provided by the company PT+A (Dresden, Germany) with the starting powder composition illustrated in Table 1. Powder size distribution was described by the provider as “Fine” grade powder size and was guaranteed for the 17–4PH and 316L powder that were used for the filament production. Fig. 1 shows an SEM image of the debonded filament showing the particles of 17–4PH and 316L, with a powder size in the range of 2 μm to 15 μm respectively. The particles for both alloys are predominantly spherical, exhibiting minimal elliptical necks. In comparison to the reference [36] the particle sizes fall within the range of typical MIM feedstock.

For the investigations of multi-alloy steels in two different constellations, cubic samples with dimensions 10 mm in each space direction

Table 1
Starting powder composition for PT+A 316-L (1.4404) and 17–4PH (1.4542) filaments.

Material	316L	17-4PH
Cr	16.8	16
Ni	10.1	4.2
Mo	2.2	0.24
Mn	1.38	0.6
Si	0.74	0.6
P	0.02	0.022
S	0.018	0.003
C	0.004	0.027
Fe	bal.	bal.
Cu	-	4.1
N	-	0.08
O	-	0.06

were produced and are depicted in Fig. 2. The sintering temperatures and deposition sequences for the samples with the labels S1 to S32 are listed in Table 2. Eight cubes were printed with 50 layers of 316L and 50 layers of 17–4PH. The other eight samples were first printed with 50 layers 17–4PH and then with 50 layers 316L. The slicing and printing parameters were chosen using Simplify 3D software (Simplify 3D, Cincinnati, USA).

Eight parts of each constellation were printed using the Renkforce RF 2000 3D printer (Conrad Electronic SE, Hirschau, Germany) with two optimized extruders for direct drive and constant filament feeding. One extruder is equipped with 316L and the other with 17–4PH filament for printing with both filaments to achieve optimal layer connection. Both extruders are driven with the same process parameters which are shown in Table 3. Solvent-based chemical debinding with acetone as solvent was applied in a volume ratio of 8 to 1 between sample and solvent for 12 h at 40 °C as recommended by the provider. An average mass loss after the chemical debinding of the green parts was around 5% as expected from the supplier's advice.

All specimens were thermally debound and sintered in an ExSO90 sinter oven (Aim3D GmbH, Rostock, Germany). Two samples for each of the four different sintering cycles were used, which are illustrated in Fig. 3. The maximum temperature of the sinter cycles is 1286 °C, 1310 °C, 1320 °C, and 1330 °C. The selected temperatures were chosen to achieve a density higher than 95% (MIM standard). The utilization of a 45° alternate filling pattern during the green part printing of the specimen results in a notable reduction in anisotropic shrinkage behavior when the sintering temperature is lowered below that of the single material [37]. Simultaneously, it is known from [19] that the sintering temperature around 1300 °C gives the desired austenitic for 316L, nickel martensite and delta ferrite microstructure for 17–4PH. Furthermore, this sintering temperature also exhibited better ferromagnetic properties [19]. 99,9% Argon was used as the sintering atmosphere and the flow rate for the thermal debinding and sintering step was 1 liter per minute. Thermal debinding and sintering were carried out in one process step. Thermal debinding was performed at two holding times at 325 °C and 600 °C. The holding time was 1 h at 325 °C and 3 h at 600 °C. The heating rate for the first holding time amounted to 300 K/h and for the second holding time to 150 K/h. The heating rate to the sintering temperatures was performed at 150 K/h. Subsequently, two samples from each constellation (316L at the bottom layers and 17–4PH at the top layers as well as 17–4PH at the bottom and 316L to the top) were sintered at four different maximum temperatures for two hours (see Fig. 3).

To quantify the shrinkage, a Digital Vernier caliper was used to measure the dimensions of the cubes in X, Y, and Z. The shrinkage was calculated as stated in the Eq. (1).

$$\text{Percent of shrinkage} = \frac{l_o - l_i}{l_o} \times 100\% \tag{1}$$

Where l_o is the length of interest in the green part and l_i is the length

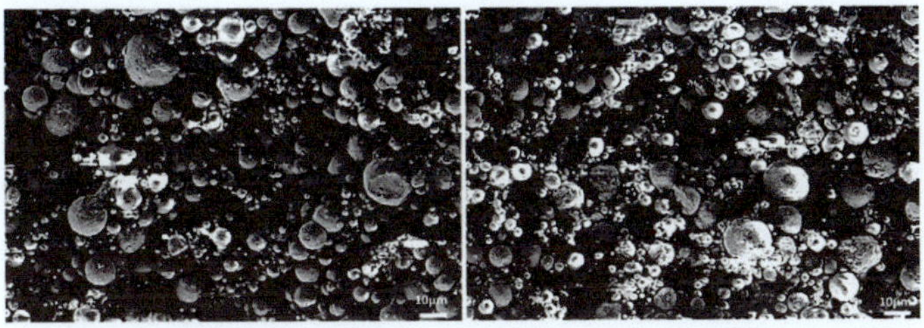

Fig. 1. SEM images of debinded PT+A GmbH 316L filament (left) and 17–4PH filament (right).

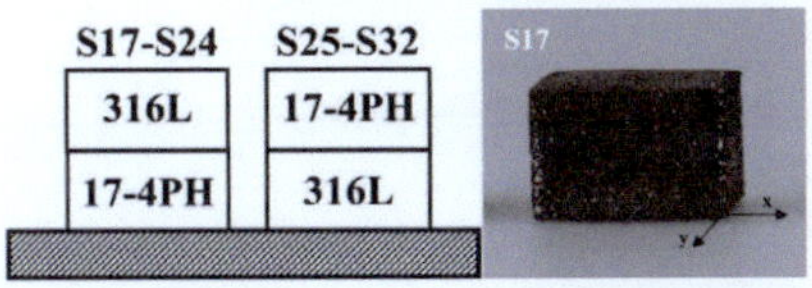

Fig. 2. Representation of selected test cubes after sintering manufactured with different sintering profiles.

Table 2
Sintering temperatures and deposition orders for the samples labeled S1 to S32.

Ref.	Sintering T. (°C)	Deposition order (Top/Bottom)
S1, S2	1286	316L
S3, S4	1310	316L
S5, S6	1320	316L
S7, S8	1330	316L
S9, S10	1286	17-4PH
S11, S12	1310	17-4PH
S13, S14	1320	17-4PH
S15, S16	1330	17-4PH
S17, S18	1286	316L/17-4PH
S19, S20	1310	316L/17-4PH
S21,S22	1320	316L/17-4PH
S23,S24	1330	316L/17-4PH
S25,S26	1286	17-4PH/316L
S27,S28	1310	17-4PH/316L
S29,S30	1320	17-4PH/316L
S31,S32	1330	17-4PH/316L

Table 3
Printing parameters for manufacturing MM stainless steel green parts using a Renkforce RF 2000 3D printer.

Extrusion temperature (°C)	140
Heat bed temperature (°C)	60
Printing speed (mm/min)	2100
Extrusion multiplier	1.35
Skirt layer	3
Retraction distance (mm)	4.5
Layer height (mm)	0.2
Infill (%)	100
Filament Diameter (mm)	2.85
Nozzle Diameter (mm)	0.4

of interest after sintering. The Measurements were conducted twice, both at the edges and in the center of the specimen. Following Buehler guidelines, samples were, embedded, and then ground and polished with progressive sanding papers and polishing solutions indicated for Fe-based alloys using a Vector Power Head (Buehler GmbH, Düsseldorf, Germany). Optical microscopy examinations were carried out using a

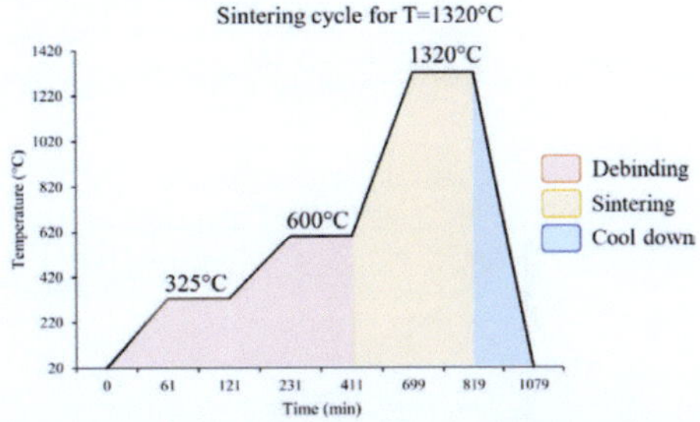

Fig. 3. Representation of the sintering cycle performed at 1320 °C.

MBL 3300 microscope (A.KRÜSS Optronic GmbH, Hamburg, Germany) to obtain further information about the internal structure of the material. Cross-section images of the samples were taken and treated using the image processing software ImageJ to determine density. The images were changed to 8-bit type and adjusted through threshold until the different microstructures were eliminated and only very small to big pores were visible. This methodology guaranteed to obtain a contrast image that includes all the areas occupied by the pores including little to null regions from grain borders. The bulk density of each region was determined through area measurement of each side of the constellation. Before deeper microscopy analysis, the samples were etched to Fry's (5 g CuCl 2, 40 mL HCl, 30 mL H_2O, and 25 mL ethanol) chemical reagent was applied for 15 s. Scanning Electron Microscopy (SEM) using a Zeiss Supra 55 VP FEG-SEM with Variable Pressure Mode (Carl Zeiss AG, Jena, Germany) was used to study the surface structure in the interface of the materials in more detail. Energy-dispersive X-ray spectroscopy (EDS) analysis was performed to investigate the chemical composition of the materials in the interface region.

3. Results and discussion

3.1. Shrinkage

The specimens tested in each constellation didn't show any significant deformation of the components as it was observed in the reference [38]. This indicates that the fabrication method and the co-sintering were successful. Fig. 4 displays the measured shrinkage in each space direction as a function of the sintering temperature.

The shrinkage in 316L specimens is very coherent in both X and Y showing a similar pattern that reaches its peak at 1310 °C with 15.88% and 16.22% in X and Y respectively. The shrinkage drops at 1320 °C to 11.76% in X and 12.09% in Y to rise again to more distinct to 12.80% and 13.94% at 1330 °C. The shrinkage experimented in the Z axis seems

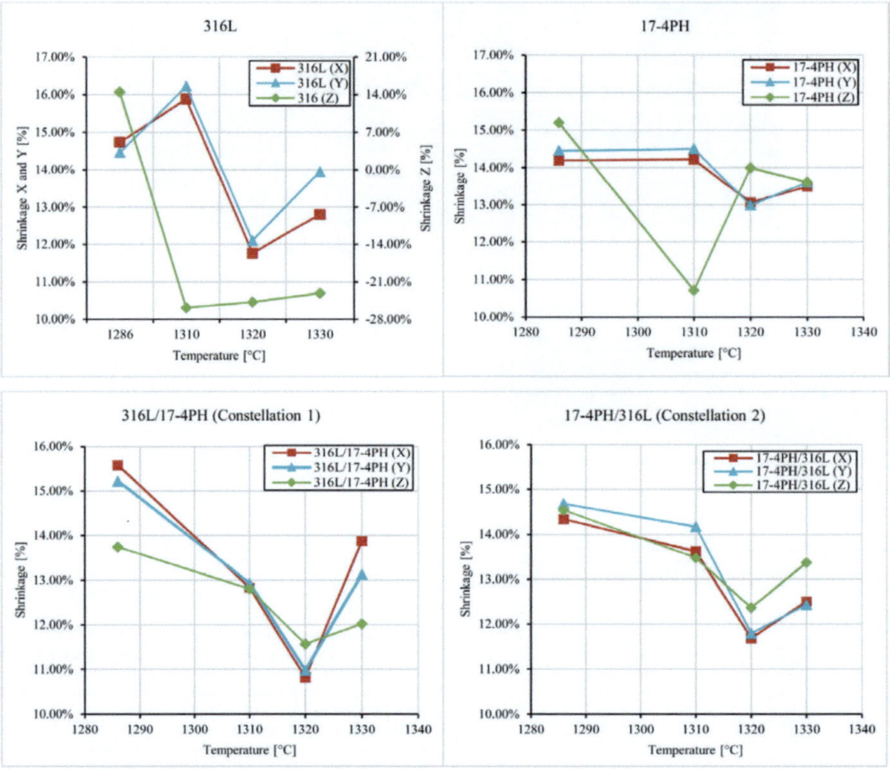

Fig. 4. Shrinkage evolution over temperature for a) 316L, b) 17–4PH, c) 316L/17–4PH and d) 17–4PH/316L samples.

to not follow the trend as the samples tend to grow in the Z axis (Fig. 4 top left, right axis). This effect manifested itself in the samples as a swelling (negative shrinkage) in the Z-axis and only occurred in the 316L samples at 1310 °C (−27%), decreasing at 1320 °C (−25%) and 1330 °C (−22%) respectively. The rounded top and bottom surface suggest that the sample displaced the material as a result of extreme swelling in X and Y. These samples' density was very low which might indicate that the part was not as dense as expected leading to the filling of this gaps during sintering and a lose of geometrical proportions in the Z axis as it was also observed in the reference [37].

17–4PH samples experimented a similar trend to the 316L samples existing with just a 1.77%, 1.91%, 0.54%, and 0.89% difference between X and Y at 1286 °C, 1310 °C, 1320 °C and 1330 °C respectively. The shrinkage in Z experiments showed an abrupt change at 1310 °C had a lower value of 10.71%, followed by a rise to 14% at 1320 °C and maintained at 13.61% at 1330 °C.

The Multimaterial samples follow a mirrored trend to the 17–4PH behavior in X and Y, having the lowest shrinkage value at 1320 °C of 10.98% and 11.79% for 316L/17–4PH and 17–4PH/316L respectively. Both constellations raise their shrinkage to 13.87% and 12.50% in X and 13.13% and 12.43% in Y for 316 L/17–4PH and 17–4PH/316L respectively. The behavior in the Y dimension shows a much similar trend to the one in X and Y contrary to the single material samples. The positioning of the material appears to play a negligible role when compared to the influence of the sintering temperature. The values obtained by the MM samples are the lowest in the graphs suggesting that the MM are less affected by the shrinkage tendency. The values obtained for the 17–4PH/316L sample are slightly higher than their counterpart. One hypothesis on this could be that the material standing on top of the sample might be receiving more heat than the one below which could make sense since 17–4PH has a lower melting point than 316L. A prolonged exposure time to heat might have resulted in increased deformation, however, the 17–4PH CTE is lower than 316L CTE.

Overall, coherent values between X and Y deformation are found in single material printing, while Multimaterial samples show slight variation. More interestingly, MM samples experiment with different shrinkage values for the two constellations proposed. This slight difference could be associated with a particle compensation of the shrinkage behavior as both materials deform in different amount. Interestingly, all values obtained for 1310 °C, 1320 °C, and 1330 °C in both constellations are lower than any of their single material counterparts, which means that an average CTE is not ruling the samples, but one that is actually closer to 17–4Ph or lower is. Nevertheless, the values only find proportion agreement at 1300 °C, therefore a more intensive dimensional study should be carried out.

3.2. Optical density measurement

The optical density measurements of the samples were within the expected range of conventional 316L and 17–4PH parts. The density was measured using SEM images and calculated from the number of pores. Optical measurements a predominantly used for process monitoring applications and allow an exact observation of specific criterias and indicators [39]. In contrast to Archimedes' principle, the current measurement method offers a more nuanced approach by taking into account the specific section under consideration. However, it is essential to acknowledge that Archimedes' principle comes with certain significant drawbacks. One notable limitation is its tendency to measure open porosities as dense, thereby potentially distorting the overall value. This flaw becomes particularly pronounced when dealing with multimaterial components, as the method is unable to measure the density of individual materials separately. This allows for a more precise characterization, circumventing the limitations associated with Archimedes' principle, especially in scenarios involving complex material compositions and transitions.In the section shown in Fig. 5, the calculated density for 316L is 97.4% and for 17–4PH 97.6%. These measurements need a secondary and morecomprehensive image based characterization method as these values correspond to some specific regions and the density of MEX/M samples could vary dramatically along the print direction due to printing and/or extrusion voids [40].

The samples show small signs of printing voids at the interface between the two materials, showing good deposition between one material onto the other. Besides those, the samples show good bonding and a small warpage indicating that the fabrication method was successful, and the materials were of high density reaching MIM standard according to ISO 22068:2012.

3.3. Microstructure

Optical microscopy (OM) was performed on the not-etched and etched specimens. No cracks nor wrecks are visible indicating no severe thermal stresses were generated during the solidification. Only in the case of 17–4PH as the top layers and 316L as the bottom layers, as in Fig. 6(a), are isolated larger pores visible at the sintering temperature of 1310 °C in the zone of material change. Likewise, a more defined grain boundary can be appreciated at 1310 °C. This region, which initially indicated the limit between the two materials shows the increased formation of expected carbides and σ-phase as a result of diffusion. The bigger pores found in this region may have originated during the printing step as the deposition of one material onto the other could lead to voids or gaps generated during the printing. A less probable justification for the large pores close to the interface at higher sintering temperatures could be due to increased stress between the materials due to thermal expansion and sintering shrinkage mismatch, however no cracks are visible. It can be assumed that in this case the diffusion process at higher temperatures as a driving force was amplified has caused more precipitations at the interface. On the contrary, these observations are not as evident as in the other material constellations at the material transition (e.g. Fig. 6(b)).

For 17–4PH the characteristic martensitic matrix with small δ-ferrite islands at the grain boundary can be identified. The 316L side shows an austenitic matrix with δ-ferrite in the grain borders. Both sides show in-grain precipitation of carbides. More pores are observed in the 316L layers than in the 17–4PH layers regardless of the sintered temperature, which is interesting as the martensitic alloy has a lower sintering temperature than the austenitic alloy. A typical grain growth at higher sintering temperatures can be observed in both stainless-steel alloys.

Fig. 7 shows the interfaces of the different constellations at two sintering temperatures and etching times. It is clear that no larger pores occur in the material transition independent of the sintering temperature, confirming that larger pores at the interface might have originated as printing voids. Furthermore, it can be observed that components of 17–4PH have diffused into the austenitic phase of 316L so that there is no clear boundary between the layers. The diffusion path achieved at 1310 °C is greater than at 1286 °C by about 50 micrometers (also compare Fig. 6). The constellation does not affect this.

3.4. SEM and EDS Analysis

As can be appreciated in Fig. 8, the material transition for S24 is inconspicuous. Ni concentration is higher in the 316L region at 6.59 wt % than in the 17–4PH area at 5.03 wt%. According to [35], the concentration difference shows that Ni from the 316 L region diffused towards the 17–4PH region as expected. In addition, the copper content is 2.92 wt% in the 316L and 5.03 wt% in the 17–4PH region, leading to two assumptions: the location of the Cu in the 316L region attends to no other explanation but the etchant containing CuCl, copper oxide may have formed on the surface during SEM preparation, or Cu diffused deep into the 316 region. The second assumption is more likely true since molybdenum was detected at 1.57 wt% in the 316L area and 1.44 wt% in the 17–4PH region. Besides the Cu contamination from etching, a significant diffusion between both materials can be denoted for the rest of the alloying elements (Mo, Ni, Cr). Also visible in this image are increased pores in the region of 316L.

4. Conclusion and outlook

With regards to MM components, MEX/M technology provides an advantageous platform to produce said components, without any major adaptations to existing equipment. Unlike the MIM process, the MEX/M process does not require a powder layering technique or co-injection molding process [30], making functional prototypes easy and cost-effective to manufacture. The main conclusions are:

- Within this work, it was shown that MM steel cubes made of 316L and 17-4PH in two different constellations were manufactured using commercial equipment and filament feedstock. The samples S7 and S8 at a sintering temperature of 1330 °C showed the best results with

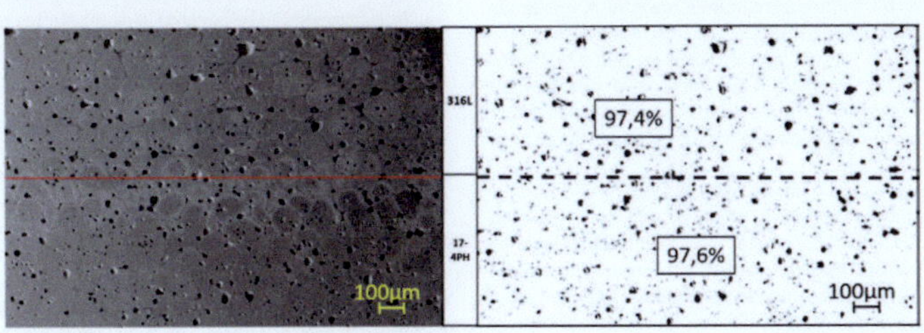

Fig. 5. SEM image of S24 (316 L/17–4PH sintered at 1330 °C) (left), and calculated density using ImageJ software (right).

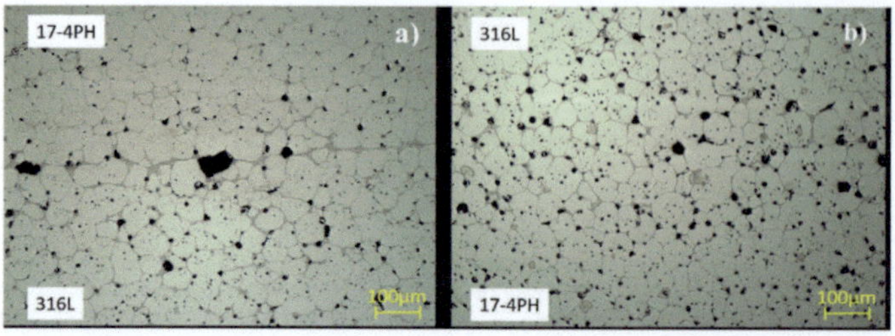

Fig. 6. Optical microscopy image of a) S27 (17–4PH/316 L sintered at 1310 °C); and b) S17 (316 L/17–4PH sintered at 1286 °C).

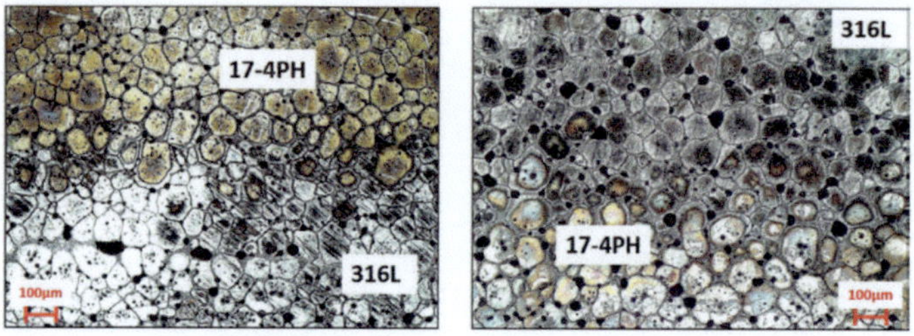

Fig. 7. Optical microscopy of the etched specimens a) S28 (17–4PH/316L sintered at 1310 °C) and etchingtime 15 s, and b) S18 (316 L/17–4PH sintered at 1286 °C) and etching time 20 s.

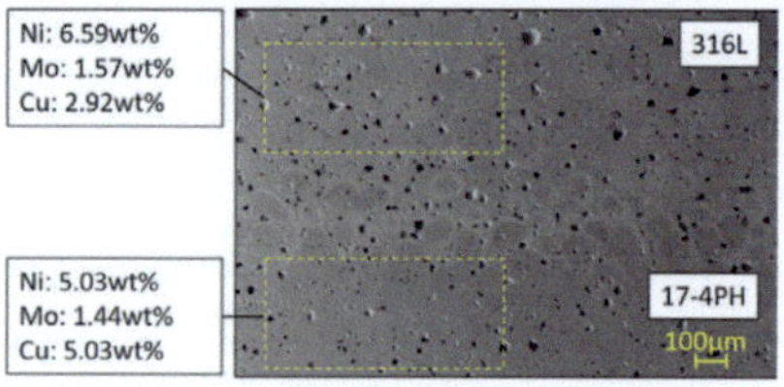

Fig. 8. SEM image of S24 (316L/17–4PH sintered at 1330 °C) and the EDS wt% measurements of Cu, Ni, and Mo at the areas highlighted.

average densities of 97.4% and 97.6%, respectively. The findings also confirm that the sintering temperature, rather than the deposition order of the materials, appears to be the key factor controlling the shrinkage of these samples. However, the shrinkage behaviour was slightly different for both constellations reaching a maximum difference of 1.44% in Z axis for 1300 °C. The difference between the values obtained for single-material and MM printing might be caused by the disposition of the samples in the furnace or due to CTEs mismatch, leading to partial compensation between them. Further in-deep investigations on this matter are needed to support the hypothesis.

- The calculated density from the image analysis also showed that the MIM standard according to ISO 22068:2012 could be met. However, a more comprehensive optical density measurement such as layer by layer x-ray image analysis must be carried out, so that an overall statement about the density of the investigated samples can be made. Further in-depth study of the density will also be covered to explore the influence of printing voids from the extrusion process into the sintered samples, exploring also the horizontal cross-section of both regions.
- As opposed to other works [27], the parts produced showed no major deformation, wherein functional parts without the need for post-processing seem to be a possible task. The measured shrinkage in each coordinate direction was consistently smaller than what might have been expected based on the individual properties of 316L and 17-4PH stainless steels. Although precise expected values were not available, the results obtained cannot provide a rigorous quantitative statement about this observation. Therefore, a statistical study such as ANOVA should be carried out in the future. Further analysis of the shrinkage can be done on complex geometries such as cylindrical structures, overhangs, walls, different infill patterns, etc. Additionally, a machine learning-based methodic approach can be conducted to identify the optimal infill pattern as in to avoid anisotropic shrinkage behaviour.

In the future, the interdiffusion of alloying elements as well as the generation of undesired intermetallic compounds for both materials at the interface need to be covered in deep. Aiming to improve the bonding of the MM samples, the mechanical properties need to be studied as well. Additionally, it is necessary to explore various multimaterial design features to establish design guidelines for the successful manufacturing of multimaterial components such as in the case of single materials [41, 42]. The results obtained in the forthcoming studies will reveal how the parameters involved in all the stages of the chain value (extrusion, debinding, and sintering) affect the mechanical strength at the interface, and therefore, shed light on their limitations for applications. In the future, different types of infill structures have to be investigated to achieve the desired mechanical properties depending on the constellations [43]. This combination of materials can potentially be used for application areas where magnetic and non-magnetic regions are required in a component such as sensors and actuators for automotive or aerospace applications. Ultimately, topology-optimized multimaterial components can also be additively manufactured [44], allowing locally advantageous material properties to be incorporated into a single unit.

CRediT authorship contribution statement

Karim Asami: Conceptualization, Data curation, Formal analysis, Investigation, Methodology Writing – original draft, Writing – review & editing. **José M. Crego Lozares:** Data curation, Formal analysis, Investigation, Validation, Visualization, Methodology, Writing – review & editing. **Abid Ullah:** Data curation, Formal analysis, Investigation, Validation, Visualization, Methodology, Writing – review & editing. **Bastian Bossen:** Resources, Validation, Visualization, Writing – review & editing. **Leighton Clague:** Resources, Validation, Visualization, Methodology, Writing – original draft, Writing – review & editing. **Claus Emmelmann:** Resources, Validation, Visualization, Project administration, Supervision.

Declaration of Competing Interest

The authors declare that they have no known competing financial interests or personal relationships that could have appeared to influence the work reported in this paper.

Data Availability

Data will be made available on request.

Acknowledgments

The authors would like to acknowledge the support and affiliation with the Hamburg University of Technology (TUHH). The authors would also like to acknowledge the support from the REDI Program, a project that has received funding from the European Union's Horizon 2020 research and innovation program under the Marie Skłodowska-Curie grant agreement no. 101034328.This paper reflects only the author's view and the Research Executive Agency is not responsible for any use that may be made of the information it contains.

Author Agreement Statement and Contributions

We the undersigned declare that this manuscript is original, has not been published before, and is not currently being considered for publication elsewhere.

We confirm that the manuscript has been read and approved by all named authors and that there are no other persons who satisfied the criteria for authorship but are not listed. We further confirm that the order of authors listed in the manuscript has been approved by all of us.

We understand that the Corresponding Author is the sole contact for the Editorial process. He/she is responsible for communicating with the other authors about progress, submissions of revisions, and final approval of proofs.

References

[1] M. Khorasani, A. Ghasemi, B. Rolfe, I. Gibson, Additive manufacturing a powerful tool for the aerospace industry, Rapid Prototyp. J. 28 (2022) 87–100.

[2] A. Ur Rehman, A. Ullah, T. Liu, R. Ur Rehman, M.U. Salamci, Additive manufacturing of Al2O3 ceramics with MgO/SiC contents by laser powder bed fusion process, Front. Chem. 11 (2023) 1034473.

[3] O.A. Mohamed, S.H. Masood, J.L. Bhowmik, Experimental investigations of process parameters influence on rheological behavior and dynamic mechanical properties of FDM manufactured parts, Mater. Manuf. Process. 31 (2016) 1983–1994.

[4] C.S. Lee, S.G. Kim, H.J. Kim, S.H. Ahn, Measurement of anisotropic compressive strength of rapid prototyping parts, J. Mater. Process. Technol. 187-188 (2007) 627–630.

[5] M. Sadaf, M. Bragaglia, F. Nanni, A simple route for additive manufacturing of 316L stainless steel via fused filament fabrication, J. Manuf. Process. 67 (2021) 141–150.

[6] C. Tosto, J. Tirillò, F. Sarasini, G. Cicala, Hybrid metal/polymer filaments for fused filament fabrication (FFF) to print metal parts, Appl. Sci. 11 (2021) 1444.

[7] Y. Thompson, J. Gonzalez-Gutierrez, C. Kukla, P. Felfer, Fused filament fabrication, debinding and sintering as a low cost additive manufacturing method of 316L stainless steel, Addit. Manuf. 30 (2019), 100861.

[8] D. Jiang, F. Ning, Additive manufacturing of 316L stainless steel by a printing-debinding-sintering method: effects of microstructure on fatigue property, J. Manuf. Sci. Eng. 143 (2021).

[9] D.F. Heaney, Handbook of metal injection molding, Woodhead Publishing.

[10] C. Suwanpreecha, A. Manonukul, A review on material extrusion additive manufacturing of metal and how it compares with metal injection moulding, Metals 12 (2022) 429.

[11] S. Jadhav, M.S. Bajestani, S. Islam, M.A. Karim, C.J. Kim, H.-J. Lee, Y.T. Cho, et al., Materials characterization of Ti6Al4V to NbZr1 bimetallic structure fabricated by wire arc additive manufacturing, Mater. Today Commun. (2023), 106934.

[12] Y. Thompson, K. Zissel, A. Förner, J. Gonzalez-Gutierrez, C. Kukla, S. Neumeier, P. Felfer, Metal fused filament fabrication of the nickel-base superalloy IN 718, J. Mater. Sci. 57 (2022) 9541–9555.

[13] J. Abel, U. Scheithauer, T. Janics, S. Hampel, S. Cano, A. Müller-Köhn, A. Günther, et al., Fused filament fabrication (FFF) of metal-ceramic components, J. Vis. Exp. JoVE (2019).

[14] J. Wang, Some critical issues for injection molding, BoD–Books on Demand, 2012.

[15] J. Gonzalez-Gutierrez, S. Cano, S. Schuschnigg, C. Kukla, J. Sapkota, C. Holzer, Additive manufacturing of metallic and ceramic components by the material extrusion of highly-filled polymers: a review and future perspectives, Mater. (Basel, Switz.) 11 (2018).

[16] P. Imgrund, A. Rota, M. Wiegmann, Powder Injection Moulding Micro MIM: Investigation of Material Interfaces Obtained by Two-Component-Micro-MIM and Co-Sintering.

[17] V. Piotter, E. Honza, A. Klein, T. Mueller, K. Plewa, Powder injection moulding of multi-material devices, Powder Metall. 58 (2015) 344–348.

[18] G.B. Dutra, M. Mulser, R. Calixto, F. Petzoldt, Investigation of material combinations processed via two-component metal injection moulding (2C-MIM), MSF 727-728 (2012) 248–253.

[19] P. Imgrund, A. Rota, F. Petzoldt, A. Simchi, Manufacturing of multi-functional micro parts by two-component metal injection moulding, Int J. Adv. Manuf. Technol. 33 (2007) 176–186.

[20] C. Pascal, J.-M. Chaix, F. Doré, C.H. Allibert, Design of multimaterial processed by powder metallurgy: processing of a (steel/cemented carbides) bilayer material, J. Mater. Process. Technol. 209 (2009) 1254–1261.

[21] J. Cheng, L. Wan, Y. Cai, J. Zhu, P. Song, J. Dong, Fabrication of W-20wt%Cu alloys by powder injection molding, J. Mater. Process. Technol. 210 (2010) 137–142.

[22] M. Mulser, T. Hartwig, F. Petzoldt, Tungsten-copper/stainless steel Bi-material parts by 2C-MIM (2012).

[23] T. Mueller, V. Piotter, K. Plewa, J. Prokop, H.J. Ritzhaupt-Kleissl, J. Hausselt, Complex shaped micro components produced by powder injection molding, in: Proceedings of 5th 4 M Conference, pp. 429–432.

[24] T. Moritz, Two-component CIM parts for the automotive and railway sectors, Powder Inject. Mould. Int. 2 (2008) 38–39.

[25] A. Ruh, V. Piotter, K. Plewa, H.-J. Ritzhaupt-Kleissl, J. Hausselt, Development of two-component micropowder injection molding (2C-MicroPIM)-process development, Int. J. Appl. Ceram. Technol. 8 (2011) 610–616.

[26] A. García-Collado, J.M. Blanco, M.K. Gupta, R. Dorado-Vicente, Advances in polymers based multi-material additive-manufacturing techniques: state-of-art review on properties and applications, Addit. Manuf. 50 (2022), 102577.

[27] B. Tomar, S. Shiva, T. Nath, A review on wire arc additive manufacturing: processing parameters, defects, quality improvement and recent advances, Mater. Today Commun. 31 (2022), 103739.

[28] R. Freund, H. Watschke, J. Heubach, T. Vietor, Determination of influencing factors on interface strength of additively manufactured multi-material parts by material extrusion, Appl. Sci. 9 (2019) 1782.

[29] S. Hasanov, A. Gupta, A. Nasirov, I. Fidan, Mechanical characterization of functionally graded materials produced by the fused filament fabrication process, J. Manuf. Process. 58 (2020) 923–935.

[30] H. Watschke, L. Waalkes, C. Schumacher, T. Vietor, Development of novel test specimens for characterization of multi-material parts manufactured by material extrusion, Appl. Sci. 8 (2018) 1220.

[31] J. Yin, C. Lu, J. Fu, Y. Huang, Y. Zheng, Interfacial bonding during multi-material fused deposition modeling (FDM) process due to inter-molecular diffusion, Mater. Des. 150 (2018) 104–112.

[32] T. Rosnitschek, A. Seefeldt, B. Alber-Laukant, T. Neumeyer, V. Altstädt, S. Tremmel, Correlations of geometry and infill degree of extrusion additively manufactured 316L stainless steel components, Mater. (Basel, Switz.) 14 (2021).

[33] M. Mousapour, M. Salmi, L. Klemettinen, J. Partanen, Feasibility study of producing multi-metal parts by fused filament fabrication (FFF) technique, J. Manuf. Process. 67 (2021) 438–446.

[34] J. Abel, U. Scheithauer, T. Janics, S. Hampel, S. Cano, A. Müller-Köhn, A. Günther, et al., Fused filament fabrication (FFF) of metal-ceramic components, JoVE (J. Vis. Exp.) (2019), e57693.

[35] A. Simchi, A. Rota, P. Imgrund, An investigation on the sintering behavior of 316L and 17-4PH stainless steel powders for graded composites, Mater. Sci. Eng.: A 424 (2006) 282–289.

[36] V. Martin, J.-F. Witz, F. Gillon, D. Najjar, P. Quaegebeur, A. Benabou, M. Hecquet, et al., Low cost 3D printing of metals using filled polymer pellets, HardwareX 11 (2022), e00292.

[37] S. You, D. Jiang, F. Wang, F. Ning, Anisotropic sintering shrinkage behavior of stainless steel fabricated by extrusion-based metal additive manufacturing, J. Manuf. Process. 101 (2023) 1508–1520.
[38] B. Tomar, S. Shiva, T. Nath, A review on wire arc additive manufacturing: processing parameters, defects, quality improvement and recent advances, Materials Today, Communications 31 (2022), 103739.
[39] K. Rane, K. Castelli, M. Strano, Rapid surface quality assessment of green 3D printed metal-binder parts, J. Manuf. Process. 38 (2019) 290–297.
[40] G. Singh, J.-M. Missiaen, D. Bouvard, J.-M. Chaix, Copper additive manufacturing using MIM feedstock: adjustment of printing, debinding, and sintering parameters for processing dense and defectless parts, Int J. Adv. Manuf. Technol. 115 (2021) 449–462.
[41] D. Herzog, K. Asami, C. Scholl, C. Ohle, C. Emmelmann, A. Sharma, N. Markovic, et al., Design guidelines for laser powder bed fusion in Inconel 718, J. Laser Appl. 34 (2022).
[42] Design Guidelines For Green Parts Manufactured With Stainless Steel In The Filament Based Material Extrusion Process For Metals (MEX/M), World PM 2022 Congress Proceedings, 2022.
[43] K. Asami, S. Roth, M. Krukenberg, T. Röver, D. Herzog, C. Emmelmann, Predictive modeling of lattice structure design for 316L stainless steel using machine learning in the L-PBF process, J. Laser Appl. 35 (2023).
[44] T. Röver, M. Bader, K. Asami, C. Emmelmann, I. Kelbassa, Development and assessment of a methodology for abstraction of topology optimization results to enable the substitution of optimized beams, J. Laser Appl. 35 (2023).

MIX
Papier aus verantwortungsvollen Quellen
Paper from responsible sources
FSC® C105338

If you have any concerns about our products,
you can contact us on
ProductSafety@springernature.com

In case Publisher is established outside the EU,
the EU authorized representative is:
Springer Nature Customer Service Center GmbH
Europaplatz 3, 69115 Heidelberg, Germany

Printed by Libri Plureos GmbH
in Hamburg, Germany